Concise Inorganic Chemistry

Concise Inorganic Chemistry

Adam Kecskemeti
Lawrence Baker

Kruger Brentt
Publishers
2025

Kruger Brentt Publishers UK. LTD.
Company Number 9728962

Regd. Office: 68 St Margarets Road, Edgware, Middlesex HA8 9UU

ISBN: 978-1-78715-438-4

For information on all our publications visit our website at http://krugerbrentt.com/

PREFACE

Concise Inorganic Chemistry serves as a vital resource in the field of chemistry, providing a streamlined approach to understanding the principles, concepts, and applications of inorganic chemistry. Inorganic chemistry itself is a branch that focuses on the study of inorganic compounds, which encompass a vast array of substances, including metals, minerals, and coordination complexes. The term "concise" implies that the material is presented in a clear, succinct manner, making complex topics more accessible to students and practitioners alike. This approach is particularly beneficial in educational settings, where students often grapple with the vastness of inorganic chemistry.

The foundation of concise inorganic chemistry lies in its emphasis on fundamental principles such as atomic structure, periodic trends, bonding theories, and molecular geometry. Understanding atomic structure is crucial, as it provides insights into how elements interact and form compounds. The periodic table, a central tool in chemistry, organizes elements based on their atomic number, electron configuration, and recurring chemical properties.

Bonding theories, including ionic, covalent, and metallic bonding, form another cornerstone of inorganic chemistry. A concise approach to these theories allows for a better understanding of how atoms combine to form various compounds, including the distinction between polar and nonpolar molecules, coordination complexes, and organometallic compounds.

Concise Inorganic Chemistry also delves into the diverse applications of inorganic compounds in various fields, including materials science, catalysis, and medicine. For instance, transition metals and their complexes are widely used as catalysts in industrial processes, enhancing reaction rates and selectivity. Coordination chemistry, a significant area within inorganic chemistry, explores the behavior of metal ions when bonded to ligands, leading to a better understanding of complex formation, stability, and reactivity. Additionally, the study of organometallic chemistry highlights the significance of carbon-metal bonds in catalysis and synthetic applications, revealing the intricate relationships between inorganic and organic chemistry.

Moreover, the text often incorporates problem-solving strategies, examples, and visual aids to enhance comprehension. By integrating practice problems and real-world examples, concise resources encourage active learning and application of theoretical knowledge. Illustrations and diagrams further aid in visualizing concepts, such as electron distribution, molecular shapes, and reaction mechanisms, making them more tangible for students.

Concise Inorganic Chemistry is a crucial resource that distills complex information into essential concepts, making it an invaluable tool for students, educators, and practitioners in the field. By focusing on foundational principles, applications, and problem-solving strategies, it equips learners with the knowledge and skills necessary to navigate the diverse landscape of inorganic chemistry. Whether in academic settings or research environments, a concise understanding of inorganic chemistry facilitates deeper exploration of the subject and its practical applications, ultimately contributing to advancements in science and technology.

Adam Kecskemeti

Lawrence Baker

CONTENTS

PART ONE

THEORETICAL CONCEPTS AND HYDROGEN

CHAPTER 1

ATOMIC STRUCTURE AND THE PERIODIC TABLE

1.1 ATOMIC STRUCTURE

Atoms are the basic particles of the chemical elements. An atom consists of a nucleus of protons and generally neutrons, surrounded by an electromagnetically bound swarm of electrons. The chemical elements are distinguished from each other by the number of protons that are in their atoms. For example, any atom that contains 11 protons is sodium, and any atom that contains 29 protons is copper. Atoms with the same number of protons but a different number of neutrons are called isotopes of the same element.

Atoms are extremely small, typically around 100 picometers across. A human hair is about a million carbon atoms wide. Atoms are smaller than the shortest wavelength of visible light, which means humans cannot see atoms with conventional microscopes. They are so small that accurately predicting their behavior using classical physics is not possible due to quantum effects.

More than 99.94% of an atom's mass is in the nucleus. Protons have a positive electric charge and neutrons have no charge, so the nucleus is positively charged. The electrons are negatively charged, and this opposing charge is what binds them to the nucleus.

If the numbers of protons and electrons are equal, as they normally are, then the atom is electrically neutral as a whole. If an atom has more electrons than protons, then it has an overall negative charge, and is called a negative ion (or anion). Conversely, if it has more protons than electrons, it has a positive charge, and is called a positive ion (or cation).

The electrons of an atom are attracted to the protons in an atomic nucleus by the electromagnetic force. The protons and neutrons in the nucleus are attracted to each other by the nuclear force. This force is usually stronger than the electromagnetic force that repels the positively charged protons from one another. Under certain circumstances, the repelling electromagnetic force becomes stronger than the nuclear force. In this case, the nucleus splits and leaves behind different elements. This is a form of nuclear decay.

Atoms can attach to one or more other atoms by chemical bonds to form chemical compounds such as molecules or crystals. The ability of atoms to attach and detach from each other is responsible for most of the physical changes observed in nature. Chemistry is the science that studies these changes.

1.1.1 History of Atomic Theory

In Philosophy

The basic idea that matter is made up of tiny indivisible particles is an old idea that appeared in many ancient cultures. The word *atom* is derived from the ancient Greek word *atomos*, which means "uncuttable". But this ancient idea was based in philosophical reasoning rather than scientific reasoning. Modern atomic theory is not based on these old concepts. In the early 19th century, the scientist John Dalton (1766–1844) found evidence that matter really is composed of discrete units, and so applied the word *atom* to those units.

Dalton's Law of Multiple Proportions

In the early 1800s, John Dalton compiled experimental data gathered by him and other scientists and discovered a pattern now known as the "law of multiple proportions". He noticed that in any group of chemical compounds which all contain two particular chemical elements, the amount of Element A per measure of Element B will differ across these compounds by ratios of small whole numbers. This pattern suggested that each element combines with other elements in multiples of a basic unit of weight, with each element having a unit of unique weight. Dalton decided to call these units "atoms".

For example, there are two types of tin oxide: one is a grey powder that is 88.1% tin and 11.9% oxygen, and the other is a white powder that is 78.7% tin and 21.3% oxygen. Adjusting these figures, in the grey powder there is about 13.5 g of oxygen for every 100 g of tin, and in the white powder there is about 27 g of oxygen for every 100 g of tin. 13.5 and 27 form a ratio of 1:2. Dalton concluded that in the grey oxide there is one atom of oxygen for every atom of tin, and in the white oxide there are two atoms of oxygen for every atom of tin (SnO and SnO_2).

Dalton also analyzed iron oxides. There is one type of iron oxide that is a black powder which is 78.1% iron and 21.9% oxygen; and there is another iron oxide that is a red powder which is 70.4% iron and 29.6% oxygen.

Adjusting these figures, in the black powder there is about 28 g of oxygen for every 100 g of iron, and in the red powder there is about 42 g of oxygen for every 100 g of iron. 28 and 42 form a ratio of 2:3. Dalton concluded that in these oxides, for every two atoms of iron, there are two or three atoms of oxygen respectively (Fe_2O_2 and Fe_2O_3).

As a final example: nitrous oxide is 63.3% nitrogen and 36.7% oxygen, nitric oxide is 44.05% nitrogen and 55.95% oxygen, and nitrogen dioxide is 29.5% nitrogen and 70.5% oxygen. Adjusting these figures, in nitrous oxide there is 80 g of oxygen for every 140 g of nitrogen, in nitric oxide there is about 160 g of oxygen for every 140 g of nitrogen, and in nitrogen dioxide there is 320 g of oxygen for every 140 g of nitrogen. 80, 160, and 320 form a ratio of 1:2:4. The respective formulas for these oxides are N_2O, NO, and NO_2.

Discovery of the Electron

In 1897, J. J. Thomson discovered that cathode rays are not a form of light but made of negatively charged particles because they can be deflected by electric and magnetic fields. He measured these particles to be at least a thousand times lighter than hydrogen (the lightest atom). He called these new particles *corpuscles* but they were later renamed *electrons* since these are the particles that carry electricity. Thomson also showed that electrons were identical to particles given off by photoelectric and radioactive materials. Thomson explained that an electric current is the passing of electrons from one atom to the next, and when there was no current the electrons embedded themselves in the atoms. This in turn meant that atoms were not indivisible as scientists thought. The atom was composed of electrons whose negative charge was balanced out by some source of positive charge to create an electrically neutral atom. Ions, Thomson explained, must be atoms which have an excess or shortage of electrons.

Discovery of the Nucleus

The electrons in the atom logically had to be balanced out by a commensurate amount of positive charge, but Thomson had no idea where this positive charge came from, so he tentatively proposed that this positive charge was everywhere in the atom, the atom being in the shape of a sphere. Following from this, he imagined the balance of electrostatic forces would distribute the electrons throughout the sphere in a more or less even manner. Thomson's model is popularly known as the plum pudding model, though neither Thomson nor his colleagues used this analogy. Thomson's model was incomplete, it was unable to predict any other properties of the elements such as emission spectra and valencies. It was soon rendered obsolete by the discovery of the atomic nucleus.

Between 1908 and 1913, Ernest Rutherford and his colleagues Hans Geiger and Ernest Marsden performed a series of experiments in which they bombarded thin foils of metal with a beam of alpha particles. They did this to measure the scattering patterns of the alpha particles. They spotted a small number of alpha particles being deflected by angles greater than 90°. This shouldn't have been possible according to the Thomson model of the atom, whose charges were too diffuse to produce a sufficiently strong electric field. The deflections should have all been negligible. Rutherford proposed that the positive charge of the atom along with most of the atom's mass is concentrated in a tiny nucleus at the center of the atom. Only such an intense concentration of positive charge, anchored by its high mass and separated from the negative charge, could produce an electric field that could deflect the alpha particles so strongly.

Bohr Model

A problem in classical mechanics is that an accelerating charged particle radiates electromagnetic radiation, causing the particle to lose kinetic energy. Circular motion counts as acceleration, which means that an electron orbiting a central charge should spiral down into the nucleus as it loses speed. In 1913, the physicist Niels Bohr proposed a new model in which the electrons of an atom were assumed to orbit the nucleus but could only do so in a finite set of orbits, and could jump between these orbits only in discrete changes of energy corresponding to absorption or radiation of a photon. This quantization was used to explain why the electrons' orbits are stable and why elements absorb and emit electromagnetic radiation in discrete spectra. Bohr's model could only predict the emission spectra of hydrogen, not atoms with more than one electron.

Discovery of Protons and Neutrons

Back in 1815, William Prout observed that the atomic weights of many elements were multiples of hydrogen's atomic weight, which is in fact true for all of them if one takes isotopes into account. In 1898, J. J. Thomson found that the positive charge of a hydrogen ion is equal to the negative charge of an electron, and these were then the smallest known charged particles. Thomson later found that the positive charge in an atom is a positive multiple of an electron's negative charge. In 1913, Henry Moseley discovered that the frequencies of X-ray emissions from an excited atom were a mathematical function of its atomic number and hydrogen's nuclear charge. In 1917 Rutherford bombarded nitrogen gas with alpha particles and detected hydrogen ions being emitted from the gas, and concluded that they were produced by alpha particles hitting and splitting the nitrogen atoms.

These observations led Rutherford to conclude that the hydrogen nucleus is a singular particle with a positive charge equal to the electron's negative charge. He named this particle "proton" in 1920. An element's atomic number, which is defined as the element's position on the periodic table, is also the number of protons it has in its nucleus.

The atomic weight of each element is higher than its proton number, so Rutherford hypothesized that the surplus weight was carried by unknown particles with no electric charge and a mass equal to that of the proton.

In 1928, Walter Bothe observed that beryllium emitted a highly penetrating, electrically neutral radiation when bombarded with alpha particles.

It was later discovered that this radiation could knock hydrogen atoms out of paraffin wax. Initially it was thought to be high-energy gamma radiation, since gamma radiation had a similar effect on electrons in metals, but James Chadwick found that the ionization effect was too strong for it to be due to electromagnetic radiation, so long as energy and momentum were conserved in the interaction.

In 1932, Chadwick exposed various elements, such as hydrogen and nitrogen, to the mysterious "beryllium radiation", and by measuring the energies of the recoiling charged particles, he deduced that the radiation was actually composed of electrically neutral particles which could not be massless like the gamma ray, but instead were required to have a mass similar to that of a proton. Chadwick now claimed these particles as Rutherford's neutrons.

The Current Consensus Model

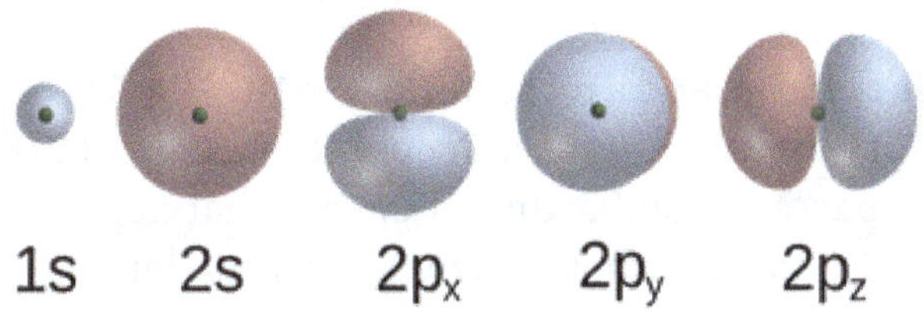

Fig 1: The Modern Model of Atomic Orbitals Draws Zones where an Electron is Most Likely to be Found at Any Moment

In 1925, Werner Heisenberg published the first consistent mathematical formulation of quantum mechanics (matrix mechanics). One year earlier, Louis de Broglie had proposed that all particles behave like waves to some extent, and in 1926 Erwin Schroedinger used this idea to develop the Schroedinger equation, which describes electrons as three-dimensional waveforms rather than points in space. A consequence of using waveforms to describe particles is that it is mathematically impossible to obtain precise values for both the position and momentum of a particle at a given point in time.

This became known as the uncertainty principle, formulated by Werner Heisenberg in 1927. In this concept, for a given accuracy in measuring a position one could only obtain a range of probable values for momentum, and vice versa.

Thus, the planetary model of the atom was discarded in favor of one that described atomic orbital zones around the nucleus where a given electron is most likely to be found. This model was able to explain observations of atomic behavior that previous models could not, such as certain structural and spectral patterns of atoms larger than hydrogen.

1.1.2 Structure

Subatomic Particles

Though the word *atom* originally denoted a particle that cannot be cut into smaller particles, in modern scientific usage the atom is composed of various subatomic particles. The constituent particles of an atom are the electron, the proton and the neutron.

The electron is the least massive of these particles by four orders of magnitude at 9.11×10^{-31} kg, with a negative electrical charge and a size that is too small to be measured using available techniques. It was the lightest particle with a positive rest mass measured, until the discovery of neutrino mass. Under ordinary conditions, electrons are bound to the positively charged nucleus by the attraction created from opposite electric charges.

If an atom has more or fewer electrons than its atomic number, then it becomes respectively negatively or positively charged as a whole; a charged atom is called an ion.

Protons have a positive charge and a mass of 1.6726×10^{-27} kg. The number of protons in an atom is called its atomic number. Ernest Rutherford (1919) observed that nitrogen under alpha-particle bombardment ejects what appeared to be hydrogen nuclei. By 1920 he had accepted that the hydrogen nucleus is a distinct particle within the atom and named it proton.

Neutrons have no electrical charge and have a mass of 1.6749×10^{-27} kg. Neutrons are the heaviest of the three constituent particles, but their mass can be reduced by the nuclear binding energy. Neutrons and protons (collectively known as nucleons) have comparable dimensions—on the order of 2.5×10^{-15} m—although the 'surface' of these particles is not sharply defined. The neutron was discovered in 1932 by the English physicist James Chadwick.

In the Standard Model of physics, electrons are truly elementary particles with no internal structure, whereas protons and neutrons are composite particles composed of elementary particles called quarks. There are two types of quarks in atoms, each having a fractional electric charge. Protons are composed of two up quarks (each with charge +2/3) and one down quark (with a charge of −1/3). Neutrons consist of one up quark and two down quarks. This distinction accounts for the difference in mass and charge between the two particles.

The quarks are held together by the strong interaction (or strong force), which is mediated by gluons. The protons and neutrons, in turn, are held to each other in the nucleus by the nuclear force, which is a residuum of the strong force that has somewhat different range-properties. The gluon is a member of the family of gauge bosons, which are elementary particles that mediate physical forces.

Nucleus

All the bound protons and neutrons in an atom make up a tiny atomic nucleus, and are collectively called nucleons. The radius of a nucleus is approximately equal to $1.07\sqrt[3]{A}$ femtometres, where A is the total number of nucleons. This is much smaller than the radius of the atom, which is on the order of 10^5 fm. The nucleons are bound together by a short-ranged attractive potential called the residual strong force. At distances smaller than 2.5 fm this force is much more powerful than the electrostatic force that causes positively charged protons to repel each other.

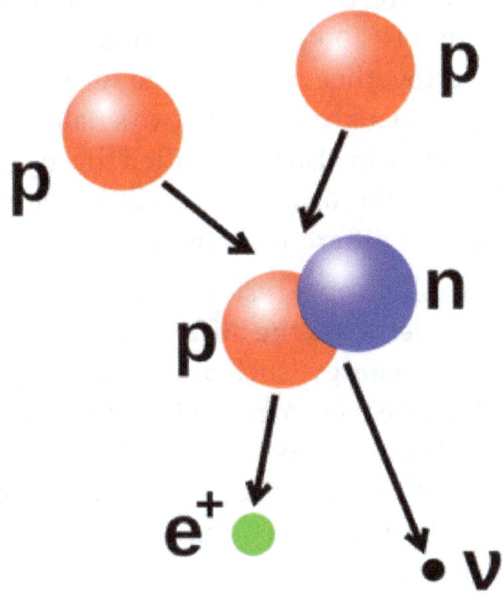

Fig 2: Illustration of a nuclear fusion process that forms a deuterium nucleus, consisting of a proton and a neutron, from two protons. A positron (e^+)—an antimatter electron—is emitted along with an electron neutrino.

Atoms of the same element have the same number of protons, called the atomic number. Within a single element, the number of neutrons may vary, determining the isotope of that element. The total number of protons and neutrons determine the nuclide. The number of neutrons relative to the protons determines the stability of the nucleus, with certain isotopes undergoing radioactive decay.

The proton, the electron, and the neutron are classified as fermions. Fermions obey the Pauli exclusion principle which prohibits *identical* fermions, such as multiple protons, from occupying the same quantum state at the same time. Thus, every proton in the nucleus must occupy a quantum state different from all other protons, and the same applies to all neutrons of the nucleus and to all electrons of the electron cloud.

A nucleus that has a different number of protons than neutrons can potentially drop to a lower energy state through a radioactive decay that causes the number of protons and neutrons to more closely match. As a result, atoms with matching numbers of protons and neutrons are more stable against decay, but with increasing atomic number, the mutual repulsion of the protons requires an increasing proportion of neutrons to maintain the stability of the nucleus.

The number of protons and neutrons in the atomic nucleus can be modified, although this can require very high energies because of the strong force. Nuclear fusion occurs when multiple atomic particles join to form a heavier nucleus, such as through the energetic collision of two nuclei. For example, at the core of the Sun protons require energies of 3 to 10 keV to overcome their mutual repulsion—the coulomb barrier—and fuse together into a single nucleus. Nuclear fission is the opposite process, causing a nucleus to split into two smaller nuclei—usually through radioactive decay. The nucleus can also be modified through bombardment by high energy subatomic particles or photons. If this modifies the number of protons in a nucleus, the atom changes to a different chemical element.

If the mass of the nucleus following a fusion reaction is less than the sum of the masses of the separate particles, then the difference between these two values can be emitted as a type of usable energy (such as a gamma ray, or the kinetic energy of a beta particle), as described by Albert Einstein's mass-energy equivalence formula, $e=mc^2$, where m is the mass loss and c is the speed of light. This deficit is part of the binding energy of the new nucleus, and it is the non-recoverable loss of the energy that causes the fused particles to remain together in a state that requires this energy to separate.

The fusion of two nuclei that create larger nuclei with lower atomic numbers than iron and nickel—a total nucleon number of about 60—is usually an exothermic process that releases more energy than is required to bring them together. It is this energy-releasing process that makes nuclear fusion in stars a self-sustaining reaction. For heavier nuclei, the binding energy per nucleon begins to decrease. That means that a fusion process producing a nucleus that has an atomic number higher than about 26, and a mass number higher than about 60, is an endothermic process. Thus, more massive nuclei cannot undergo an energy-producing fusion reaction that can sustain the hydrostatic equilibrium of a star.

Electron Cloud

The electrons in an atom are attracted to the protons in the nucleus by the electromagnetic force. This force binds the electrons inside an electrostatic potential well surrounding the smaller nucleus, which means that an external source of energy is needed for the electron to escape. The closer an electron is to the nucleus, the greater the attractive force. Hence electrons bound near the center of the potential well require more energy to escape than those at greater separations.

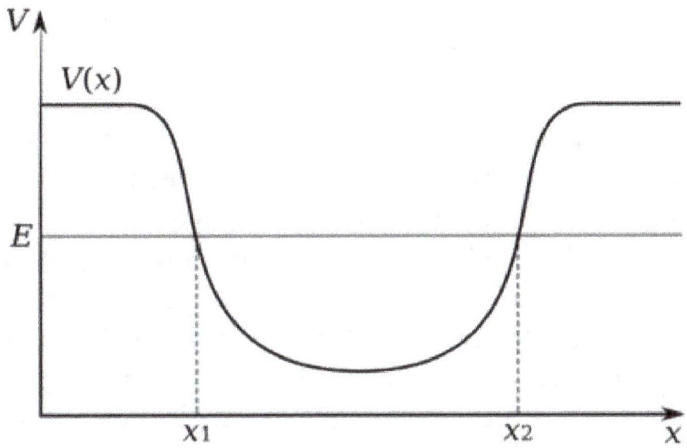

Fig 3: A potential well, showing, according to classical mechanics, the minimum energy $V(x)$ needed to reach each position x. Classically, a particle with energy E is constrained to a range of positions between x_1 and x_2.

Electrons, like other particles, have properties of both a particle and a wave. The electron cloud is a region inside the potential well where each electron forms a type of three-dimensional standing wave—a wave form that does not move relative to the nucleus. This behavior is defined by an atomic orbital, a mathematical function that characterises the probability that an electron appears to be at a particular location when its position is measured. Only a discrete (or quantized) set of these orbitals exist around the nucleus, as other possible wave patterns rapidly decay into a more stable form. Orbitals can have one or more ring or node structures, and differ from each other in size, shape and orientation.

Each atomic orbital corresponds to a particular energy level of the electron. The electron can change its state to a higher energy level by absorbing a photon with sufficient energy to boost it into the new quantum state. Likewise, through spontaneous emission, an electron in a higher energy state can drop to a lower energy state while radiating the excess energy as a photon. These characteristic energy values, defined by the differences in the energies of the quantum states, are responsible for atomic spectral lines.

The amount of energy needed to remove or add an electron—the electron binding energy—is far less than the binding energy of nucleons. For example, it requires only 13.6 eV to strip a ground-state electron from a hydrogen atom, compared to 2.23 *million* eV for splitting a deuterium nucleus. Atoms are electrically neutral if they have an equal number of protons and electrons. Atoms that have either a deficit or a surplus of electrons are called ions. Electrons that are farthest from the nucleus may be transferred to other nearby atoms or shared between atoms. By this mechanism, atoms are able to bond into molecules and other types of chemical compounds like ionic and covalent network crystals.

1.1.3 Properties

Nuclear Properties

By definition, any two atoms with an identical number of *protons* in their nuclei belong to the same chemical element. Atoms with equal numbers of protons but a different number of *neutrons* are different isotopes of the same element. For example, all hydrogen atoms admit exactly one proton, but isotopes exist with no neutrons (hydrogen-1, by far the most common form, also called protium), one neutron (deuterium), two neutrons (tritium) and more than two neutrons. The known elements form a set of atomic numbers, from the single-proton element hydrogen up to the 118-proton element oganesson. All known isotopes of elements with atomic numbers greater than 82 are radioactive, although the radioactivity of element 83 (bismuth) is so slight as to be practically negligible.

About 339 nuclides occur naturally on Earth, of which 251 (about 74%) have not been observed to decay, and are referred to as "stable isotopes". Only 90 nuclides are stable theoretically, while another 161 (bringing the total to 251) have not been observed to decay, even though in theory it is energetically possible. These are also formally classified as "stable". An additional 35 radioactive nuclides have half-lives longer than 100 million years, and are long-lived enough to have been present since the birth of the Solar System. This collection of 286 nuclides are known as primordial nuclides. Finally, an additional 53 short-lived nuclides are known to occur naturally, as daughter products of primordial nuclide decay (such as radium from uranium), or as products of natural energetic processes on Earth, such as cosmic ray bombardment (for example, carbon-14).

For 80 of the chemical elements, at least one stable isotope exists. As a rule, there is only a handful of stable isotopes for each of these elements, the average being 3.1 stable isotopes per element. Twenty-six "monoisotopic elements" have only a single stable isotope, while the largest number of stable isotopes observed for any element is ten, for the element tin. Elements 43, 61, and all elements numbered 83 or higher have no stable isotopes.

Stability of isotopes is affected by the ratio of protons to neutrons, and also by the presence of certain "magic numbers" of neutrons or protons that represent closed and filled quantum shells. These quantum shells correspond to a set of energy levels within the shell model of the nucleus; filled shells, such as the filled shell of 50 protons for tin, confers unusual stability on the nuclide. Of the 251 known stable nuclides, only four have both an odd number of protons *and* odd number of neutrons: hydrogen-2 (deuterium), lithium-6, boron-10, and nitrogen-14. (Tantalum-180m is odd-odd and observationally stable, but is predicted to decay with a very long half-life.) Also, only four naturally occurring, radioactive odd-odd nuclides have a half-life over a billion years: potassium-40, vanadium-50, lanthanum-138, and lutetium-176. Most odd-odd nuclei are highly unstable with respect to beta decay, because the decay products are even-even, and are therefore more strongly bound, due to nuclear pairing effects.

Mass

The large majority of an atom's mass comes from the protons and neutrons that make it up. The total number of these particles (called "nucleons") in a given atom is called the mass number. It is a positive integer and dimensionless (instead of having dimension of mass), because it expresses a count. An example of use of a mass number is "carbon-12," which has 12 nucleons (six protons and six neutrons).

The actual mass of an atom at rest is often expressed in daltons (Da), also called the unified atomic mass unit (u). This unit is defined as a twelfth of the mass of a free neutral atom of carbon-12, which is approximately 1.66×10^{-27} kg. Hydrogen-1 (the lightest isotope of hydrogen which is also the nuclide with the lowest mass) has an atomic weight of 1.007825 Da. The value of this number is called the atomic mass. A given atom has an atomic mass approximately equal (within 1%) to its mass number times the atomic mass unit (for example the mass of a nitrogen-14 is roughly 14 Da), but this number will not be exactly an integer except (by definition) in the case of carbon-12. The heaviest stable atom is lead-208, with a mass of 207.9766521 Da.

As even the most massive atoms are far too light to work with directly, chemists instead use the unit of moles. One mole of atoms of any element always has the same number of atoms (about 6.022×10^{23}). This number was chosen so that if an element has an atomic mass of 1 u, a mole of atoms of that element has a mass close to one gram. Because of the definition of the unified atomic mass unit, each carbon-12 atom has an atomic mass of exactly 12 Da, and so a mole of carbon-12 atoms weighs exactly 0.012 kg.

Shape and Size

Atoms lack a well-defined outer boundary, so their dimensions are usually described in terms of an atomic radius. This is a measure of the distance out to which the electron cloud extends from the nucleus. This assumes the atom to exhibit a spherical shape, which is only obeyed for atoms in vacuum or free space. Atomic radii may be derived from the distances between two nuclei when the two atoms are joined in a chemical bond. The radius varies with the location of an atom on the atomic chart, the type of chemical bond, the number of neighboring atoms (coordination number) and a quantum mechanical property known as spin. On the periodic table of the elements, atom size tends to increase when moving down columns, but decrease when moving across rows (left to right). Consequently, the smallest atom is helium with a radius of 32 pm, while one of the largest is caesium at 225 pm.

When subjected to external forces, like electrical fields, the shape of an atom may deviate from spherical symmetry. The deformation depends on the field magnitude and the orbital type of outer shell electrons, as shown by group-theoretical considerations. Aspherical deviations might be elicited for instance in crystals, where large crystal-electrical fields may occur at low-symmetry lattice sites. Significant ellipsoidal deformations have been shown to occur for sulfur ions and chalcogen ions in pyrite-type compounds.

Atomic dimensions are thousands of times smaller than the wavelengths of light (400–700 nm) so they cannot be viewed using an optical microscope, although individual atoms can be observed using a scanning tunneling microscope. To visualize the minuteness of the atom, consider that a typical human hair is about 1 million carbon atoms in width. A single drop of water contains about 2 sextillion (2×10^{21}) atoms of oxygen, and twice the number of hydrogen atoms. A single carat diamond with a mass of 2×10^{-4} kg contains about 10 sextillion (10^{22}) atoms of carbon. If an apple were magnified to the size of the Earth, then the atoms in the apple would be approximately the size of the original apple.

Radioactive Decay

Every element has one or more isotopes that have unstable nuclei that are subject to radioactive decay, causing the nucleus to emit particles or electromagnetic radiation. Radioactivity can occur when the radius of a nucleus is large compared with the radius of the strong force, which only acts over distances on the order of 1 fm. The most common forms of radioactive decay are:

- **Alpha Decay:** This process is caused when the nucleus emits an alpha particle, which is a helium nucleus consisting of two protons and two neutrons. The result of the emission is a new element with a lower atomic number.
- **Beta Decay (Electron Capture):** These processes are regulated by the weak force, and result from a transformation of a neutron into a proton, or a proton into a neutron. The neutron to proton transition is accompanied by the emission of an electron and an antineutrino, while proton to neutron transition (except in electron capture) causes the emission of a positron and a neutrino. The electron or positron emissions are called beta particles. Beta decay either increases or decreases the atomic number of the nucleus by one. Electron capture is more common than positron emission, because it requires less energy.

In this type of decay, an electron is absorbed by the nucleus, rather than a positron emitted from the nucleus. A neutrino is still emitted in this process, and a proton changes to a neutron.

- **Gamma Decay:** This process results from a change in the energy level of the nucleus to a lower state, resulting in the emission of electromagnetic radiation. The excited state of a nucleus which results in gamma emission usually occurs following the emission of an alpha or a beta particle. Thus, gamma decay usually follows alpha or beta decay.

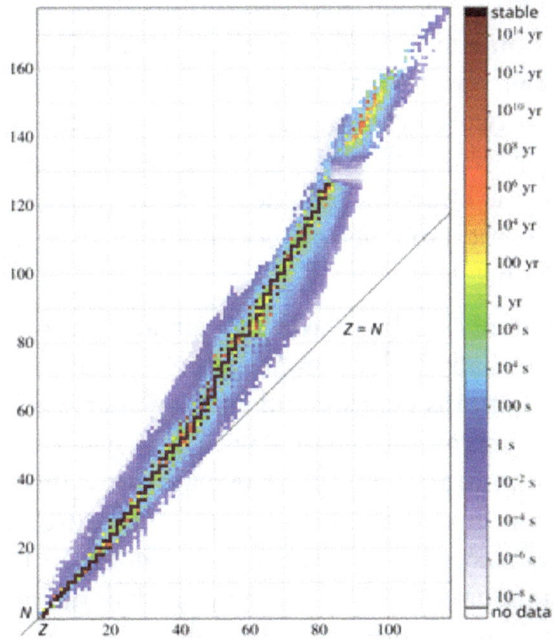

Fig 4: This Diagram Shows the Half-Life ($T_{1/2}$) of Various Isotopes with Z Protons and N Neutrons

Other more rare types of radioactive decay include ejection of neutrons or protons or clusters of nucleons from a nucleus, or more than one beta particle. An analog of gamma emission which allows excited nuclei to lose energy in a different way, is internal conversion—a process that produces high-speed electrons that are not beta rays, followed by production of high-energy photons that are not gamma rays. A few large nuclei explode into two or more charged fragments of varying masses plus several neutrons, in a decay called spontaneous nuclear fission. Each radioactive isotope has a characteristic decay time period—the half-life—that is determined by the amount of time needed for half of a sample to decay. This is an exponential decay process that steadily decreases the proportion of the remaining isotope by 50% every half-life. Hence after two half-lives have passed only 25% of the isotope is present, and so forth.

Magnetic Moment

Elementary particles possess an intrinsic quantum mechanical property known as spin. This is analogous to the angular momentum of an object that is spinning around its center of mass, although strictly speaking these particles are believed to be point-like and cannot be said to be rotating. In an atom, electrons in motion around the nucleus possess orbital angular momentum in addition to their spin, while the nucleus itself possesses angular momentum due to its nuclear spin.

The magnetic field produced by an atom—its magnetic moment—is determined by these various forms of angular momentum, just as a rotating charged object classically produces a magnetic field, but the most dominant contribution comes from electron spin. Due to the nature of electrons to obey the Pauli exclusion principle, in which no two electrons may be found in the same quantum state, bound electrons pair up with each other, with one member of each pair in a spin up state and the other in the opposite, spin down state. Thus these spins cancel each other out, reducing the total magnetic dipole moment to zero in some atoms with even number of electrons.

In ferromagnetic elements such as iron, cobalt and nickel, an odd number of electrons leads to an unpaired electron and a net overall magnetic moment. The orbitals of neighboring atoms overlap and a lower energy state is achieved when the spins of unpaired electrons are aligned with each other, a spontaneous process known as an exchange interaction. When the magnetic moments of ferromagnetic atoms are lined up, the material can produce a measurable macroscopic field. Paramagnetic materials have atoms with magnetic moments that line up in random directions when no magnetic field is present, but the magnetic moments of the individual atoms line up in the presence of a field.

The nucleus of an atom will have no spin when it has even numbers of both neutrons and protons, but for other cases of odd numbers, the nucleus may have a spin. Normally nuclei with spin are aligned in random directions because of thermal equilibrium, but for certain elements (such as xenon-129) it is possible to polarize a significant proportion of the nuclear spin states so that they are aligned in the same direction—a condition called hyperpolarization. This has important applications in magnetic resonance imaging.

Energy Levels

The potential energy of an electron in an atom is negative relative to when the distance from the nucleus goes to infinity; its dependence on the electron's position reaches the minimum inside the nucleus, roughly in inverse proportion to the distance. In the quantum-mechanical model, a bound electron can occupy only a set of states centered on the nucleus, and each state corresponds to a specific energy level. An energy level can be measured by the amount of energy needed to unbind the electron from the atom, and is usually given in units of electronvolts (eV). The lowest energy state of a bound electron is called the ground state, i.e. stationary state, while an electron transition to a higher level results in an excited state. The electron's energy increases along with *n* because the (average) distance to the nucleus increases. Dependence of the energy on ℓ is caused not by the electrostatic potential of the nucleus, but by interaction between electrons.

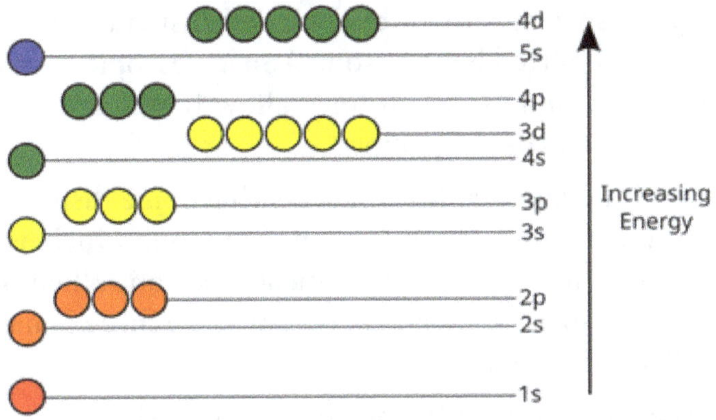

Fig 5: These Electron's Energy Levels (Not to Scale) are Sufficient for Ground States of Atoms Up to Cadmium ($5s^2$ $4d^{10}$) Inclusively. Do not Forget that Even the Top of the Diagram is Lower than an Unbound Electron State

For an electron to transition between two different states, e.g. ground state to first excited state, it must absorb or emit a photon at an energy matching the difference in the potential energy of those levels, according to the Niels Bohr model, what can be precisely calculated by the Schrödinger equation. Electrons jump between orbitals in a particle-like fashion.

For example, if a single photon strikes the electrons, only a single electron changes states in response to the photon.

The energy of an emitted photon is proportional to its frequency, so these specific energy levels appear as distinct bands in the electromagnetic spectrum. Each element has a characteristic spectrum that can depend on the nuclear charge, subshells filled by electrons, the electromagnetic interactions between the electrons and other factors.

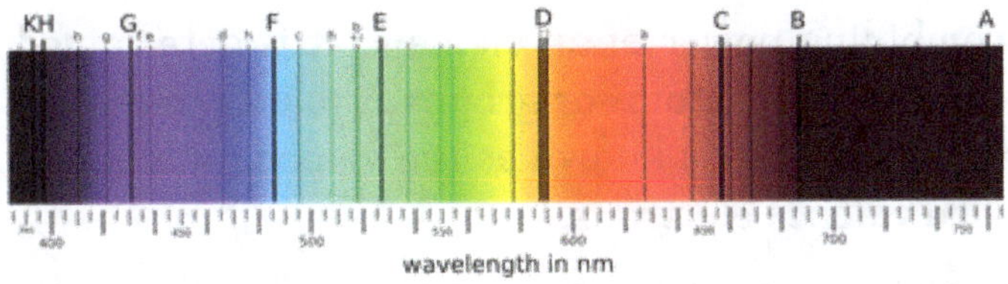

Fig 6: An Example of Absorption Lines in a Spectrum

When a continuous spectrum of energy is passed through a gas or plasma, some of the photons are absorbed by atoms, causing electrons to change their energy level. Those excited electrons that remain bound to their atom spontaneously emit this energy as a photon, traveling in a random direction, and so drop back to lower energy levels.

Thus the atoms behave like a filter that forms a series of dark absorption bands in the energy output. (An observer viewing the atoms from a view that does not include the continuous spectrum in the background, instead sees a series of emission lines from the photons emitted by the atoms.) Spectroscopic measurements of the strength and width of atomic spectral lines allow the composition and physical properties of a substance to be determined.

Close examination of the spectral lines reveals that some display a fine structure splitting. This occurs because of spin–orbit coupling, which is an interaction between the spin and motion of the outermost electron. When an atom is in an external magnetic field, spectral lines become split into three or more components; a phenomenon called the Zeeman effect.

This is caused by the interaction of the magnetic field with the magnetic moment of the atom and its electrons. Some atoms can have multiple electron configurations with the same energy level, which thus appear as a single spectral line. The interaction of the magnetic field with the atom shifts these electron configurations to slightly different energy levels, resulting in multiple spectral lines. The presence of an external electric field can cause a comparable splitting and shifting of spectral lines by modifying the electron energy levels, a phenomenon called the Stark effect.

If a bound electron is in an excited state, an interacting photon with the proper energy can cause stimulated emission of a photon with a matching energy level. For this to occur, the electron must drop to a lower energy state that has an energy difference matching the energy of the interacting photon. The emitted photon and the interacting photon then move off in parallel and with matching phases. That is, the wave patterns of the two photons are synchronized. This physical property is used to make lasers, which can emit a coherent beam of light energy in a narrow frequency band.

Valence and Bonding Behavior

Valency is the combining power of an element. It is determined by the number of bonds it can form to other atoms or groups. The outermost electron shell of an atom in its uncombined state is known as the valence shell, and the electrons in that shell are called valence electrons.

The number of valence electrons determines the bonding behavior with other atoms. Atoms tend to chemically react with each other in a manner that fills (or empties) their outer valence shells. For example, a transfer of a single electron between atoms is a useful approximation for bonds that form between atoms with one-electron more than a filled shell, and others that are one-electron short of a full shell, such as occurs in the compound sodium chloride and other chemical ionic salts.

Many elements display multiple valences, or tendencies to share differing numbers of electrons in different compounds. Thus, chemical bonding between these elements takes many forms of electron-sharing that are more than simple electron transfers. Examples include the element carbon and the organic compounds.

The chemical elements are often displayed in a periodic table that is laid out to display recurring chemical properties, and elements with the same number of valence electrons form a group that is aligned in the same column of the table. (The horizontal rows correspond to the filling of a quantum shell of electrons.) The elements at the far right of the table have their outer shell completely filled with electrons, which results in chemically inert elements known as the noble gases.

States

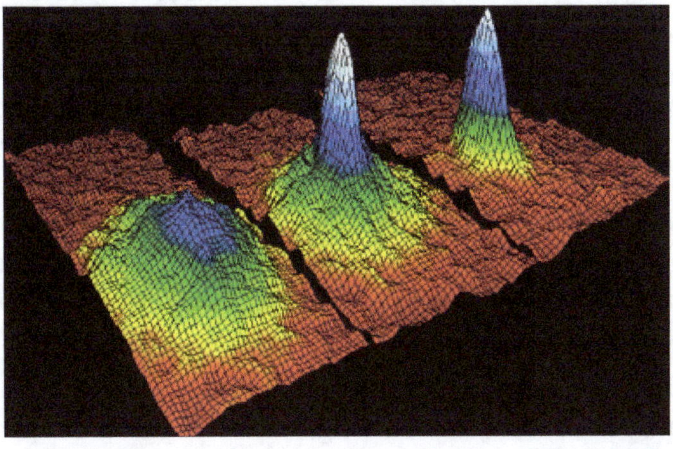

Fig 7: Graphic Illustrating the Formation of a Bose–Einstein Condensate

Quantities of atoms are found in different states of matter that depend on the physical conditions, such as temperature and pressure. By varying the conditions, materials can transition between solids, liquids, gases and plasmas. Within a state, a material can also exist in different allotropes. An example of this is solid carbon, which can exist as graphite or diamond. Gaseous allotropes exist as well, such as dioxygen and ozone. At temperatures close to absolute zero, atoms can form a Bose–Einstein condensate, at which point quantum mechanical effects, which are normally only observed at the atomic scale, become apparent on a macroscopic scale. This super-cooled collection of atoms then behaves as a single super atom, which may allow fundamental checks of quantum mechanical behavior.

1.1.4 Identification

While atoms are too small to be seen, devices such as the scanning tunneling microscope (STM) enable their visualization at the surfaces of solids. The microscope uses the quantum tunneling phenomenon, which allows particles to pass through a barrier that would be insurmountable in the classical perspective. Electrons tunnel through the vacuum between two biased electrodes, providing a tunneling current that is exponentially dependent on their separation. One electrode is a sharp tip ideally ending with a single atom. At each point of the scan of the surface the tip's height is adjusted so as to keep the tunneling current at a set value. How much the tip moves to and away from the surface is interpreted as the height profile. For low bias, the microscope images the averaged electron orbitals across closely packed energy levels—the local density of the electronic states near the Fermi level. Because of the distances involved, both electrodes need to be extremely stable; only then periodicities can be observed that correspond to individual atoms. The method alone is not chemically specific, and cannot identify the atomic species present at the surface.

Atoms can be easily identified by their mass. If an atom is ionized by removing one of its electrons, its trajectory when it passes through a magnetic field will bend. The radius by which the trajectory of a moving ion is turned by the magnetic field is determined by the mass of the atom. The mass spectrometer uses this principle to measure the mass-to-charge ratio of ions. If a sample contains multiple isotopes, the mass spectrometer can determine the proportion of each isotope in the sample by measuring the intensity of the different beams of ions. Techniques to vaporize atoms include inductively coupled plasma atomic emission spectroscopy and inductively coupled plasma mass spectrometry, both of which use a plasma to vaporize samples for analysis.

The atom-probe tomograph has sub-nanometer resolution in 3-D and can chemically identify individual atoms using time-of-flight mass spectrometry.

Electron emission techniques such as X-ray photoelectron spectroscopy (XPS) and Auger electron spectroscopy (AES), which measure the binding energies of the core electrons, are used to identify the atomic species present in a sample in a non-destructive way. With proper focusing both can be made area-specific. Another such method is electron energy loss spectroscopy (EELS), which measures the energy loss of an electron beam within a transmission electron microscope when it interacts with a portion of a sample.

Spectra of excited states can be used to analyze the atomic composition of distant stars. Specific light wavelengths contained in the observed light from stars can be separated out and related to the quantized transitions in free gas atoms. These colors can be replicated using a gas-discharge lamp containing the same element.

Helium was discovered in this way in the spectrum of the Sun 23 years before it was found on Earth.

1.1.5 Origin and Current State

Baryonic matter forms about 4% of the total energy density of the observable universe, with an average density of about 0.25 particles/m^3 (mostly protons and electrons). Within a galaxy such as the Milky Way, particles have a much higher concentration, with the density of matter in the interstellar medium (ISM) ranging from 10^5 to 10^9 atoms/m^3. The Sun is believed to be inside the Local Bubble, so the density in the solar neighborhood is only about 10^3 atoms/m^3. Stars form from dense clouds in the ISM, and the evolutionary processes of stars result in the steady enrichment of the ISM with elements more massive than hydrogen and helium.

Up to 95% of the Milky Way's baryonic matter are concentrated inside stars, where conditions are unfavorable for atomic matter. The total baryonic mass is about 10% of the mass of the galaxy; the remainder of the mass is an unknown dark matter. High temperature inside stars makes most "atoms" fully ionized, that is, separates *all* electrons from the nuclei. In stellar remnants—with exception of their surface layers—an immense pressure make electron shells impossible.

Formation

Electrons are thought to exist in the Universe since early stages of the Big Bang. Atomic nuclei forms in nucleosynthesis reactions. In about three minutes Big Bang nucleosynthesis produced most of the helium, lithium, and deuterium in the Universe, and perhaps some of the beryllium and boron.

Ubiquitousness and stability of atoms relies on their binding energy, which means that an atom has a lower energy than an unbound system of the nucleus and electrons. Where the temperature is much higher than ionization potential, the matter exists in the form of plasma—a gas of positively charged ions (possibly, bare nuclei) and electrons. When the temperature drops below the ionization potential, atoms become statistically favorable. Atoms (complete with bound electrons) became to dominate over charged particles 380,000 years after the Big Bang—an epoch called recombination, when the expanding Universe cooled enough to allow electrons to become attached to nuclei.

Since the Big Bang, which produced no carbon or heavier elements, atomic nuclei have been combined in stars through the process of nuclear fusion to produce more of the element helium, and (via the triple-alpha process) the sequence of elements from carbon up to iron.

Isotopes such as lithium-6, as well as some beryllium and boron are generated in space through cosmic ray spallation. This occurs when a high-energy proton strikes an atomic nucleus, causing large numbers of nucleons to be ejected.

Elements heavier than iron were produced in supernovae and colliding neutron stars through the r-process, and in AGB stars through the s-process, both of which involve the capture of neutrons by atomic nuclei. Elements such as lead formed largely through the radioactive decay of heavier elements.

Earth

Most of the atoms that make up the Earth and its inhabitants were present in their current form in the nebula that collapsed out of a molecular cloud to form the Solar System. The rest are the result of radioactive decay, and their relative proportion can be used to determine the age of the Earth through radiometric dating. Most of the helium in the crust of the Earth (about 99% of the helium from gas wells, as shown by its lower abundance of helium-3) is a product of alpha decay.

There are a few trace atoms on Earth that were not present at the beginning (i.e., not "primordial"), nor are results of radioactive decay. Carbon-14 is continuously generated by cosmic rays in the atmosphere. Some atoms on Earth have been artificially generated either deliberately or as by-products of nuclear reactors or explosions. Of the transuranic elements—those with atomic numbers greater than 92—only plutonium and neptunium occur naturally on Earth. Transuranic elements have radioactive lifetimes shorter than the current age of the Earth and thus identifiable quantities of these elements have long since decayed, with the exception of traces of plutonium-244 possibly deposited by cosmic dust. Natural deposits of plutonium and neptunium are produced by neutron capture in uranium ore.

The Earth contains approximately 1.33×10^{50} atoms. Although small numbers of independent atoms of noble gases exist, such as argon, neon, and helium, 99% of the atmosphere is bound in the form of molecules, including carbon dioxide and diatomic oxygen and nitrogen. At the surface of the Earth, an overwhelming majority of atoms combine to form various compounds, including water, salt, silicates and oxides. Atoms can also combine to create materials that do not consist of discrete molecules, including crystals and liquid or solid metals. This atomic matter forms networked arrangements that lack the particular type of small-scale interrupted order associated with molecular matter.

Rare and Theoretical Forms

Superheavy Elements

All nuclides with atomic numbers higher than 82 (lead) are known to be radioactive. No nuclide with an atomic number exceeding 92 (uranium) exists on Earth as a primordial nuclide, and heavier elements generally have shorter half-lives. Nevertheless, an "island of stability" encompassing relatively long-lived isotopes of superheavy elements with atomic numbers 110 to 114 might exist. Predictions for the half-life of the most stable nuclide on the island range from a few minutes to millions of years. In any case, superheavy elements (with $Z > 104$) would not exist due to increasing Coulomb repulsion (which results in spontaneous fission with increasingly short half-lives) in the absence of any stabilizing effects.

Exotic Matter

Each particle of matter has a corresponding antimatter particle with the opposite electrical charge. Thus, the positron is a positively charged antielectron and the antiproton is a negatively charged equivalent of a proton. When a matter and corresponding antimatter particle meet, they annihilate each other. Because of this, along with an imbalance between the number of matter and antimatter particles, the latter are rare in the universe. The first causes of this imbalance are not yet fully understood, although theories of baryogenesis may offer an explanation. As a result, no antimatter atoms have been discovered in nature. In 1996, the antimatter counterpart of the hydrogen atom (antihydrogen) was synthesized at the CERN laboratory in Geneva.

Other exotic atoms have been created by replacing one of the protons, neutrons or electrons with other particles that have the same charge. For example, an electron can be replaced by a more massive muon, forming a muonic atom. These types of atoms can be used to test fundamental predictions of physics.

1.2 PERIODIC TABLE

The periodic table, also known as the periodic table of the elements, is an ordered arrangement of the chemical elements into rows ("periods") and columns ("groups"). It is an icon of chemistry and is widely used in physics and other sciences. It is a depiction of the periodic law, which states that when the elements are arranged in order of their atomic numbers an approximate recurrence of their properties is evident. The table is divided into four roughly rectangular areas called blocks. Elements in the same group tend to show similar chemical characteristics.

Vertical, horizontal and diagonal trends characterize the periodic table. Metallic character increases going down a group and from right to left across a period. Nonmetallic character increases going from the bottom left of the periodic table to the top right.

The first periodic table to become generally accepted was that of the Russian chemist Dmitri Mendeleev in 1869; he formulated the periodic law as a dependence of chemical properties on atomic mass. As not all elements were then known, there were gaps in his periodic table, and Mendeleev successfully used the periodic law to predict some properties of some of the missing elements. The periodic law was recognized as a fundamental discovery in the late 19th century. It was explained early in the 20th century, with the discovery of atomic numbers and associated pioneering work in quantum mechanics, both ideas serving to illuminate the internal structure of the atom. A recognisably modern form of the table was reached in 1945 with Glenn T. Seaborg's discovery that the actinides were in fact f-block rather than d-block elements. The periodic table and law are now a central and indispensable part of modern chemistry.

The periodic table continues to evolve with the progress of science. In nature, only elements up to atomic number 94 exist; to go further, it was necessary to synthesize new elements in the laboratory. By 2010, the first 118 elements were known, thereby completing the first seven rows of the table; however, chemical characterization is still needed for the heaviest elements to confirm that their properties match their positions. New discoveries will extend the table beyond these seven rows, though it is not yet known how many more elements are possible; moreover, theoretical calculations suggest that this unknown region will not follow the patterns of the known part of the table. Some scientific discussion also continues regarding whether some elements are correctly positioned in today's table. Many alternative representations of the periodic law exist, and there is some discussion as to whether there is an optimal form of the periodic table.

1.2.1 Structure

Each chemical element has a unique atomic number (Z— for "Zahl", German for "number") representing the number of protons in its nucleus. Each distinct atomic number therefore corresponds to a class of atom: these classes are called the chemical elements.

The chemical elements are what the periodic table classifies and organizes. Hydrogen is the element with atomic number 1; helium, atomic number 2; lithium, atomic number 3; and so on. Each of these names can be further abbreviated by a one- or two-letter chemical symbol; those for hydrogen, helium, and lithium are respectively H, He, and Li. Neutrons do not affect the atom's chemical identity, but do affect its weight.

Atoms with the same number of protons but different numbers of neutrons are called isotopes of the same chemical element. Naturally occurring elements usually occur as mixes of different isotopes; since each isotope usually occurs with a characteristic abundance, naturally occurring elements have well-defined atomic weights, defined as the average mass of a naturally occurring atom of that element.

All elements have multiple isotopes, variants with the same number of protons but different numbers of neutrons. For example, carbon has three naturally occurring isotopes: all of its atoms have six protons and most have six neutrons as well, but about one per cent have seven neutrons, and a very small fraction have eight neutrons. Isotopes are never separated in the periodic table; they are always grouped together under a single element. When atomic mass is shown, it is usually the weighted average of naturally occurring isotopes; but if no isotopes occur naturally in significant quantities, the mass of the most stable isotope usually appears, often in parentheses.

In the standard periodic table, the elements are listed in order of increasing atomic number. A new row (*period*) is started when a new electron shell has its first electron. Columns (*groups*) are determined by the electron configuration of the atom; elements with the same number of electrons in a particular subshell fall into the same columns (e.g. oxygen, sulfur, and selenium are in the same column because they all have four electrons in the outermost p-subshell). Elements with similar chemical properties generally fall into the same group in the periodic table, although in the f-block, and to some respect in the d-block, the elements in the same period tend to have similar properties, as well. Thus, it is relatively easy to predict the chemical properties of an element if one knows the properties of the elements around it.

Today, 118 elements are known, the first 94 of which are known to occur naturally on Earth at present. The remaining 24, americium to oganesson (95–118), occur only when synthesized in laboratories. Of the 94 naturally occurring elements, 83 are primordial and 11 occur only in decay chains of primordial elements. A few of the latter are so rare that they were not discovered in nature, but were synthesized in the laboratory before it was determined that they do exist in nature after all: technetium (element 43), promethium (element 61), astatine (element 85), neptunium (element 93), and plutonium (element 94). No element heavier than einsteinium (element 99) has ever been observed in macroscopic quantities in its pure form, nor has astatine; francium (element 87) has been only photographed in the form of light emitted from microscopic quantities (300,000 atoms). Of the 94 natural elements, eighty have a stable isotope and one more (bismuth) has an almost-stable isotope (with a half-life of 2.01×10^{19} years, over a billion times the age of the universe).

Two more, thorium and uranium, have isotopes undergoing radioactive decay with a half-life comparable to the age of the Earth. The stable elements plus bismuth, thorium, and uranium make up the 83 primordial elements that survived from the Earth formation. The remaining eleven natural elements decay quickly enough that their continued trace occurrence rests primarily on being constantly regenerated as intermediate products of the decay of thorium and uranium. All 24 known artificial elements are radioactive.

Group Names and Numbers

Under an international naming convention, the groups are numbered numerically from 1 to 18 from the leftmost column (the alkali metals) to the rightmost column (the noble gases). The f-block groups are ignored in this numbering. Groups can also be named by their first element, e.g. the "scandium group" for group 3. Previously, groups were known by Roman numerals. In the United States, the Roman numerals were followed by either an "A" if the group was in the s- or p-block, or a "B" if the group was in the d-block. The Roman numerals used correspond to the last digit of today's naming convention (e.g. the group 4 elements were group IVB, and the group 14 elements were group IVA). In Europe, the lettering was similar, except that "A" was used for groups 1 through 7, and "B" was used for groups 11 through 17. In addition, groups 8, 9 and 10 used to be treated as one triple-sized group, known collectively in both notations as group VIII. In 1988, the new IUPAC (International Union of Pure and Applied Chemistry) naming system (1–18) was put into use, and the old group names (I–VIII) were deprecated.

Presentation Forms

For reasons of space, the periodic table is commonly presented with the f-block elements cut out and positioned as a distinct part below the main body. This reduces the number of element columns from 32 to 18.

Both forms represent the same periodic table. The form with the f-block included in the main body is sometimes called the 32-column or long form; the form with the f-block cut out the 18-column or medium-long form. The 32-column form has the advantage of showing all elements in their correct sequence, but it has the disadvantage of requiring more space. The form chosen is an editorial choice, and does not imply any change of scientific claim or statement. For example, when discussing the composition of group 3, the options can be shown equally (unprejudiced) in both forms.

Periodic tables usually at least show the elements' symbols; many also provide supplementary information about the elements, either via colour-coding or as data in the cells. The above table shows the names and atomic numbers of the elements, and also their blocks, natural occurrences and standard atomic weights. For the short-lived elements without standard atomic weights, the mass number of the most stable known isotope is used instead. Other tables may include properties such as state of matter, melting and boiling points, densities, as well as provide different classifications of the elements.

Electron Configurations

The periodic table is a graphic description of the periodic law, which states that the properties and atomic structures of the chemical elements are a periodic function of their atomic number. Elements are placed in the periodic table according to their electron configurations, the periodic recurrences of which explain the trends in properties across the periodic table.

An electron can be thought of as inhabiting an atomic orbital, which characterizes the probability it can be found in any particular region around the atom. Their energies are quantised, which is to say that they can only take discrete values. Furthermore, electrons obey the Pauli exclusion principle: different electrons must always be in different states.

This allows classification of the possible states an electron can take in various energy levels known as shells, divided into individual subshells, which each contain one or more orbitals. Each orbital can contain up to two electrons: they are distinguished by a quantity known as spin, conventionally labelled "up" or "down". In a cold atom (one in its ground state), electrons arrange themselves in such a way that the total energy they have is minimized by occupying the lowest-energy orbitals available. Only the outermost electrons (so-called valence electrons) have enough energy to break free of the nucleus and participate in chemical reactions with other atoms. The others are called core electrons.

$\ell =$	0	1	2	3	4	5	6	**Shell capacity ($2n^2$)**
Orbital	s	p	d	f	g	h	i	
$n = 1$	1s							2
$n = 2$	2s	2p						8
$n = 3$	3s	3p	3d					18
$n = 4$	4s	4p	4d	4f				32
$n = 5$	5s	5p	5d	5f	5g			50
$n = 6$	6s	6p	6d	6f	6g	6h		72
$n = 7$	7s	7p	7d	7f	7g	7h	7i	98
Subshell capacity ($4\ell+2$)	2	6	10	14	18	22	26	

Elements are known with up to the first seven shells occupied. The first shell contains only one orbital, a spherical s orbital. As it is in the first shell, this is called the 1s orbital. This can hold up to two electrons. The second shell similarly contains a 2s orbital, and it also contains three dumbbell-shaped 2p orbitals, and can thus fill up to eight electrons ($2\times1 + 2\times3 = 8$). The third shell contains one 3s orbital, three 3p orbitals, and five 3d orbitals, and thus has a capacity of $2\times1 + 2\times3 + 2\times5 = 18$. The fourth shell contains one 4s orbital, three 4p orbitals, five 4d orbitals, and seven 4f orbitals, thus leading to a capacity of $2\times1 + 2\times3 + 2\times5 + 2\times7 = 32$. Higher shells contain more types of orbitals that continue the pattern, but such types of orbitals are not filled in the ground states of known elements. The subshell types are characterized by the quantum numbers. Four numbers describe an orbital in an atom completely: the principal quantum number n, the azimuthal quantum number l (the orbital type), the orbital magnetic quantum number m_l, and the spin magnetic quantum number m_s.

Order of Subshell Filling

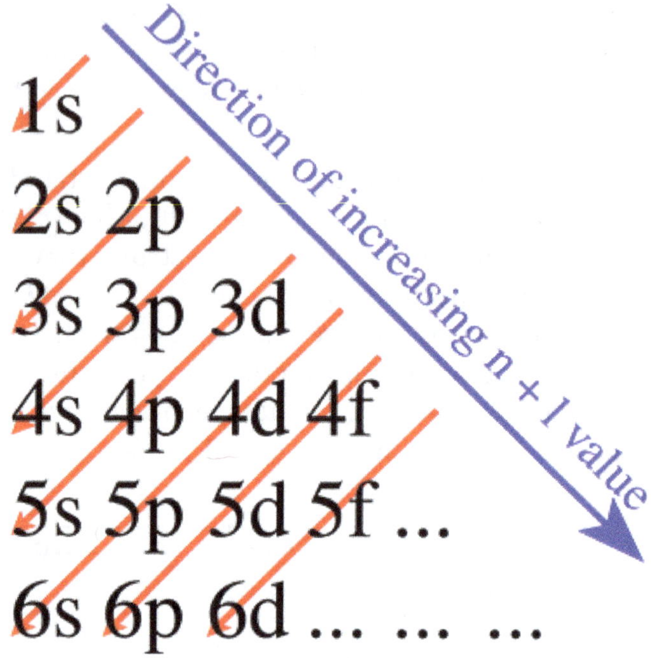

Fig 8: Idealized Order of Subshell Filling According to the Madelung Rule

The sequence in which the subshells are filled is given in most cases by the Aufbau principle, also known as the Madelung or Klechkovsky rule (after Erwin Madelung and Vsevolod Klechkovsky respectively). This rule was first observed empirically by Madelung, and Klechkovsky and later authors gave it theoretical justification. The shells overlap in energies, and the Madelung rule specifies the sequence of filling according to:

- 1s ≪ 2s < 2p ≪ 3s < 3p ≪ 4s < 3d < 4p ≪ 5s < 4d < 5p ≪ 6s < 4f < 5d < 6p ≪ 7s < 5f < 6d < 7p ≪ ...

Here the sign ≪ means "much less than" as opposed to < meaning just "less than". Phrased differently, electrons enter orbitals in order of increasing $n + \ell$, and if two orbitals are available with the same value of $n + \ell$, the one with lower n is occupied first. In general, orbitals with the same value of $n + \ell$ are similar in energy, but in the case of the s-orbitals (with $\ell = 0$), quantum effects raise their energy to approach that of the next $n + \ell$ group. Hence the periodic table is usually drawn to begin each row (often called a period) with the filling of a new s-orbital, which corresponds to the beginning of a new shell. Thus, with the exception of the first row, each period length appears twice:

- 2, 8, 8, 18, 18, 32, 32, ...

The overlaps get quite close at the point where the d-orbitals enter the picture, and the order can shift slightly with atomic number and atomic charge.

Starting from the simplest atom, this lets us build up the periodic table one at a time in order of atomic number, by considering the cases of single atoms. In hydrogen, there is only one electron, which must go in the lowest-energy orbital 1s. This electron configuration is written $1s^1$, where the superscript indicates the number of electrons in the subshell. Helium adds a second electron, which also goes into 1s, completely filling the first shell and giving the configuration $1s^2$.

Starting from the third element, lithium, the first shell is full, so its third electron occupies a 2s orbital, giving a $1s^2\ 2s^1$ configuration. The 2s electron is lithium's only valence electron, as the 1s subshell is now too tightly bound to the nucleus to participate in chemical bonding to other atoms: such a shell is called a "core shell". The 1s subshell is a core shell for all elements from lithium onward. The 2s subshell is completed by the next element beryllium ($1s^2\ 2s^2$). The following elements then proceed to fill the 2p subshell. Boron ($1s^2\ 2s^2\ 2p^1$) puts its new electron in a 2p orbital; carbon ($1s^2\ 2s^2\ 2p^2$) fills a second 2p orbital; and with nitrogen ($1s^2\ 2s^2\ 2p^3$) all three 2p orbitals become singly occupied. This is consistent with Hund's rule, which states that atoms usually prefer to singly occupy each orbital of the same type before filling them with the second electron. Oxygen ($1s^2\ 2s^2\ 2p^4$), fluorine ($1s^2\ 2s^2\ 2p^5$), and neon ($1s^2\ 2s^2\ 2p^6$) then complete the already singly filled 2p orbitals; the last of these fills the second shell completely.

Starting from element 11, sodium, the second shell is full, making the second shell a core shell for this and all heavier elements. The eleventh electron begins the filling of the third shell by occupying a 3s orbital, giving a configuration of $1s^2\ 2s^2\ 2p^6\ 3s^1$ for sodium. This configuration is abbreviated [Ne] $3s^1$, where [Ne] represents neon's configuration. Magnesium ([Ne] $3s^2$) finishes this 3s orbital, and the following six elements aluminium, silicon, phosphorus, sulfur, chlorine, and argon fill the three 3p orbitals ([Ne] $3s^2\ 3p^1$ through [Ne] $3s^2\ 3p^6$). This creates an analogous series in which the outer shell structures of sodium through argon are analogous to those of lithium through neon, and is the basis for the periodicity of chemical properties that the periodic table illustrates: at regular but changing intervals of atomic numbers, the properties of the chemical elements approximately repeat.

The first eighteen elements can thus be arranged as the start of a periodic table. Elements in the same column have the same number of valence electrons and have analogous valence electron configurations: these columns are called groups. The single exception is helium, which has two valence electrons like beryllium and magnesium, but is typically placed in the column of neon and argon to emphasise that its outer shell is full. (Some contemporary authors question even this single exception, preferring to consistently follow the valence configurations and place helium over beryllium.) There are eight columns in this periodic table fragment, corresponding to at most eight outer-shell electrons. A period begins when a new shell starts filling. Finally, the colouring illustrates the blocks: the elements in the s-block (coloured red) are filling s-orbitals, while those in the p-block (coloured yellow) are filling p-orbitals.

1 H							2 He	2×1 = 2 elements 1s
3 Li	4 Be	5 B	6 C	7 N	8 O	9 F	10 Ne	2×(1+3) = 8 elements 2s 2p
11 Na	12 Mg	13 Al	14 Si	15 P	16 S	17 Cl	18 Ar	2×(1+3) = 8 elements 3s 3p

Starting the next row, for potassium and calcium the 4s subshell is the lowest in energy, and therefore they fill it. Potassium adds one electron to the 4s shell ([Ar] $4s^1$), and calcium then completes it ([Ar] $4s^2$). However, starting from scandium ([Ar] $3d^1$ $4s^2$) the 3d subshell becomes the next highest in energy. The 4s and 3d subshells have approximately the same energy and they compete for filling the electrons, and so the occupation is not quite consistently filling the 3d orbitals one at a time.

The precise energy ordering of 3d and 4s changes along the row, and also changes depending on how many electrons are removed from the atom. For example, due to the repulsion between the 3d electrons and the 4s ones, at chromium the 4s energy level becomes slightly higher than 3d, and so it becomes more profitable for a chromium atom to have a [Ar] $3d^5$ $4s^1$ configuration than an [Ar] $3d^4$ $4s^2$ one. A similar anomaly occurs at copper, whose atom has a [Ar] $3d^{10}$ $4s^1$ configuration rather than the expected [Ar] $3d^9$ $4s^2$.

These are violations of the Madelung rule. Such anomalies, however, do not have any chemical significance: most chemistry is not about isolated gaseous atoms, and the various configurations are so close in energy to each other that the presence of a nearby atom can shift the balance. Therefore, the periodic table ignores them and considers only idealized configurations.

At zinc ([Ar] $3d^{10}$ $4s^2$), the 3d orbitals are completely filled with a total of ten electrons. Next come the 4p orbitals, completing the row, which are filled progressively by gallium ([Ar] $3d^{10}$ $4s^2$ $4p^1$) through krypton ([Ar] $3d^{10}$ $4s^2$ $4p^6$), in a manner analogous to the previous p-block elements.

From gallium onwards, the 3d orbitals form part of the electronic core, and no longer participate in chemistry. The s- and p-block elements, which fill their outer shells, are called main-group elements; the d-block elements (coloured blue below), which fill an inner shell, are called transition elements (or transition metals, since they are all metals).

The next eighteen elements fill the 5s orbitals (rubidium and strontium), then 4d (yttrium through cadmium, again with a few anomalies along the way), and then 5p (indium through xenon). Again, from indium onward the 4d orbitals are in the core. Hence the fifth row has the same structure as the fourth.

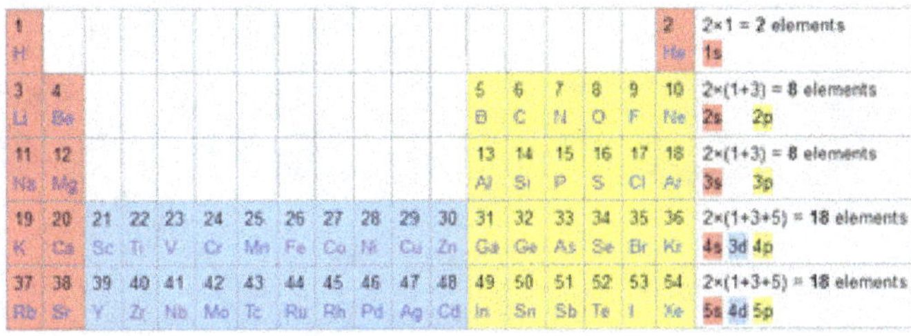

The sixth row of the table likewise starts with two s-block elements: caesium and barium. After this, the first f-block elements (coloured green below) begin to appear, starting with lanthanum. These are sometimes termed inner transition elements. As there are now not only 4f but also 5d and 6s subshells at similar energies, competition occurs once again with many irregular configurations; this resulted in some dispute about where exactly the f-block is supposed to begin, but most who study the matter agree that it starts at lanthanum in accordance with the Aufbau principle.

Even though lanthanum does not itself fill the 4f subshell as a single atom, because of repulsion between electrons, its 4f orbitals are low enough in energy to participate in chemistry. At ytterbium, the seven 4f orbitals are completely filled with fourteen electrons; thereafter, a series of ten transition elements (lutetium through mercury) follows, and finally six main-group elements (thallium through radon) complete the period. From lutetium onwards the 4f orbitals are in the core, and from thallium onwards so are the 5d orbitals.

The seventh row is analogous to the sixth row: 7s fills (francium and radium), then 5f (actinium to nobelium), then 6d (lawrencium to copernicium), and finally 7p (nihonium to oganesson). Starting from lawrencium the 5f orbitals are in the core, and probably the 6d orbitals join the core starting from nihonium. Again there are a few anomalies along the way: for example, as single atoms neither actinium nor thorium actually fills the 5f subshell, and lawrencium does not fill the 6d shell, but all these subshells can still become filled in chemical environments. For a very long time, the seventh row was incomplete as most of its elements do not occur in nature. The missing elements beyond uranium started to be synthesized in the laboratory in 1940, when neptunium was made. (However, the first element to be discovered by synthesis rather than in nature was technetium in 1937.) The row was completed with the synthesis of tennessine in 2010 (the last element oganesson had already been made in 2002), and the last elements in this seventh row were given names in 2016.

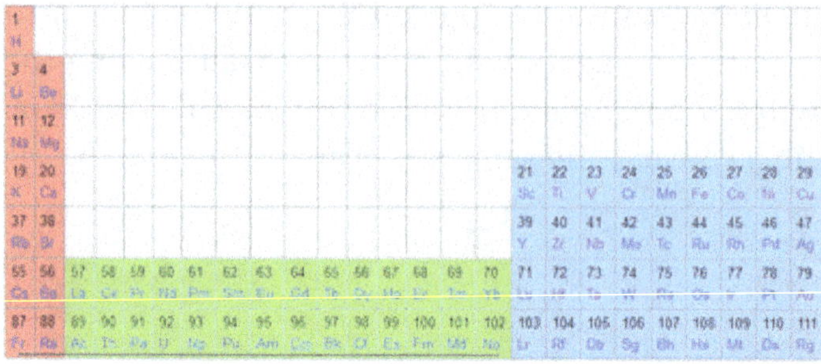

1.2.2 Variations

Period 1

Although the modern periodic table is standard today, the placement of the period 1 elements hydrogen and helium remains an open issue under discussion, and some variation can be found. Following their respective s^1 and s^2 electron configurations, hydrogen would be placed in group 1, and helium would be placed in group 2. The group 1 placement of hydrogen is common, but helium is almost always placed in group 18 with the other noble gases. The debate has to do with conflicting understandings of the extent to which chemical or electronic properties should decide periodic table placement.

Like the group 1 metals, hydrogen has one electron in its outermost shell and typically loses its only electron in chemical reactions. Hydrogen has some metal-like chemical properties, being able to displace some metals from their salts. But it forms a diatomic nonmetallic gas at standard conditions, unlike the alkali metals which are reactive solid metals. This and hydrogen's formation of hydrides, in which it gains an electron, brings it close to the properties of the halogens which do the same (though it is rarer for hydrogen to form H^- than H^+). Moreover, the lightest two halogens (fluorine and chlorine) are gaseous like hydrogen at standard conditions. Some properties of hydrogen are not a good fit for either group: hydrogen is neither highly oxidizing nor highly reducing and is not reactive with water. Hydrogen thus has properties corresponding to both those of the alkali metals and the halogens, but matches neither group perfectly, and is thus difficult to place by its chemistry. Therefore, while the electronic placement of hydrogen in group 1 predominates, some rarer arrangements show either hydrogen in group 17, duplicate hydrogen in both groups 1 and 17, or float it separately from all groups. This last option has nonetheless been criticized by the chemist and philosopher of science Eric Scerri on the grounds that it appears to imply that hydrogen is above the periodic law altogether, unlike all the other elements.

Helium is the only element that routinely occupies a position in the periodic table that is not consistent with its electronic structure. It has two electrons in its outermost shell, whereas the other noble gases have eight; and it is an s-block element, whereas all other noble gases are p-block elements. However it is unreactive at standard conditions, and has a full outer shell: these properties are like the noble gases in group 18, but not at all like the reactive alkaline earth metals of group 2. For these reasons helium is nearly universally placed in group 18 which its properties best match; a proposal to move helium to group 2 was rejected by IUPAC in 1988 for these reasons.

Nonetheless, helium is still occasionally placed in group 2 today, and some of its physical and chemical properties are closer to the group 2 elements and support the electronic placement. Solid helium crystallises in a hexagonal close-packed structure, which matches beryllium and magnesium in group 2, but not the other noble gases in group 18.

Recent theoretical developments in noble gas chemistry, in which helium is expected to show slightly less inertness than neon and to form $(HeO)(LiF)_2$ with a structure similar to the analogous beryllium compound (but with no expected neon analogue), have resulted in more chemists advocating a placement of helium in group 2. This relates to the electronic argument, as the reason for neon's greater inertness is repulsion from its filled p-shell that helium lacks, though realistically it is unlikely that helium-containing molecules will be stable outside extreme low-temperature conditions (around 10 K).

The first-row anomaly in the periodic table has additionally been cited to support moving helium to group 2. It arises because the first orbital of any type is unusually small, since unlike its higher analogues, it does not experience interelectronic repulsion from a smaller orbital of the same type. This makes the first row of elements in each block unusually small, and such elements tend to exhibit characteristic kinds of anomalies for their group.

Some chemists arguing for the repositioning of helium have pointed out that helium exhibits these anomalies if it is placed in group 2, but not if it is placed in group 18: on the other hand, neon, which would be the first group 18 element if helium was removed from that spot, does exhibit those anomalies. The relationship between helium and beryllium is then argued to resemble that between hydrogen and lithium, a placement which is much more commonly accepted. For example, because of this trend in the sizes of orbitals, a large difference in atomic radii between the first and second members of each main group is seen in groups 1 and 13–17: it exists between neon and argon, and between helium and beryllium, but not between helium and neon.

This similarly affects the noble gases' boiling points and solubilities in water, where helium is too close to neon, and the large difference characteristic between the first two elements of a group appears only between neon and argon. Moving helium to group 2 makes this trend consistent in groups 2 and 18 as well, by making helium the first group 2 element and neon the first group 18 element: both exhibit the characteristic properties of a kainosymmetric first element of a group. The group 18 placement of helium nonetheless remains near-universal due to its extreme inertness. Additionally, tables that float both hydrogen and helium outside all groups may rarely be encountered.

Group 3

In many periodic tables, the f-block is shifted one element to the right, so that lanthanum and actinium become d-block elements in group 3, and Ce–Lu and Th–Lr form the f-block. Thus the d-block is split into two very uneven portions. This is a holdover from early mistaken measurements of electron configurations; modern measurements are more consistent with the form with lutetium and lawrencium in group 3, and with La–Yb and Ac–No as the f-block.

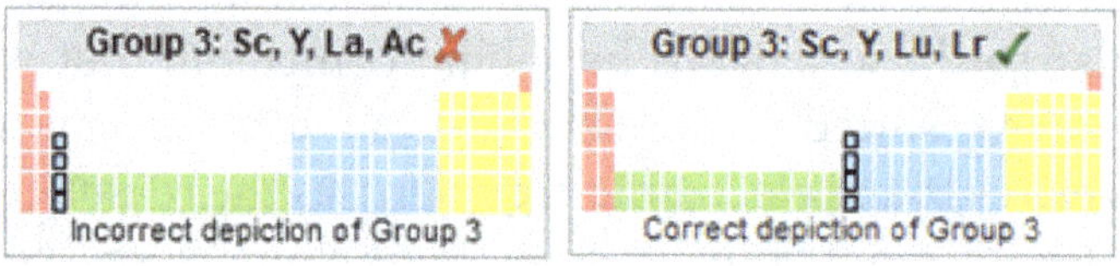

The 4f shell is completely filled at ytterbium, and for that reason Lev Landau and Evgeny Lifshitz in 1948 considered it incorrect to group lutetium as an f-block element. They did not yet take the step of removing lanthanum from the d-block as well, but Jun Kondō realized in 1963 that lanthanum's low-temperature superconductivity implied the activity of its 4f shell. In 1965, David C. Hamilton linked this observation to its position in the periodic table, and argued that the f-block should be composed of the elements La–Yb and Ac–No. Since then, physical, chemical, and electronic evidence has supported this assignment. The issue was brought to wide attention by William B. Jensen in 1982, and the reassignment of lutetium and lawrencium to group 3 was supported by IUPAC reports dating from 1988 (when the 1–18 group numbers were recommended) and 2021. The variation nonetheless still exists because most textbook writers are not aware of the issue.

A third form can sometimes be encountered in which the spaces below yttrium in group 3 are left empty, such as the table appearing on the IUPAC web site, but this creates an inconsistency with quantum mechanics by making the f-block 15 elements wide (La–Lu and Ac–Lr) even though only 14 electrons can fit in an f-subshell.

There is moreover some confusion in the literature on which elements are then implied to be in group 3. While the 2021 IUPAC report noted that 15-element-wide f-blocks are supported by some practitioners of a specialized branch of relativistic quantum mechanics focusing on the properties of superheavy elements, the project's opinion was that such interest-dependent concerns should not have any bearing on how the periodic table is presented to "the general chemical and scientific community". Other authors focusing on superheavy elements since clarified that the "15th entry of the f-block represents the first slot of the d-block which is left vacant to indicate the place of the f-block inserts", which would imply that this form still has lutetium and lawrencium (the 15th entries in question) as d-block elements in group 3. Indeed, when IUPAC publications expand the table to 32 columns, they make this clear and place lutetium and lawrencium under yttrium in group 3.

Several arguments in favour of Sc-Y-La-Ac can be encountered in the literature, but they have been challenged as being logically inconsistent. For example, it has been argued that lanthanum and actinium cannot be f-block elements because as individual gas-phase atoms, they have not begun to fill the f-subshells. But the same is true of thorium which is never disputed as an f-block element, and this argument overlooks the problem on the other end: that the f-shells complete filling at ytterbium and nobelium, matching the Sc-Y-Lu-Lr form, and not at lutetium and lawrencium as the Sc-Y-La-Ac form would have it.

Not only are such exceptional configurations in the minority, but they have also in any case never been considered as relevant for positioning any other elements on the periodic table: in gaseous atoms, the d-shells complete their filling at copper, palladium, and gold, but it is universally accepted by chemists that these configurations are exceptional and that the d-block really ends in accordance with the Madelung rule at zinc, cadmium, and mercury. The relevant fact for placement is that lanthanum and actinium (like thorium) have valence f-orbitals that can become occupied in chemical environments, whereas lutetium and lawrencium do not: their f-shells are in the core, and cannot be used for chemical reactions. Thus the relationship between yttrium and lanthanum is only a secondary relationship between elements with the same number of valence electrons but different kinds of valence orbitals, such as that between chromium and uranium; whereas the relationship between yttrium and lutetium is primary, sharing both valence electron count and valence orbital type.

1.2.3 Periodic Trends

As chemical reactions involve the valence electrons, elements with similar outer electron configurations may be expected to react similarly and form compounds with similar proportions of elements in them. Such elements are placed in the same group, and thus there tend to be clear similarities and trends in chemical behaviour as one proceeds down a group. As analogous configurations occur at regular intervals, the properties of the elements thus exhibit periodic recurrences, hence the name of the periodic table and the periodic law. These periodic recurrences were noticed well before the underlying theory that explains them was developed.

Atomic Radius

Historically, the physical size of atoms was unknown until the early 20th century. The first calculated estimate of the atomic radius of hydrogen was published by physicist Arthur Haas in 1910 to within an order of magnitude (a factor of 10) of the accepted value, the Bohr radius (~0.529 Å). In his model, Haas used a single-electron configuration based on the classical atomic model proposed by J. J. Thomson in 1904, often called the plum-pudding model.

Atomic radii (the size of atoms) are dependent on the sizes of their outermost orbitals. They generally decrease going left to right along the main-group elements, because the nuclear charge increases but the outer electrons are still in the same shell. However, going down a column, the radii generally increase, because the outermost electrons are in higher shells that are thus further away from the nucleus. The first row of each block is abnormally small, due to an effect called kainosymmetry or primogenic repulsion: the 1s, 2p, 3d, and 4f subshells have no inner analogues.

For example, the 2p orbitals do not experience strong repulsion from the 1s and 2s orbitals, which have quite different angular charge distributions, and hence are not very large; but the 3p orbitals experience strong repulsion from the 2p orbitals, which have similar angular charge distributions. Thus higher s-, p-, d-, and f-subshells experience strong repulsion from their inner analogues, which have approximately the same angular distribution of charge, and must expand to avoid this.

This makes significant differences arise between the small 2p elements, which prefer multiple bonding, and the larger 3p and higher p-elements, which do not. Similar anomalies arise for the 1s, 2p, 3d, 4f, and the hypothetical 5g elements: the degree of this first-row anomaly is highest for the s-block, is moderate for the p-block, and is less pronounced for the d- and f-blocks.

In the transition elements, an inner shell is filling, but the size of the atom is still determined by the outer electrons. The increasing nuclear charge across the series and the increased number of inner electrons for shielding somewhat compensate each other, so the decrease in radius is smaller.

The 4p and 5d atoms, coming immediately after new types of transition series are first introduced, are smaller than would have been expected, because the added core 3d and 4f subshells provide only incomplete shielding of the nuclear charge for the outer electrons.

Hence for example gallium atoms are slightly smaller than aluminium atoms. Together with kainosymmetry, this results in an even-odd difference between the periods (except in the s-block) that is sometimes known as secondary periodicity:

elements in even periods have smaller atomic radii and prefer to lose fewer electrons, while elements in odd periods (except the first) differ in the opposite direction. Thus for example many properties in the p-block show a zigzag rather than a smooth trend along the group. For example, phosphorus and antimony in odd periods of group 15 readily reach the +5 oxidation state, whereas nitrogen, arsenic, and bismuth in even periods prefer to stay at +3. A similar situation holds for the d-block, with lutetium through tungsten atoms being slightly smaller than yttrium through molybdenum atoms respectively.

Thallium and lead atoms are about the same size as indium and tin atoms respectively, but from bismuth to radon the 6p atoms are larger than the analogous 5p atoms. This happens because when atomic nuclei become highly charged, special relativity becomes needed to gauge the effect of the nucleus on the electron cloud. These relativistic effects result in heavy elements increasingly having differing properties compared to their lighter homologues in the periodic table. Spin-orbit interaction splits the p-subshell: one p-orbital is relativistically stabilized and shrunken (it fills in thallium and lead), but the other two (filling in bismuth through radon) are relativistically destabilized and expanded. Relativistic effects also explain why gold is golden and mercury is a liquid at room temperature. They are expected to become very strong in the late seventh period, potentially leading to a collapse of periodicity.

Electron configurations are only clearly known until element 108 (hassium), and experimental chemistry beyond 108 has only been done for 112 (copernicium), 113 (nihonium), and 114 (flerovium), so the chemical characterization of the heaviest elements remains a topic of current research.

The trend that atomic radii decrease from left to right is also present in ionic radii, though it is more difficult to examine because the most common ions of consecutive elements normally differ in charge. Ions with the same electron configuration decrease in size as their atomic number rises, due to increased attraction from the more positively charged nucleus: thus for example ionic radii decrease in the series Se^{2-}, Br^{-}, Rb^{+}, Sr^{2+}, Y^{3+}, Zr^{4+}, Nb^{5+}, Mo^{6+}, Tc^{7+}. Ions of the same element get smaller as more electrons are removed, because the attraction from the nucleus begins to outweigh the repulsion between electrons that causes electron clouds to expand: thus for example ionic radii decrease in the series V^{2+}, V^{3+}, V^{4+}, V^{5+}.

Ionisation Energy

The first ionisation energy of an atom is the energy required to remove an electron from it. This varies with the atomic radius: ionisation energy increases left to right and down to up, because electrons that are closer to the nucleus are held more tightly and are more difficult to remove. Ionisation energy thus is minimized at the first element of each period – hydrogen and the alkali metals – and then generally rises until it reaches the noble gas at the right edge of the period. There are some exceptions to this trend, such as oxygen, where the electron being removed is paired and thus interelectronic repulsion makes it easier to remove than expected.

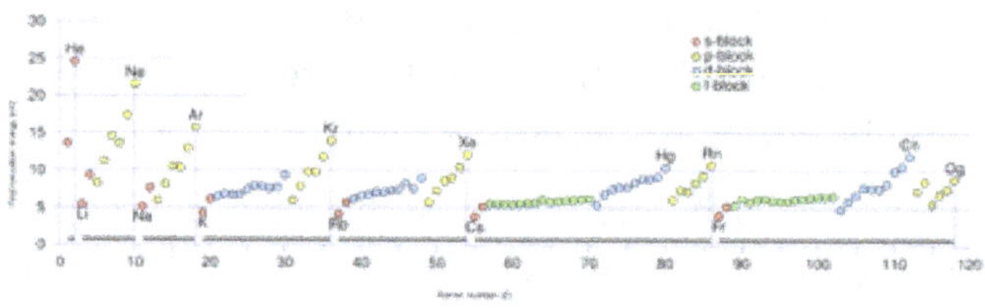

Fig 9: Graph of First Ionisation Energies of the Elements in Electronvolts (Predictions Used for Elements 109–118)

In the transition series, the outer electrons are preferentially lost even though the inner orbitals are filling. For example, in the 3d series, the 4s electrons are lost first even though the 3d orbitals are being filled. The shielding effect of adding an extra 3d electron approximately compensates the rise in nuclear charge, and therefore the ionisation energies stay mostly constant, though there is a small increase especially at the end of each transition series.

As metal atoms tend to lose electrons in chemical reactions, ionisation energy is generally correlated with chemical reactivity, although there are other factors involved as well.

Electron Affinity

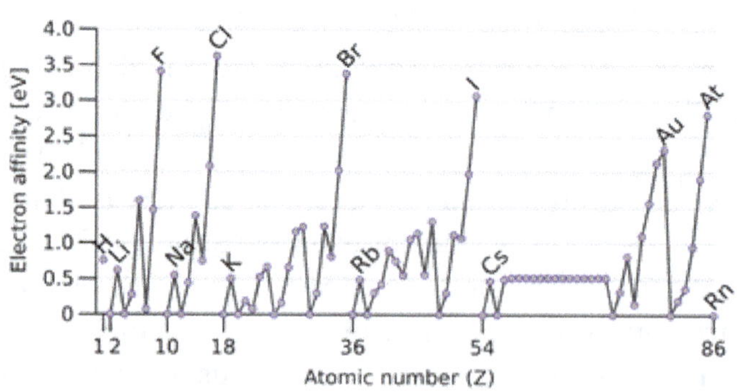

Fig 10: Trend in Electron Affinities

The opposite property to ionisation energy is the electron affinity, which is the energy released when adding an electron to the atom. A passing electron will be more readily attracted to an atom if it feels the pull of the nucleus more strongly, and especially if there is an available partially filled outer orbital that can accommodate it. Therefore, electron affinity tends to increase down to up and left to right. The exception is the last column, the noble gases, which have a full shell and have no room for another electron. This gives the halogens in the next-to-last column the highest electron affinities.

Some atoms, like the noble gases, have no electron affinity: they cannot form stable gas-phase anions. (They can form metastable resonances if the incoming electron arrives with enough kinetic energy, but these inevitably and rapidly autodetach: for example, the lifetime of the most long-lived He^- level is about 359 microseconds.) The noble gases, having high ionisation energies and no electron affinity, have little inclination towards gaining or losing electrons and are generally unreactive.

Some exceptions to the trends occur: oxygen and fluorine have lower electron affinities than their heavier homologues sulfur and chlorine, because they are small atoms and hence the newly added electron would experience significant repulsion from the already present ones. For the nonmetallic elements, electron affinity likewise somewhat correlates with reactivity, but not perfectly since other factors are involved. For example, fluorine has a lower electron affinity than chlorine (because of extreme interelectronic repulsion for the very small fluorine atom), but is more reactive.

Valence and Oxidation States

The valence of an element can be defined either as the number of hydrogen atoms that can combine with it to form a simple binary hydride, or as twice the number of oxygen atoms that can combine with it to form a simple binary oxide (that is, not a peroxide or a superoxide).

The valences of the main-group elements are directly related to the group number: the hydrides in the main groups 1–2 and 13–17 follow the formulae MH, MH_2, MH_3, MH_4, MH_3, MH_2, and finally MH.

The highest oxides instead increase in valence, following the formulae M_2O, MO, M_2O_3, MO_2, M_2O_5, MO_3, M_2O_7. Today the notion of valence has been extended by that of the oxidation state, which is the formal charge left on an element when all other elements in a compound have been removed as their ions.

The electron configuration suggests a ready explanation from the number of electrons available for bonding; indeed, the number of valence electrons starts at 1 in group 1, and then increases towards the right side of the periodic table, only resetting at 3 whenever each new block starts.

Thus in period 6, Cs–Ba have 1–2 valence electrons; La–Yb have 3–16; Lu–Hg have 3–12; and Tl–Rn have 3–8. However, towards the right side of the d- and f-blocks, the theoretical maximum corresponding to using all valence electrons is not achievable at all; the same situation affects oxygen, fluorine, and the light noble gases up to krypton.

Table 1: Number of Valence Electrons

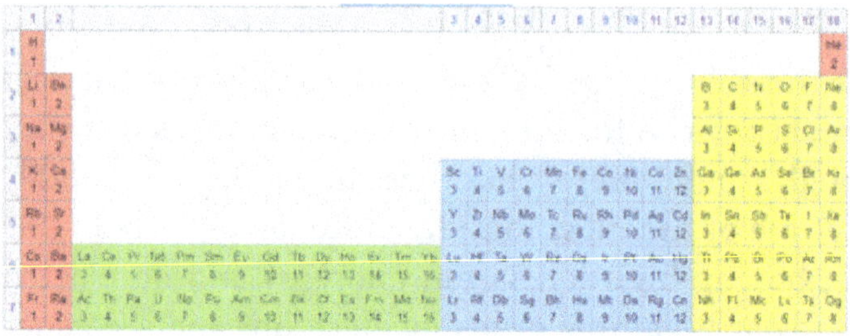

A full explanation requires considering the energy that would be released in forming compounds with different valences rather than simply considering electron configurations alone. For example, magnesium forms Mg^{2+} rather than Mg^{+} cations when dissolved in water, because the latter would spontaneously disproportionate into Mg^{0} and Mg^{2+} cations. This is because the enthalpy of hydration (surrounding the cation with water molecules) increases in magnitude with the charge and radius of the ion. In Mg^{+}, the outermost orbital (which determines ionic radius) is still 3s, so the hydration enthalpy is small and insufficient to compensate the energy required to remove the electron; but ionizing again to Mg^{2+} uncovers the core 2p subshell, making the hydration enthalpy large enough to allow magnesium(II) compounds to form. For similar reasons, the common oxidation states of the heavier p-block elements (where the ns electrons become lower in energy than the np) tend to vary by steps of 2, because that is necessary to uncover an inner subshell and decrease the ionic radius (e.g. Tl^{+} uncovers 6s, and Tl^{3+} uncovers 5d, so once thallium loses two electrons it tends to lose the third one as well). Analogous arguments based on orbital hybridization can be used for the less electronegative p-block elements.

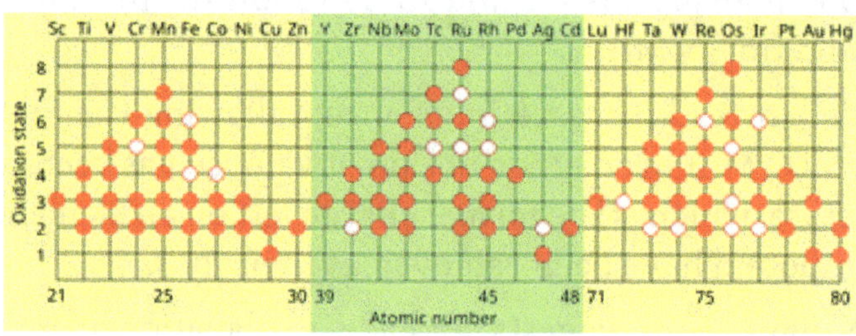

Fig 11: Oxidation States of the Transition Metals. The Solid Dots Show Common Oxidation States, and the Hollow Dots Show Possible but Unlikely States

For transition metals, common oxidation states are nearly always at least +2 for similar reasons (uncovering the next subshell); this holds even for the metals with anomalous $d^{x+1}s^{1}$ or $d^{x+2}s^{0}$ configurations (except for silver), because repulsion between d-electrons means that the movement of the second electron from the s- to the d-subshell does not appreciably change its ionisation energy. Because ionizing the transition metals further does not uncover any new inner subshells, their oxidation states tend to vary by steps of 1 instead. The lanthanides and late actinides generally show a stable +3 oxidation state, removing the outer s-electrons and then (usually) one electron from the (n-2)f-orbitals, that are similar in energy to ns. The common and maximum oxidation states of the d- and f-block elements tend to depend on the ionisation energies. As the energy difference between the (n-1)d and ns orbitals rises along each transition series, it becomes less energetically favourable to ionize further electrons.

Thus, the early transition metal groups tend to prefer higher oxidation states, but the +2 oxidation state becomes more stable for the late transition metal groups. The highest formal oxidation state thus increases from +3 at the beginning of each d-block row, to +7 or +8 in the middle (e.g. OsO_4), and then decrease to +2 at the end. The lanthanides and late actinides usually have high fourth ionisation energies and hence rarely surpass the +3 oxidation state, whereas early actinides have low fourth ionisation energies and so for example neptunium and plutonium can reach +7. The very last actinides go further than the lanthanides towards low oxidation states: mendelevium is more easily reduced to the +2 state than thulium or even europium (the lanthanide with the most stable +2 state, on account of its half-filled f-shell), and nobelium outright favours +2 over +3, in contrast to ytterbium.

As elements in the same group share the same valence configurations, they usually exhibit similar chemical behaviour. For example, the alkali metals in the first group all have one valence electron, and form a very homogeneous class of elements: they are all soft and reactive metals. However, there are many factors involved, and groups can often be rather heterogeneous. For instance, hydrogen also has one valence electron and is in the same group as the alkali metals, but its chemical behaviour is quite different. The stable elements of group 14 comprise a nonmetal (carbon), two semiconductors (silicon and germanium), and two metals (tin and lead); they are nonetheless united by having four valence electrons. This often leads to similarities in maximum and minimum oxidation states (e.g. sulfur and selenium in group 16 both have maximum oxidation state +6, as in SO_3 and SeO_3, and minimum oxidation state -2, as in sulfides and selenides); but not always (e.g. oxygen is not known to form oxidation state +6, despite being in the same group as sulfur and selenium).

Electronegativity

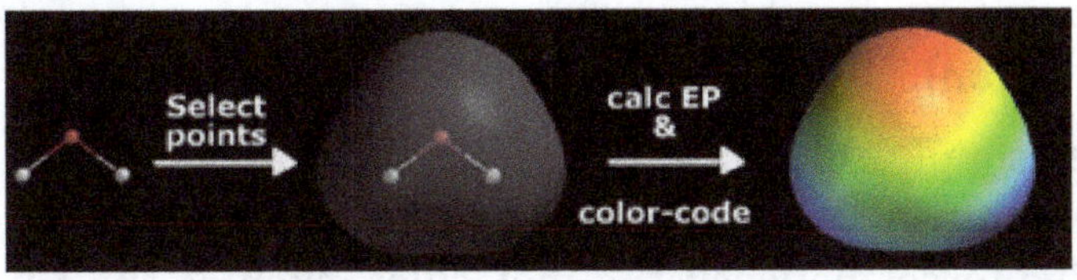

Fig 12: Electrostatic Potential Map of a Water Molecule, where the Oxygen Atom has a More Negative Charge (Red) than the Positive (Blue) Hydrogen Atoms

Another important property of elements is their electronegativity. Atoms can form covalent bonds to each other by sharing electrons in pairs, creating an overlap of valence orbitals. The degree to which each atom attracts the shared electron pair depends on the atom's electronegativity—the tendency of an atom towards gaining or losing electrons.

The more electronegative atom will tend to attract the electron pair more, and the less electronegative (or more electropositive) one will attract it less. In extreme cases, the electron can be thought of as having been passed completely from the more electropositive atom to the more electronegative one, though this is a simplification. The bond then binds two ions, one positive (having given up the electron) and one negative (having accepted it), and is termed an ionic bond.

Electronegativity depends on how strongly the nucleus can attract an electron pair, and so it exhibits a similar variation to the other properties already discussed: electronegativity tends to fall going up to down, and rise going left to right. The alkali and alkaline earth metals are among the most electropositive elements, while the chalcogens, halogens, and noble gases are among the most electronegative ones.

Electronegativity is generally measured on the Pauling scale, on which the most electronegative reactive atom (fluorine) is given electronegativity 4.0, and the least electronegative atom (caesium) is given electronegativity 0.79. In fact neon is the most electronegative element, but the Pauling scale cannot measure its electronegativity because it does not form covalent bonds with most elements.

An element's electronegativity varies with the identity and number of the atoms it is bonded to, as well as how many electrons it has already lost: an atom becomes more electronegative when it has lost more electrons. This sometimes makes a large difference: lead in the +2 oxidation state has electronegativity 1.87 on the Pauling scale, while lead in the +4 oxidation state has electronegativity 2.33.

Metallicity

A simple substance is a substance formed from atoms of one chemical element. The simple substances of the more electronegative atoms tend to share electrons (form covalent bonds) with each other. They form either small molecules (like hydrogen or oxygen, whose atoms bond in pairs) or giant structures stretching indefinitely (like carbon or silicon).

The noble gases simply stay as single atoms, as they already have a full shell. Substances composed of discrete molecules or single atoms are held together by weaker attractive forces between the molecules, such as the London dispersion force: as electrons move within the molecules, they create momentary imbalances of electrical charge, which induce similar imbalances on nearby molecules and create synchronized movements of electrons across many neighbouring molecules. The more electropositive atoms, however, tend to instead lose electrons, creating a "sea" of electrons engulfing cations. The outer orbitals of one atom overlap to share electrons with all its neighbours, creating a giant structure of molecular orbitals extending over all the atoms.

This negatively charged "sea" pulls on all the ions and keeps them together in a metallic bond. Elements forming such bonds are often called metals; those which do not are often called nonmetals. Some elements can form multiple simple substances with different structures: these are called allotropes. For example, diamond and graphite are two allotropes of carbon.

The metallicity of an element can be predicted from electronic properties. When atomic orbitals overlap during metallic or covalent bonding, they create both bonding and antibonding molecular orbitals of equal capacity, with the antibonding orbitals of higher energy. Net bonding character occurs when there are more electrons in the bonding orbitals than there are in the antibonding orbitals. Metallic bonding is thus possible when the number of electrons delocalized by each atom is less than twice the number of orbitals contributing to the overlap. This is the situation for elements in groups 1 through 13; they also have too few valence electrons to form giant covalent structures where all atoms take equivalent positions, and so almost all of them metallise. The exceptions are hydrogen and boron, which have too high an ionisation energy. Hydrogen thus forms a covalent H_2 molecule, and boron forms a giant covalent structure based on icosahedral B_{12} clusters. In a metal, the bonding and antibonding orbitals have overlapping energies, creating a single band that electrons can freely flow through, allowing for electrical conduction.

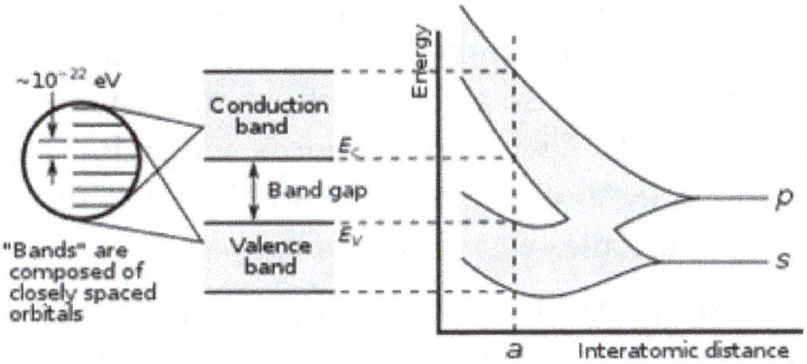

Fig 13: Graph of Carbon Atoms Being Brought Together to Form a Diamond Crystal, Demonstrating Formation of the Electronic Band Structure and Band Gap

The right graph shows the energy levels as a function of the spacing between atoms. When far apart *(right side of graph)* all the atoms have discrete valence orbitals *p* and *s* with the same energies. However, when the atoms come closer *(left side)*, their electron orbitals begin to spatially overlap. The Pauli exclusion principle prohibits them from having the same energy, so the orbitals hybridize into *N* molecular orbitals each with a different energy, where *N* is the number of atoms in the crystal. Since *N* is such a large number, adjacent orbitals are extremely close together in energy so the orbitals can be considered a continuous energy band. At the actual diamond crystal cell size (denoted by *a*), two bands are formed, called the valence and conduction bands, separated by a 5.5 eV band gap. (Here only the valence 2s and 2p electrons have been illustrated; the 1s orbitals do not significantly overlap, so the bands formed from them are much narrower.)

In group 14, both metallic and covalent bonding become possible. In a diamond crystal, covalent bonds between carbon atoms are strong, because they have a small atomic radius and thus the nucleus has more of a hold on the electrons. Therefore, the bonding orbitals that result are much lower in energy than the antibonding orbitals, and there is no overlap, so electrical conduction becomes impossible: carbon is a nonmetal. However, covalent bonding becomes weaker for larger atoms and the energy gap between the bonding and antibonding orbitals decreases. Therefore, silicon and germanium have smaller band gaps and are semiconductors at ambient conditions: electrons can cross the gap when thermally excited. (Boron is also a semiconductor at ambient conditions.) The band gap disappears in tin, so that tin and lead become metals. As the temperature rises, all nonmetals develop some semiconducting properties, to a greater or lesser extent depending on the size of the band gap. Thus metals and nonmetals may be distinguished by the temperature dependence of their electrical conductivity: a metal's conductivity lowers as temperature rises (because thermal motion makes it more difficult for the electrons to flow freely), whereas a nonmetal's conductivity rises (as more electrons may be excited to cross the gap).

Elements in groups 15 through 17 have too many electrons to form giant covalent molecules that stretch in all three dimensions. For the lighter elements, the bonds in small diatomic molecules are so strong that a condensed phase is disfavoured: thus nitrogen (N_2), oxygen (O_2), white phosphorus and yellow arsenic (P_4 and As_4), sulfur and red selenium (S_8 and Se_8), and the stable halogens (F_2, Cl_2, Br_2, and I_2) readily form covalent molecules with few atoms.

The heavier ones tend to form long chains (e.g. red phosphorus, grey selenium, tellurium) or layered structures (e.g. carbon as graphite, black phosphorus, grey arsenic, antimony, bismuth) that only extend in one or two rather than three dimensions. Both kinds of structures can be found as allotropes of phosphorus, arsenic, and selenium, although the long-chained allotropes are more stable in all three. As these structures do not use all their orbitals for bonding, they end up with bonding, nonbonding, and antibonding bands in order of increasing energy. Similarly to group 14, the band gaps shrink for the heavier elements and free movement of electrons between the chains or layers becomes possible. Thus for example black phosphorus, black arsenic, grey selenium, tellurium, and iodine are semiconductors; grey arsenic, antimony, and bismuth are semimetals (exhibiting quasi-metallic conduction, with a very small band overlap); and polonium and probably astatine are true metals. Finally, the natural group 18 elements all stay as individual atoms.

The dividing line between metals and nonmetals is roughly diagonal from top left to bottom right, with the transition series appearing to the left of this diagonal (as they have many available orbitals for overlap). This is expected, as metallicity tends to be correlated with electropositivity and the willingness to lose electrons, which increases right to left and up to down. Thus the metals greatly outnumber the nonmetals. Elements near the borderline are difficult to classify: they tend to have properties that are intermediate between those of metals and nonmetals, and may have some properties characteristic of both. They are often termed semimetals or metalloids. The term "semimetal" used in this sense should not be confused with its strict physical sense having to do with band structure: bismuth is physically a semimetal, but is generally considered a metal by chemists.

Table 2: Bonding of Simple Substances in the Periodic Table

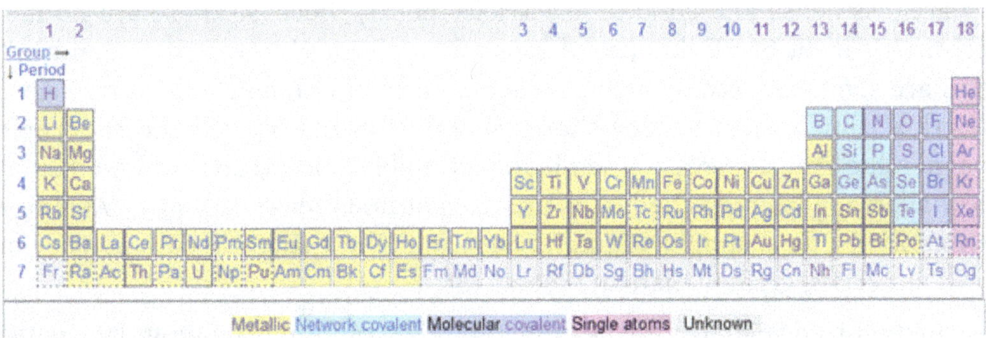

The table considers the most stable allotropes at standard conditions. The elements coloured yellow form simple substances that are well-characterised by metallic bonding. Elements coloured light blue form giant network covalent structures, whereas those coloured dark blue form small covalently bonded molecules that are held together by weaker van der Waals forces. The noble gases are coloured in violet: their molecules are single atoms and no covalent bonding occurs. Greyed-out cells are for elements which have not been prepared in sufficient quantities for their most stable allotropes to have been characterized in this way. Theoretical considerations and current experimental evidence suggest that all of those elements would metallise if they could form condensed phases, except perhaps for oganesson. Generally, metals are shiny and dense. They usually have high melting and boiling points due to the strength of the metallic bond, and are often malleable and ductile (easily stretched and shaped) because the atoms can move relative to each other without breaking the metallic bond. They conduct electricity because their electrons are free to move in all three dimensions. Similarly, they conduct heat, which is transferred by the electrons as extra kinetic energy: they move faster. These properties persist in the liquid state, as although the crystal structure is destroyed on melting, the atoms still touch and the metallic bond persists, though it is weakened. Metals tend to be reactive towards nonmetals. Some exceptions can be found to these generalizations: for example, beryllium, chromium, manganese, antimony, bismuth, and uranium are brittle (not an exhaustive list); chromium is extremely hard; gallium, rubidium, caesium, and mercury are liquid at or close to room temperature; and noble metals such as gold are chemically very inert.

Nonmetals exhibit different properties. Those forming giant covalent crystals exhibit high melting and boiling points, as it takes considerable energy to overcome the strong covalent bonds. Those forming discrete molecules are held together mostly by dispersion forces, which are more easily overcome; thus they tend to have lower melting and boiling points, and many are liquids or gases at room temperature. Nonmetals are often dull-looking. They tend to be reactive towards metals, except for the noble gases, which are inert towards most substances. They are brittle when solid as their atoms are held tightly in place. They are less dense and conduct electricity poorly, because there are no mobile electrons. Near the borderline, band gaps are small and thus many elements in that region are semiconductors, such as silicon, germanium, and tellurium. Selenium has both a semiconducting grey allotrope and an insulating red allotrope; arsenic has a metallic grey allotrope, a semiconducting black allotrope, and an insulating yellow allotrope (though the last is unstable at ambient conditions). Again there are exceptions; for example, diamond has the highest thermal conductivity of all known materials, greater than any metal.

It is common to designate a class of metalloids straddling the boundary between metals and nonmetals, as elements in that region are intermediate in both physical and chemical properties. However, no consensus exists in the literature for precisely which elements should be so designated. When such a category is used, silicon, germanium, arsenic, and tellurium are almost always included, and boron and antimony usually are; but most sources include other elements as well, without agreement on which extra elements should be added, and some others subtract from this list instead.

For example, unlike all the other elements generally considered metalloids or nonmetals, antimony's only stable form has metallic conductivity. Moreover, the element resembles bismuth and, more generally, the other p-block metals in its physical and chemical behaviour. On this basis some authors have argued that it is better classified as a metal than as a metalloid. On the other hand, selenium has some semiconducting properties in its most stable form (though it also has insulating allotropes) and it has been argued that it should be considered a metalloid – though this situation also holds for phosphorus, which is a much rarer inclusion among the metalloids.

Further Manifestations of Periodicity

There are some other relationships throughout the periodic table between elements that are not in the same group, such as the diagonal relationships between elements that are diagonally adjacent (e.g. lithium and magnesium). Some similarities can also be found between the main groups and the transition metal groups, or between the early actinides and early transition metals, when the elements have the same number of valence electrons. Thus uranium somewhat resembles chromium and tungsten in group 6, as all three have six valence electrons.

Relationships between elements with the same number of valence electrons but different types of valence orbital have been called secondary or isodonor relationships: they usually have the same maximum oxidation states, but not the same minimum oxidation states. For example, chlorine and manganese both have +7 as their maximum oxidation state (e.g. Cl_2O_7 and Mn_2O_7), but their respective minimum oxidation states are -1 (e.g. HCl) and -3 ($K_2[Mn(CO)_4]$). Elements with the same number of valence vacancies but different numbers of valence electrons are related by a tertiary or isoacceptor relationship: they usually have similar minimum but not maximum oxidation states. For example, hydrogen and chlorine both have -1 as their minimum oxidation state (in hydrides and chlorides), but hydrogen's maximum oxidation state is +1 (e.g. H_2O) while chlorine's is +7.

Many other physical properties of the elements exhibit periodic variation in accordance with the periodic law, such as melting points, boiling points, heats of fusion, heats of vaporization, atomisation energy, and so on. Similar periodic variations appear for the compounds of the elements, which can be observed by comparing hydrides, oxides, sulfides, halides, and so on. Chemical properties are more difficult to describe quantitatively, but likewise exhibit their own periodicities. Examples include the variation in the acidic and basic properties of the elements and their compounds, the stabilities of compounds, and methods of isolating the elements. Periodicity is and has been used very widely to predict the properties of unknown new elements and new compounds, and is central to modern chemistry.

1.2.4 Classification of Elements

Many terms have been used in the literature to describe sets of elements that behave similarly. The group names *alkali metal*, *alkaline earth metal*, *triel*, *tetrel*, *pnictogen*, *chalcogen*, *halogen*, and *noble gas* are acknowledged by IUPAC; the other groups can be referred to by their number, or by their first element (e.g., group

6 is the chromium group). Some divide the p-block elements from groups 13 to 16 by metallicity, although there is neither an IUPAC definition nor a precise consensus on exactly which elements should be considered metals, nonmetals, or semi-metals (sometimes called metalloids). Neither is there a consensus on what the metals succeeding the transition metals ought to be called, with *post-transition metal* and *poor metal* being among the possibilities having been used. Some advanced monographs exclude the elements of group 12 from the transition metals on the grounds of their sometimes quite different chemical properties, but this is not a universal practice and IUPAC does not presently mention it as allowable in its *Principles of Chemical Nomenclature*.

The *lanthanides* are considered to be the elements La–Lu, which are all very similar to each other: historically they included only Ce–Lu, but lanthanum became included by common usage. The *rare earth elements* (or rare earth metals) add scandium and yttrium to the lanthanides. Analogously, the *actinides* are considered to be the elements Ac–Lr (historically Th–Lr), although variation of properties in this set is much greater than within the lanthanides. IUPAC recommends the names *lanthanoids* and *actinoids* to avoid ambiguity, as the -ide suffix typically denotes a negative ion; however *lanthanides* and *actinides* remain common. With the increasing recognition of lutetium and lawrencium as d-block elements, some authors began to define the lanthanides as La–Yb and the actinides as Ac–No, matching the f-block. The *transactinides* or *superheavy elements* are the short-lived elements beyond the actinides, starting at lawrencium or rutherfordium (depending on where the actinides are taken to end).

Many more categorizations exist and are used according to certain disciplines. In astrophysics, a metal is defined as any element with atomic number greater than 2, i.e. anything except hydrogen and helium. The term “semimetal” has a different definition in physics than it does in chemistry: bismuth is a semimetal by physical definitions, but chemists generally consider it a metal. A few terms are widely used, but without any very formal definition, such as “heavy metal”, which has been given such a wide range of definitions that it has been criticized as “effectively meaningless”.

The scope of terms varies significantly between authors. For example, according to IUPAC, the noble gases extend to include the whole group, including the very radioactive superheavy element oganesson. However, among those who specialize in the superheavy elements, this is not often done: in this case “noble gas” is typically taken to imply the unreactive behaviour of the lighter elements of the group. Since calculations generally predict that oganesson should not be particularly inert due to relativistic effects, and may not even be a gas at room temperature if it could be produced in bulk, its status as a noble gas is often questioned in this context.

Furthermore, national variations are sometimes encountered: in Japan, alkaline earth metals often do not include beryllium and magnesium as their behaviour is different from the heavier group 2 metals.

1.2.5 Future Extension Beyond the Seventh Period

The most recently named elements - nihonium (113), moscovium (115), tennessine (117), and oganesson (118)—completed the seventh row of the periodic table. Future elements would have to begin an eighth row. These elements may be referred to either by their atomic numbers (e.g. "element 164"), or by the IUPAC systematic element names adopted in 1978, which directly relate to the atomic numbers.

All attempts to synthesize such elements have failed so far. An attempt to make element 119 has been ongoing since 2018 at the Riken research institute in Japan. The LBNL in the United States, the JINR in Russia, and the Heavy Ion Research Facility in Lanzhou (HIRFL) in China also plan to make their own attempts at synthesizing the first few period 8 elements.

If the eighth period followed the pattern set by the earlier periods, then it would contain fifty elements, filling the 8s, 5g, 6f, 7d, and finally 8p subshells in that order. But by this point, relativistic effects should result in significant deviations from the Madelung rule.

Various different models have been suggested for the configurations of eighth-period elements, as well as how to show the results in a periodic table. All agree that the eighth period should begin like the previous ones with two 8s elements, 119 and 120. However, after that the massive energetic overlaps between the 5g, 6f, 7d, and 8p subshells means that they all begin to fill together, and it is not clear how to separate out specific 5g and 6f series.

Elements 121 through 156 thus do not fit well as chemical analogues of any previous group in the earlier parts of the table, although they have sometimes been placed as 5g, 6f, and other series to formally reflect their electron configurations. Eric Scerri has raised the question of whether an extended periodic table should take into account the failure of the Madelung rule in this region, or if such exceptions should be ignored. The shell structure may also be fairly formal at this point: already the electron distribution in an oganesson atom is expected to be rather uniform, with no discernible shell structure. The situation from elements 157 to 172 should return to normalcy and be more reminiscent of the earlier rows. The heavy p-shells are split by the spin–orbit interaction: one p-orbital ($p_{1/2}$) is more stabilized, and the other two ($p_{3/2}$) are destabilized. (Such shifts in the quantum numbers happen for all types of shells, but it makes the biggest difference to the order for the p-shells.) It is likely that by element 157, the filled 8s and $8p_{1/2}$ shells with four electrons in total have sunk into the core.

Beyond the core, the next orbitals are 7d and 9s at similar energies, followed by $9p_{1/2}$ and $8p_{3/2}$ at similar energies, and then a large gap. Thus, the 9s and $9p_{1/2}$ orbitals in essence replace the 8s and $8p_{1/2}$ ones, making elements 157–172 probably chemically analogous to groups 3–18: for example, element 164 would appear two places below lead in group 14 under the usual pattern, but is calculated to be very analogous to palladium in group 10 instead.

Thus, it takes fifty-four elements rather than fifty to reach the next noble element after 118. However, while these conclusions about elements 157 through 172's chemistry are generally agreed by models, there is disagreement on whether the periodic table should be drawn to reflect chemical analogies, or if it should reflect likely formal electron configurations, which should be quite different from earlier periods and are not agreed between sources. Discussion about the format of the eighth row thus continues.

Beyond element 172, calculation is complicated by the 1s electron energy level becoming imaginary. Such a situation does have physical interpretation, and does not in itself pose an electronic limit to the periodic table, but the correct way to incorporate such states into multi-electron calculations is still an open problem. This would need to be solved to continue calculating the periodic table's structure beyond this point.

Nuclear stability will likely prove a decisive factor constraining the number of possible elements. It depends on the balance between the electric repulsion between protons and the strong force binding protons and neutrons together. Protons and neutrons are arranged in shells, just like electrons, and so a closed shell can significantly increase stability: the known superheavy nuclei exist because of such a shell closure, probably at around 114–126 protons and 184 neutrons.

They are probably close to a predicted island of stability, where superheavy nuclides should be more long-lived than expected: predictions for the longest-lived nuclides on the island range from microseconds to millions of years. It should nonetheless be noted that these are essentially extrapolations into an unknown part of the chart of nuclides, and systematic model uncertainties need to be taken into account.

As the closed shells are passed, the stabilizing effect should vanish: thus, superheavy nuclides with more than 184 neutrons are expected to have much shorter lifetimes, spontaneously fissioning within 10^{-15} seconds.

If this is so, then it would not make sense to consider them chemical elements: IUPAC defines an element to exist only if the nucleus lives longer than $^{14-}10$ seconds, the time needed for it to gather an electron cloud. Nonetheless, theoretical estimates of half-lives are very model-dependent, ranging over many orders of magnitude. The extreme repulsion between protons is predicted to result in exotic nuclear topologies, with bubbles, rings, and tori expected: this further complicates extrapolation. It is not clear if any further-out shell closures exist, due to an expected smearing out of distinct nuclear shells (as is already expected for the electron shells at oganesson). Furthermore, even if later shell closures exist, it is not clear if they would allow such heavy elements to exist. As such, it may be that the periodic table practically ends around element 120, as elements become too short-lived to observe, and then too short-lived to have chemistry; the era of discovering new elements would thus be close to its end. If another proton shell closure beyond 126 does exist, then it probably occurs around 164; thus the region where periodicity fails more or less matches the region of instability between the shell closures.

Alternatively, quark matter may become stable at high mass numbers, in which the nucleus is composed of freely flowing up and down quarks instead of binding them into protons and neutrons; this would create a continent of stability instead of an island. Other effects may come into play: for example, in very heavy elements the 1s electrons are likely to spend a significant amount of time so close to the nucleus that they are actually inside it, which would make them vulnerable to electron capture.

Even if eighth-row elements can exist, producing them is likely to be difficult, and it should become even more difficult as atomic number rises. Although the 8s elements 119 and 120 are expected to be reachable with present means, the elements beyond that are expected to require new technology, if they can be produced at all. Experimentally characterizing these elements chemically would also pose a great challenge.

1.2.6 Alternative Periodic Tables

The periodic law may be represented in multiple ways, of which the standard periodic table is only one. Within 100 years of the appearance of Mendeleev's table in 1869, Edward G. Mazurs had collected an estimated 700 different published versions of the periodic table. Many forms retain the rectangular structure, including Charles Janet's left-step periodic table (pictured below), and the modernised form of Mendeleev's original 8-column layout that is still common in Russia. Other periodic table formats have been shaped much more exotically, such as spirals (Otto Theodor Benfey's pictured to the right), circles and triangles.

Alternative periodic tables are often developed to highlight or emphasize chemical or physical properties of the elements that are not as apparent in traditional periodic tables, with different ones skewed more towards emphasizing chemistry or physics at either end. The standard form, which remains by far the most common, is somewhere in the middle.

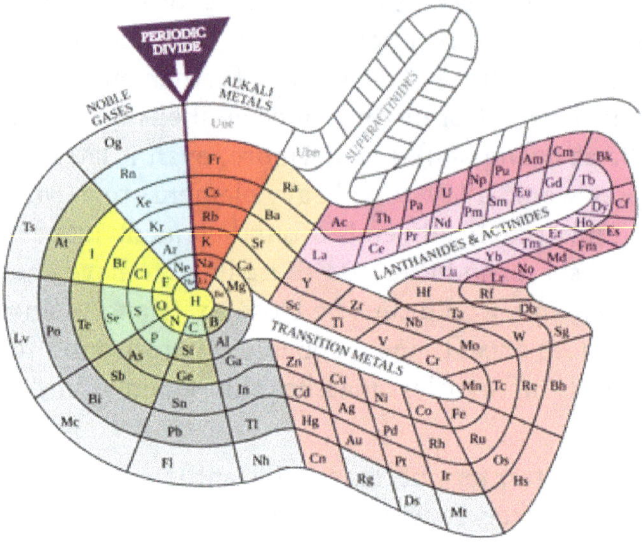

Fig 14: Otto Theodor Benfey's Spiral Periodic Table (1964)

The many different forms of the periodic table have prompted the questions of whether there is an optimal or definitive form of the periodic table, and if so, what it might be. There are no current consensus answers to either question. Janet's left-step table is being increasingly discussed as a candidate for being the optimal or most fundamental form; Scerri has written in support of it, as it clarifies helium's nature as an s-block element, increases regularity by having all period lengths repeated, faithfully follows Madelung's rule by making each period correspond to one value of $n+\ell$, and regularises atomic number triads and the first-row anomaly trend. While he notes that its placement of helium atop the alkaline earth metals can be seen a disadvantage from a chemical perspective, he counters this by appealing to the first-row anomaly, pointing out that the periodic table "fundamentally reduces to quantum mechanics", and that it is concerned with "abstract elements" and hence atomic properties rather than macroscopic properties.

Table 3: Left-Step Periodic Table (by Charles Janet)

	f^1	f^2	f^3	f^4	f^5	f^6	f^7	f^8	f^9	f^{10}	f^{11}	f^{12}	f^{13}	f^{14}	d^1	d^2	d^3	d^4	d^5	d^6	d^7	d^8	d^9	d^{10}	p^1	p^2	p^3	p^4	p^5	p^6	s^1	s^2
1s																															H	He
2s																															Li	Be
2p 3s																									B	C	N	O	F	Ne	Na	Mg
3p 4s																									Al	Si	P	S	Cl	Ar	K	Ca
3d 4p 5s															Sc	Ti	V	Cr	Mn	Fe	Co	Ni	Cu	Zn	Ga	Ge	As	Se	Br	Kr	Rb	Sr
4d 5p 6s															Y	Zr	Nb	Mo	Tc	Ru	Rh	Pd	Ag	Cd	In	Sn	Sb	Te	I	Xe	Cs	Ba
4f 5d 6p 7s	La	Ce	Pr	Nd	Pm	Sm	Eu	Gd	Tb	Dy	Ho	Er	Tm	Yb	Lu	Hf	Ta	W	Re	Os	Ir	Pt	Au	Hg	Tl	Pb	Bi	Po	At	Rn	Fr	Ra
5f 6d 7p 8s	Ac	Th	Pa	U	Np	Pu	Am	Cm	Bk	Cf	Es	Fm	Md	No	Lr	Rf	Db	Sg	Bh	Hs	Mt	Ds	Rg	Cn	Nh	Fl	Mc	Lv	Ts	Og	Uue	Ubn
	f-block														d-block										p-block						s-block	

CHAPTER 2

INTRODUCTION TO BONDING

A chemical bond is the association of atoms or ions to form molecules, crystals, and other structures. The bond may result from the electrostatic force between oppositely charged ions as in ionic bonds or through the sharing of electrons as in covalent bonds, or some combination of these effects. Chemical bonds are described as having different strengths: there are "strong bonds" or "primary bonds" such as covalent, ionic and metallic bonds, and "weak bonds" or "secondary bonds" such as dipole–dipole interactions, the London dispersion force, and hydrogen bonding.

Since opposite electric charges attract, the negatively charged electrons surrounding the nucleus and the positively charged protons within a nucleus attract each other. Electrons shared between two nuclei will be attracted to both of them. "Constructive quantum mechanical wavefunction interference" stabilizes the paired nuclei. Bonded nuclei maintain an optimal distance (the bond distance) balancing attractive and repulsive effects explained quantitatively by quantum theory.

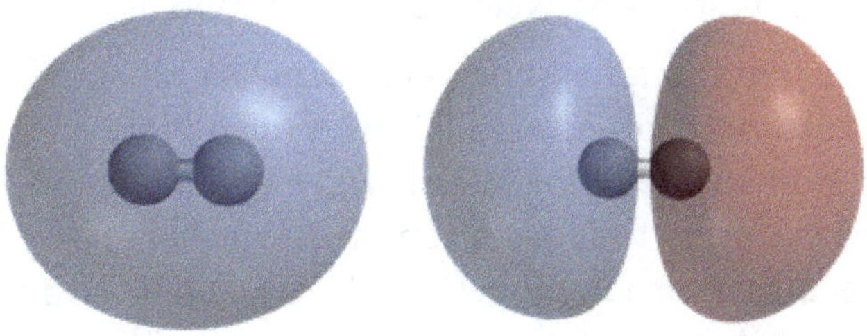

(a) (b)

Fig 1: Covalent Bonding of Two Hydrogen Atoms to form a Hydrogen Molecule, H_2

In (a) the two nuclei are surrounded by a cloud of two electrons in the bonding orbital that holds the molecule together. (b) shows hydrogen's antibonding orbital, which is higher in energy and is normally not occupied by any electrons.

The atoms in molecules, crystals, metals and other forms of matter are held together by chemical bonds, which determine the structure and properties of matter.

All bonds can be described by quantum theory, but, in practice, simplified rules and other theories allow chemists to predict the strength, directionality, and polarity of bonds. The octet rule and VSEPR theory are examples. More sophisticated theories are valence bond theory, which includes orbital hybridization and resonance, and molecular orbital theory which includes the linear combination of atomic orbitals and ligand field theory. Electrostatics are used to describe bond polarities and the effects they have on chemical substances.

2.1 MAIN TYPES OF CHEMICAL BONDS

A chemical bond is an attraction between atoms. This attraction may be seen as the result of different behaviors of the outermost or valence electrons of atoms. These behaviors merge into each other seamlessly in various circumstances, so that there is no clear line to be drawn between them. However it remains useful and customary to differentiate between different types of bond, which result in different properties of condensed matter.

In the simplest view of a covalent bond, one or more electrons (often a pair of electrons) are drawn into the space between the two atomic nuclei. Energy is released by bond formation. This is not as a result of reduction in potential energy, because the attraction of the two electrons to the two protons is offset by the electron-electron and proton-proton repulsions. Instead, the release of energy (and hence stability of the bond) arises from the reduction in kinetic energy due to the electrons being in a more spatially distributed (i.e., longer de Broglie wavelength) orbital compared with each electron being confined closer to its respective nucleus.

These bonds exist between two particular identifiable atoms and have a direction in space, allowing them to be shown as single connecting lines between atoms in drawings, or modeled as sticks between spheres in models.

In a polar covalent bond, one or more electrons are unequally shared between two nuclei. Covalent bonds often result in the formation of small collections of better-connected atoms called molecules, which in solids and liquids are bound to other molecules by forces that are often much weaker than the covalent bonds that hold the molecules internally together. Such weak intermolecular bonds give organic molecular substances, such as waxes and oils, their soft bulk character, and their low melting points (in liquids, molecules must cease most structured or oriented contact with each other). When covalent bonds link long chains of atoms in large molecules, however (as in polymers such as nylon), or when covalent bonds extend in networks through solids that are not composed of discrete molecules (such as diamond or quartz or the silicate minerals in many types of rock) then the structures that result may be both strong and tough, at least in the direction oriented correctly with networks of covalent bonds. Also, the melting points of such covalent polymers and networks increase greatly.

In a simplified view of an *ionic* bond, the bonding electron is not shared at all, but transferred. In this type of bond, the outer atomic orbital of one atom has a vacancy which allows the addition of one or more electrons. These newly added electrons potentially occupy a lower energy-state (effectively closer to more nuclear charge) than they experience in a different atom. Thus, one nucleus offers a more tightly bound position to an electron than does another nucleus, with the result that one atom may transfer an electron to the other.

This transfer causes one atom to assume a net positive charge, and the other to assume a net negative charge. The *bond* then results from electrostatic attraction between the positive and negatively charged ions. Ionic bonds may be seen as extreme examples of polarization in covalent bonds. Often, such bonds have no particular orientation in space, since they result from equal electrostatic attraction of each ion to all ions around them. Ionic bonds are strong (and thus ionic substances require high temperatures to melt) but also brittle, since the forces between ions are short-range and do not easily bridge cracks and fractures. This type of bond gives rise to the physical characteristics of crystals of classic mineral salts, such as table salt.

A less often mentioned type of bonding is *metallic* bonding. In this type of bonding, each atom in a metal donates one or more electrons to a "sea" of electrons that reside between many metal atoms. In this sea, each electron is free (by virtue of its wave nature) to be associated with a great many atoms at once. The bond results because the metal atoms become somewhat positively charged due to loss of their electrons while the electrons remain attracted to many atoms, without being part of any given atom. Metallic bonding may be seen as an extreme example of delocalization of electrons over a large system of covalent bonds, in which every atom participates. This type of bonding is often very strong (resulting in the tensile strength of metals). However, metallic bonding is more collective in nature than other types, and so they allow metal crystals to more easily deform, because they are composed of atoms attracted to each other, but not in any particularly-oriented ways. This results in the malleability of metals. The cloud of electrons in metallic bonding causes the characteristically good electrical and thermal conductivity of metals, and also their shiny lustre that reflects most frequencies of white light.

2.2 HISTORY

Early speculations about the nature of the chemical bond, from as early as the 12th century, supposed that certain types of chemical species were joined by a type of chemical affinity. In 1704, Sir Isaac Newton famously outlined his atomic bonding theory, in "Query 31" of his *Opticks*, whereby atoms attach to each other by some "force". Specifically, after acknowledging the various popular theories in vogue at the time, of how atoms were reasoned to attach to each other, i.e. "hooked atoms", "glued together by rest", or "stuck together by conspiring motions", Newton states that he would rather infer from their cohesion, that "particles attract one another by some force, which in immediate contact is exceedingly strong, at small distances performs the chemical operations, and reaches not far from the particles with any sensible effect."

In 1819, on the heels of the invention of the voltaic pile, Jöns Jakob Berzelius developed a theory of chemical combination stressing the electronegative and electropositive characters of the combining atoms. By the mid 19th century, Edward Frankland, F.A. Kekulé, A.S. Couper, Alexander Butlerov, and Hermann Kolbe, building on the theory of radicals, developed the theory of valency, originally called "combining power", in which compounds were joined owing to an attraction of positive and negative poles.

In 1904, Richard Abegg proposed his rule that the difference between the maximum and minimum valencies of an element is often eight. At this point, valency was still an empirical number based only on chemical properties.

However the nature of the atom became clearer with Ernest Rutherford's 1911 discovery that of an atomic nucleus surrounded by electrons in which he quoted Nagaoka rejected Thomson's model on the grounds that opposite charges are impenetrable.

In 1904, Nagaoka proposed an alternative planetary model of the atom in which a positively charged center is surrounded by a number of revolving electrons, in the manner of Saturn and its rings.

Nagaoka's model made two predictions:

- A very massive atomic center (in analogy to a very massive planet).
- Electrons revolving around the nucleus, bound by electrostatic forces (in analogy to the rings revolving around Saturn, bound by gravitational forces.).

Rutherford mentions Nagaoka's model in his 1911 paper in which the atomic nucleus is proposed. At the 1911 Solvay Conference, in the discussion of what could regulate energy differences between atoms, Max Planck stated: "The intermediaries could be the electrons." These nuclear models suggested that electrons determine chemical behavior.

Next came Niels Bohr's 1913 model of a nuclear atom with electron orbits. In 1916, chemist Gilbert N. Lewis developed the concept of electron-pair bonds, in which two atoms may share one to six electrons, thus forming the single electron bond, a single bond, a double bond, or a triple bond; in Lewis's own words, "An electron may form a part of the shell of two different atoms and cannot be said to belong to either one exclusively." Also in 1916, Walther Kossel put forward a theory similar to Lewis' only his model assumed complete transfers of electrons between atoms, and was thus a model of ionic bonding. Both Lewis and Kossel structured their bonding models on that of Abegg's rule (1904). Niels Bohr also proposed a model of the chemical bond in 1913. According to his model for a diatomic molecule, the electrons of the atoms of the molecule form a rotating ring whose plane is perpendicular to the axis of the molecule and equidistant from the atomic nuclei. The dynamic equilibrium of the molecular system is achieved through the balance of forces between the forces of attraction of nuclei to the plane of the ring of electrons and the forces of mutual repulsion of the nuclei. The Bohr model of the chemical bond took into account the Coulomb repulsion – the electrons in the ring are at the maximum distance from each other.

In 1927, the first mathematically complete quantum description of a simple chemical bond, i.e. that produced by one electron in the hydrogen molecular ion, H_2^+, was derived by the Danish physicist Øyvind Burrau. This work showed that the quantum approach to chemical bonds could be fundamentally and quantitatively correct, but the mathematical methods used could not be extended to molecules containing more than one electron.

A more practical, albeit less quantitative, approach was put forward in the same year by Walter Heitler and Fritz London. The Heitler–London method forms the basis of what is now called valence bond theory. In 1929, the linear combination of atomic orbitals molecular orbital method (LCAO) approximation was introduced by Sir John Lennard-Jones, who also suggested methods to derive electronic structures of molecules of F_2 (fluorine) and O_2 (oxygen) molecules, from basic quantum principles. This molecular orbital theory represented a covalent bond as an orbital formed by combining the quantum mechanical Schrödinger atomic orbitals which had been hypothesized for electrons in single atoms. The equations for bonding electrons in multi-electron atoms could not be solved to mathematical perfection (i.e., *analytically*), but approximations for them still gave many good qualitative predictions and results. Most quantitative calculations in modern quantum chemistry use either valence bond or molecular orbital theory as a starting point, although a third approach, density functional theory, has become increasingly popular in recent years.

In 1933, H. H. James and A. S. Coolidge carried out a calculation on the dihydrogen molecule that, unlike all previous calculation which used functions only of the distance of the electron from the atomic nucleus, used functions which also explicitly added the distance between the two electrons.

With up to 13 adjustable parameters they obtained a result very close to the experimental result for the dissociation energy. Later extensions have used up to 54 parameters and gave excellent agreement with experiments. This calculation convinced the scientific community that quantum theory could give agreement with experiment. However this approach has none of the physical pictures of the valence bond and molecular orbital theories and is difficult to extend to larger molecules.

2.3 BONDS IN CHEMICAL FORMULAS

Because atoms and molecules are three-dimensional, it is difficult to use a single method to indicate orbitals and bonds. In molecular formulas the chemical bonds (binding orbitals) between atoms are indicated in different ways depending on the type of discussion. Sometimes, some details are neglected. For example, in organic chemistry one is sometimes concerned only with the functional group of the molecule. Thus, the molecular formula of ethanol may be written in conformational form, three-dimensional form, full two-dimensional form (indicating every bond with no three-dimensional directions), compressed two-dimensional form (CH_3–CH_2–OH), by separating the functional group from another part of the molecule (C_2H_5OH), or by its atomic constituents (C_2H_6O), according to what is discussed. Sometimes, even the non-bonding valence shell electrons (with the two-dimensional approximate directions) are marked, e.g. for elemental carbon ˙C˙. Some chemists may also mark the respective orbitals, e.g. the hypothetical ethene^{-4} anion ($_{\backslash}{}^{/}C{=}C_{/}{}^{\backslash}$ $^{-4}$) indicating the possibility of bond formation.

2.4 STRONG CHEMICAL BONDS

Strong chemical bonds are the *intramolecular* forces that hold atoms together in molecules. A strong chemical bond is formed from the transfer or sharing of electrons between atomic centers and relies on the electrostatic attraction between the protons in nuclei and the electrons in the orbitals.

Table 1: Typical Bond Lengths in pm and Bond Energies in kJ/mol

Bond	Length (pm)	Energy (kJ/mol)
H — Hydrogen		
H–H	74	436
H–O	96	467
H–F	92	568
H–Cl	127	432
C — Carbon		
C–H	109	413
C–C	154	347
C–C=	151	
=C–C⍰	147	
=C–C=	148	
C=C	134	614
C⍰C	120	839

Bond	Length (pm)	Energy (kJ/mol)
C–N	147	308
C–O	143	358
C=O		745
C⍰O		1,072
C–F	134	488
C–Cl	177	330
N — Nitrogen		
N–H	101	391
N–N	145	170
N⍰N	110	945
O — Oxygen		
O–O	148	146
O=O	121	495
F, Cl, Br, I — Halogens		
F–F	142	158
Cl–Cl	199	243
Br–H	141	366
Br–Br	228	193
I–H	161	298
I–I	267	151

The types of strong bond differ due to the difference in electronegativity of the constituent elements. Electronegativity is the tendency for an atom of a given chemical element to attract shared electrons when forming a chemical bond, where the higher the associated electronegativity then the more it attracts electrons. Electronegativity serves as a simple way to quantitatively estimate the bond energy, which characterizes a bond along the continuous scale from covalent to ionic bonding. A large difference in electronegativity leads to more polar (ionic) character in the bond.

2.4.1 Ionic Bond

Ionic bonding is a type of electrostatic interaction between atoms that have a large electronegativity difference.

There is no precise value that distinguishes ionic from covalent bonding, but an electronegativity difference of over 1.7 is likely to be ionic while a difference of less than 1.7 is likely to be covalent. Ionic bonding leads to separate positive and negative ions. Ionic charges are commonly between -3e to +3e. Ionic bonding commonly occurs in metal salts such as sodium chloride (table salt). A typical feature of ionic bonds is that the species form into ionic crystals, in which no ion is specifically paired with any single other ion in a specific directional bond. Rather, each species of ion is surrounded by ions of the opposite charge, and the spacing between it and each of the oppositely charged ions near it is the same for all surrounding atoms of the same type. It is thus no longer possible to associate an ion with any specific other single ionized atom near it. This is a situation unlike that in covalent crystals, where covalent bonds between specific atoms are still discernible from the shorter distances between them, as measured via such techniques as X-ray diffraction.

Fig 2: Crystal Structure of Sodium Chloride (NaCl) with Sodium Cations (Na^+) in Purple and Chloride Anions (Cl^-) in Green. The Yellow Stipples Represent the Electrostatic Force between the Ions of Opposite Charge

Ionic crystals may contain a mixture of covalent and ionic species, as for example salts of complex acids such as sodium cyanide, NaCN. X-ray diffraction shows that in NaCN, for example, the bonds between sodium cations (Na^+) and the cyanide anions (CN^-) are *ionic*, with no sodium ion associated with any particular cyanide. However, the bonds between the carbon (C) and nitrogen (N) atoms in cyanide are of the *covalent* type, so that each carbon is strongly bound to *just one* nitrogen, to which it is physically much closer than it is to other carbons or nitrogens in a sodium cyanide crystal.

When such crystals are melted into liquids, the ionic bonds are broken first because they are non-directional and allow the charged species to move freely. Similarly, when such salts dissolve into water, the ionic bonds are typically broken by the interaction with water but the covalent bonds continue to hold. For example, in solution, the cyanide ions, still bound together as single CN^- ions, move independently through the solution, as do sodium ions, as Na^+. In water, charged ions move apart because each of them are more strongly attracted to a number of water molecules than to each other. The attraction between ions and water molecules in such solutions is due to a type of weak dipole-dipole type chemical bond. In melted ionic compounds, the ions continue to be attracted to each other, but not in any ordered or crystalline way.

2.4.2 Covalent Bond

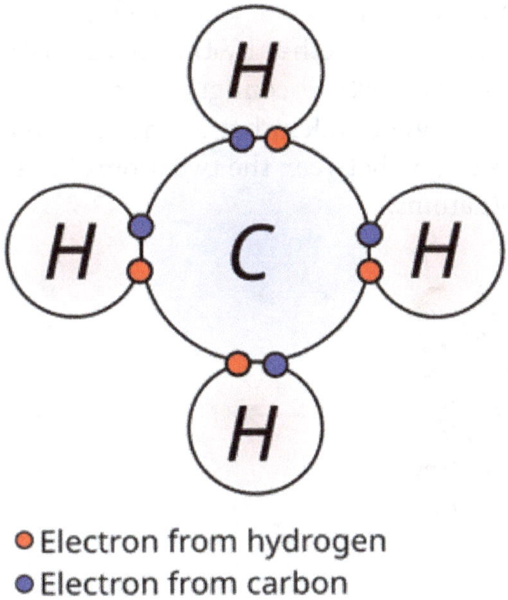

Fig 3: Non-Polar Covalent Bonds in Methane (CH_4). The Lewis Structure Shows Electrons Shared between C and H Atoms

Covalent bonding is a common type of bonding in which two or more atoms share valence electrons more or less equally. The simplest and most common type is a single bond in which two atoms share two electrons. Other types include the double bond, the triple bond, one- and three-electron bonds, the three-center two-electron bond and three-center four-electron bond.

In non-polar covalent bonds, the electronegativity difference between the bonded atoms is small, typically 0 to 0.3. Bonds within most organic compounds are described as covalent. The figure shows methane (CH_4), in which each hydrogen forms a covalent bond with the carbon.

Molecules that are formed primarily from non-polar covalent bonds are often immiscible in water or other polar solvents, but much more soluble in non-polar solvents such as hexane.

A polar covalent bond is a covalent bond with a significant ionic character. This means that the two shared electrons are closer to one of the atoms than the other, creating an imbalance of charge. Such bonds occur between two atoms with moderately different electronegativities and give rise to dipole–dipole interactions. The electronegativity difference between the two atoms in these bonds is 0.3 to 1.7.

Single and Multiple Bonds

A single bond between two atoms corresponds to the sharing of one pair of electrons. The Hydrogen (H) atom has one valence electron. Two Hydrogen atoms can then form a molecule, held together by the shared pair of electrons. Each H atom now has the noble gas electron configuration of helium (He). The pair of shared electrons forms a single covalent bond. The electron density of these two bonding electrons in the region between the two atoms increases from the density of two non-interacting H atoms.

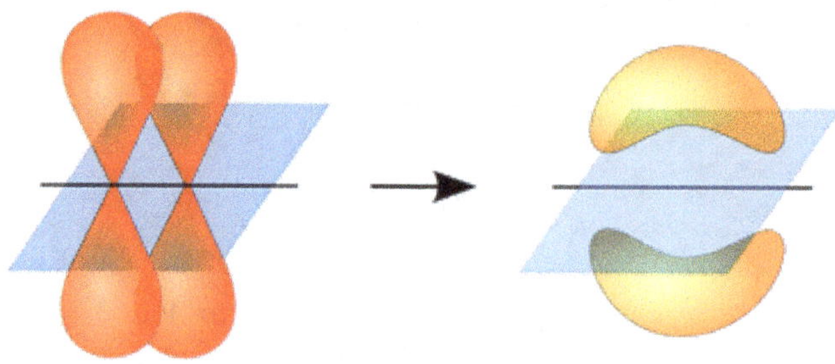

Fig 4: Two P-Orbitals Forming a Pi-Bond

A double bond has two shared pairs of electrons, one in a sigma bond and one in a pi bond with electron density concentrated on two opposite sides of the internuclear axis. A triple bond consists of three shared electron pairs, forming one sigma and two pi bonds. An example is nitrogen. Quadruple and higher bonds are very rare and occur only between certain transition metal atoms.

Coordinate Covalent Bond (Dipolar Bond)

A coordinate covalent bond is a covalent bond in which the two shared bonding electrons are from the same one of the atoms involved in the bond. For example, boron trifluoride (BF_3) and ammonia (NH_3) form an adduct or coordination complex $F_3B{\leftarrow}NH_3$ with a B–N bond in which a lone pair of electrons on N is shared with an empty atomic orbital on B. BF_3 with an empty orbital is described as an electron pair acceptor or Lewis acid, while NH_3 with a lone pair that can be shared is described as an electron-pair donor or Lewis base. The electrons are shared roughly equally between the atoms in contrast to ionic bonding. Such bonding is shown by an arrow pointing to the Lewis acid. (In the Figure, solid lines are bonds in the plane of the diagram, wedged bonds point towards the observer, and dashed bonds point away from the observer.)

Transition metal complexes are generally bound by coordinate covalent bonds. For example, the ion Ag^+ reacts as a Lewis acid with two molecules of the Lewis base NH_3 to form the complex ion $Ag(NH_3)_2^+$, which has two Ag←N coordinate covalent bonds.

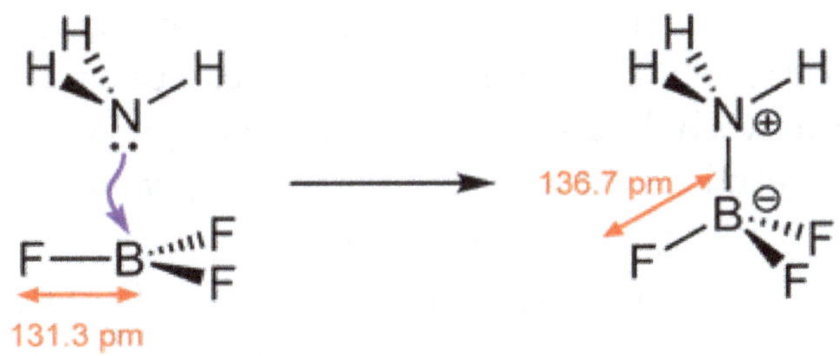

Fig 5: Adduct of Ammonia and Boron Trifluoride

2.4.3 Metallic Bonding

In metallic bonding, bonding electrons are delocalized over a lattice of atoms. By contrast, in ionic compounds, the locations of the binding electrons and their charges are static. The free movement or delocalization of bonding electrons leads to classical metallic properties such as luster (surface light reflectivity), electrical and thermal conductivity, ductility, and high tensile strength.

2.5 INTERMOLECULAR BONDING

There are several types of weak bonds that can be formed between two or more molecules which are not covalently bound. Intermolecular forces cause molecules to attract or repel each other. Often, these forces influence physical characteristics (such as the melting point) of a substance.

Van der Waals forces are interactions between closed-shell molecules. They include both Coulombic interactions between partial charges in polar molecules, and Pauli repulsions between closed electrons shells.

Keesom forces are the forces between the permanent dipoles of two polar molecules. London dispersion forces are the forces between induced dipoles of different molecules. There can also be an interaction between a permanent dipole in one molecule and an induced dipole in another molecule.

Hydrogen bonds of the form A--H•••B occur when A and B are two highly electronegative atoms (usually N, O or F) such that A forms a highly polar covalent bond with H so that H has a partial positive charge, and B has a lone pair of electrons which is attracted to this partial positive charge and forms a hydrogen bond. Hydrogen bonds are responsible for the high boiling points of water and ammonia with respect to their heavier analogues. In some cases a similar halogen bond can be formed by a halogen atom located between two electronegative atoms on different molecules. At short distances, repulsive forces between atoms also become important.

2.6 THEORIES OF CHEMICAL BONDING

In the (unrealistic) limit of "pure" ionic bonding, electrons are perfectly localized on one of the two atoms in the bond. Such bonds can be understood by classical physics. The force between the atoms depends on isotropic continuum electrostatic potentials. The magnitude of the force is in simple proportion to the product of the two ionic charges according to Coulomb's law.

Covalent bonds are better understood by valence bond (VB) theory or molecular orbital (MO) theory. The properties of the atoms involved can be understood using concepts such as oxidation number, formal charge, and electronegativity. The electron density within a bond is not assigned to individual atoms, but is instead delocalized between atoms. In valence bond theory, bonding is conceptualized as being built up from electron pairs that are localized and shared by two atoms via the overlap of atomic orbitals.

The concepts of orbital hybridization and resonance augment this basic notion of the electron pair bond. In molecular orbital theory, bonding is viewed as being delocalized and apportioned in orbitals that extend throughout the molecule and are adapted to its symmetry properties, typically by considering linear combinations of atomic orbitals (LCAO). Valence bond theory is more chemically intuitive by being spatially localized, allowing attention to be focused on the parts of the molecule undergoing chemical change. In contrast, molecular orbitals are more "natural" from a quantum mechanical point of view, with orbital energies being physically significant and directly linked to experimental ionization energies from photoelectron spectroscopy. Consequently, valence bond theory and molecular orbital theory are often viewed as competing but complementary frameworks that offer different insights into chemical systems. As approaches for electronic structure theory, both MO and VB methods can give approximations to any desired level of accuracy, at least in principle. However, at lower levels, the approximations differ, and one approach may be better suited for computations involving a particular system or property than the other.

Unlike the spherically symmetrical Coulombic forces in pure ionic bonds, covalent bonds are generally directed and anisotropic. These are often classified based on their symmetry with respect to a molecular plane as sigma bonds and pi bonds. In the general case, atoms form bonds that are intermediate between ionic and covalent, depending on the relative electronegativity of the atoms involved. Bonds of this type are known as polar covalent bonds.

CHAPTER 3

THE IONIC BOND

Ionic bonding is a type of chemical bonding that involves the electrostatic attraction between oppositely charged ions, or between two atoms with sharply different electronegativities, and is the primary interaction occurring in ionic compounds. It is one of the main types of bonding, along with covalent bonding and metallic bonding. Ions are atoms (or groups of atoms) with an electrostatic charge. Atoms that gain electrons make negatively charged ions (called anions). Atoms that lose electrons make positively charged ions (called cations). This transfer of electrons is known as electrovalence in contrast to covalence. In the simplest case, the cation is a metal atom and the anion is a nonmetal atom, but these ions can be more complex, e.g., molecular ions like NH^{+}_{4} or SO^{2-}_{4}. In simpler words, an ionic bond results from the transfer of electrons from a metal to a non-metal to obtain a full valence shell for both atoms.

Clean ionic bonding—in which one atom or molecule completely transfers an electron to another—cannot exist: all ionic compounds have some degree of covalent bonding or electron sharing. Thus, the term "ionic bonding" is given when the ionic character is greater than the covalent character – that is, a bond in which there is a large difference in electronegativity between the two atoms, causing the bonding to be more polar (ionic) than in covalent bonding where electrons are shared more equally. Bonds with partially ionic and partially covalent characters are called polar covalent bonds.

Ionic compounds conduct electricity when molten or in solution, typically not when solid. Ionic compounds generally have a high melting point, depending on the charge of the ions they consist of. The higher the charges the stronger the cohesive forces and the higher the melting point. They also tend to be soluble in water; the stronger the cohesive forces, the lower the solubility.

3.1 OVERVIEW

Atoms that have an almost full or almost empty valence shell tend to be very reactive. Strongly electronegative atoms (such as halogens) often have only one or two empty electron states in their valence shell, and frequently bond with other atoms or gain electrons to form anions. Weakly electronegative atoms (such as alkali metals) have relatively few valence electrons, which can easily be lost to strongly electronegative atoms. As a result, weakly electronegative atoms tend to distort their electron cloud and form cations.

3.1.1 Properties of Ionic Bonds

- They are considered to be among the strongest of all types of chemical bonds. This often causes ionic compounds to be very stable.
- Ionic bonds have high bond energy. Bond energy is the mean amount of energy required to break the bond in the gaseous state.
- Most ionic compounds exist in the form of a crystal structure, in which the ions occupy the corners of the crystal. Such a structure is called a crystal lattice.
- Ionic compounds lose their crystal lattice structure and break up into ions when dissolved in water or any other polar solvent. This process is called solvation. The presence of these free ions makes aqueous ionic compound solutions good conductors of electricity. The same occurs when the compounds are heated above their melting point in a process known as melting.

3.2 FORMATION

Ionic bonding can result from a redox reaction when atoms of an element (usually metal), whose ionization energy is low, give some of their electrons to achieve a stable electron configuration.

In doing so, cations are formed. An atom of another element (usually nonmetal) with greater electron affinity accepts one or more electrons to attain a stable electron configuration, and after accepting electrons an atom becomes an anion. Typically, the stable electron configuration is one of the noble gases for elements in the s-block and the p-block, and particular stable electron configurations for d-block and f-block elements.

The electrostatic attraction between the anions and cations leads to the formation of a solid with a crystallographic lattice in which the ions are stacked in an alternating fashion. In such a lattice, it is usually not possible to distinguish discrete molecular units, so that the compounds formed are not molecular. However, the ions themselves can be complex and form molecular ions like the acetate anion or the ammonium cation.

Lithium has a low ionization energy and readily gives up its lone valence electron to a fluorine atom, which has a positive electron affinity and accepts the electron that was donated by the lithium atom. The end-result is that lithium is isoelectronic with helium and fluorine is isoelectronic with neon. Electrostatic interaction occurs between the two resulting ions, but typically aggregation is not limited to two of them. Instead, aggregation into a whole lattice held together by ionic bonding is the result.

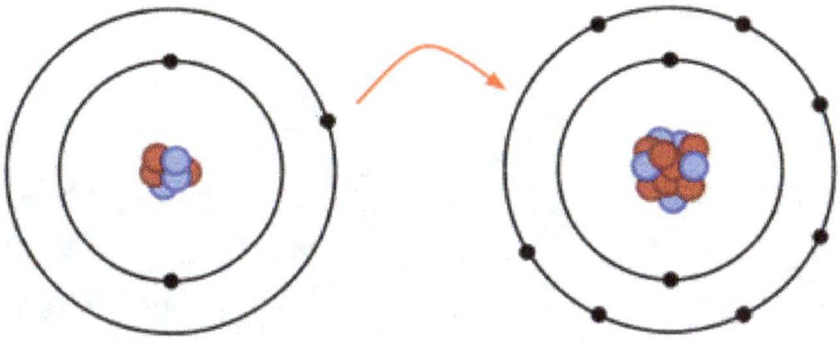

Fig. 4: Representation of Ionic Bonding between Lithium and Fluorine to form Lithium Fluoride

For example, common table salt is sodium chloride. When sodium (Na) and chlorine (Cl) are combined, the sodium atoms each lose an electron, forming cations (Na^+), and the chlorine atoms each gain an electron to form anions (Cl^-). These ions are then attracted to each other in a 1:1 ratio to form sodium chloride (NaCl).

$$Na + Cl \longrightarrow Na^+ + Cl^- \longrightarrow NaCl$$

However, to maintain charge neutrality, strict ratios between anions and cations are observed so that ionic compounds, in general, obey the rules of stoichiometry despite not being molecular compounds. For compounds that are transitional to the alloys and possess mixed ionic and metallic bonding, this may not be the case anymore. Many sulfides, e.g., do form non-stoichiometric compounds.

Many ionic compounds are referred to as salts as they can also be formed by the neutralization reaction of an Arrhenius base like NaOH with an Arrhenius acid like HCl

$$NaOH + HCl \longrightarrow NaCl + H_2O$$

The salt NaCl is then said to consist of the acid rest Cl^- and the base rest Na^+. The removal of electrons to form the cation is endothermic, raising the system's overall energy. There may also be energy changes associated with breaking of existing bonds or the addition of more than one electron to form anions. However, the action of the anion's accepting the cation's valence electrons and the subsequent attraction of the ions to each other releases (lattice) energy and, thus, lowers the overall energy of the system.

Ionic bonding will occur only if the overall energy change for the reaction is favorable. In general, the reaction is exothermic, but, e.g., the formation of mercuric oxide (HgO) is endothermic. The charge of the resulting ions is a major factor in the strength of ionic bonding, e.g. a salt C^+A^- is held together by electrostatic forces roughly four times weaker than $C^{2+}A^{2-}$ according to Coulomb's law, where C and A represent a generic cation and anion respectively. The sizes of the ions and the particular packing of the lattice are ignored in this rather simplistic argument.

3.3 STRUCTURES

Fig. 2: In the Rock Salt Lattice, Each Sodium Ion (Purple Sphere) has an Electrostatic Interaction with its Eight Nearest-Neighbour Chloride Ions (Green Spheres)

Ionic compounds in the solid state form lattice structures. The two principal factors in determining the form of the lattice are the relative charges of the ions and their relative sizes. Some structures are adopted by a number of compounds; for example, the structure of the rock salt sodium chloride is also adopted by many alkali halides, and binary oxides such as magnesium oxide. Pauling's rules provide guidelines for predicting and rationalizing the crystal structures of ionic crystals

3.4 STRENGTH OF THE BONDING

For a solid crystalline ionic compound the enthalpy change in forming the solid from gaseous ions is termed the lattice energy. The experimental value for the lattice energy can be determined using the Born–Haber cycle. It can also be calculated (predicted) using the Born–Landé equation as the sum of the electrostatic potential energy, calculated by summing interactions between cations and anions, and a short-range repulsive potential energy term.

The electrostatic potential can be expressed in terms of the interionic separation and a constant (Madelung constant) that takes account of the geometry of the crystal. The further away from the nucleus the weaker the shield. The Born–Landé equation gives a reasonable fit to the lattice energy of, e.g., sodium chloride, where the calculated (predicted) value is -756 kJ/mol, which compares to -787 kJ/

mol using the Born–Haber cycle. In aqueous solution the binding strength can be described by the Bjerrum or Fuoss equation as function of the ion charges, rather independent of the nature of the ions such as polarizability or size. The strength of salt bridges is most often evaluated by measurements of equilibria between molecules containing cationic and anionic sites, most often in solution. Equilibrium constants in water indicate additive free energy contributions for each salt bridge. Another method for the identification of hydrogen bonds also in complicated molecules is crystallography, sometimes also NMR-spectroscopy. The attractive forces defining the strength of ionic bonding can be modeled by Coulomb's Law. Ionic bond strengths are typically (cited ranges vary) between 170 and 1500 kJ/mol.

3.5 POLARIZATION POWER EFFECTS

Ions in crystal lattices of purely ionic compounds are spherical; however, if the positive ion is small and/or highly charged, it will distort the electron cloud of the negative ion, an effect summarised in Fajans' rules. This polarization of the negative ion leads to a build-up of extra charge density between the two nuclei, that is, to partial covalency. Larger negative ions are more easily polarized, but the effect is usually important only when positive ions with charges of 3+ (e.g., Al^{3+}) are involved. However, 2+ ions (Be^{2+}) or even 1+ (Li^{+}) show some polarizing power because their sizes are so small (e.g., LiI is ionic but has some covalent bonding present). Note that this is not the ionic polarization effect that refers to displacement of ions in the lattice due to the application of an electric field.

3.6 COMPARISON WITH COVALENT BONDING

In ionic bonding, the atoms are bound by attraction of oppositely charged ions, whereas, in covalent bonding, atoms are bound by sharing electrons to attain stable electron configurations. In covalent bonding, the molecular geometry around each atom is determined by valence shell electron pair repulsion VSEPR rules, whereas, in ionic materials, the geometry follows maximum packing rules. One could say that covalent bonding is more *directional* in the sense that the energy penalty for not adhering to the optimum bond angles is large, whereas ionic bonding has no such penalty. There are no shared electron pairs to repel each other, the ions should simply be packed as efficiently as possible. This often leads to much higher coordination numbers. In NaCl, each ion has 6 bonds and all bond angles are 90°. In CsCl the coordination number is 8. By comparison carbon typically has a maximum of four bonds.

Purely ionic bonding cannot exist, as the proximity of the entities involved in the bonding allows some degree of sharing electron density between them. Therefore, all ionic bonding has some covalent character. Thus, bonding is considered ionic where the ionic character is greater than the covalent character. The larger the difference in electronegativity between the two types of atoms involved in the bonding, the more ionic (polar) it is. Bonds with partially ionic and partially covalent character are called polar covalent bonds. For example, Na–Cl and Mg–O interactions have a few percent covalency, while Si–O bonds are usually ~50% ionic and ~50% covalent. Pauling estimated that an electronegativity difference of 1.7 (on the Pauling scale) corresponds to 50% ionic character, so that a difference greater than 1.7 corresponds to a bond which is predominantly ionic.

Ionic character in covalent bonds can be directly measured for atoms having quadrupolar nuclei (^{2}H, ^{14}N, $^{81,79}Br$, $^{35,37}Cl$ or ^{127}I). These nuclei are generally objects of NQR nuclear quadrupole resonance and NMR nuclear magnetic resonance studies. Interactions between the nuclear quadrupole moments Q and the electric field gradients (EFG) are characterized via the nuclear quadrupole coupling constants

$$QCC = \frac{e^2 q_{zz} Q}{h}$$

Where the eq_{zz} term corresponds to the principal component of the EFG tensor and e is the elementary charge. In turn, the electric field gradient opens the way to description of bonding modes in molecules when the QCC values are accurately determined by NMR or NQR methods.

In general, when ionic bonding occurs in the solid (or liquid) state, it is not possible to talk about a single "ionic bond" between two individual atoms, because the cohesive forces that keep the lattice together are of a more collective nature. This is quite different in the case of covalent bonding, where we can often speak of a distinct bond localized between two particular atoms. However, even if ionic bonding is combined with some covalency, the result is *not* necessarily discrete bonds of a localized character. In such cases, the resulting bonding often requires description in terms of a band structure consisting of gigantic molecular orbitals spanning the entire crystal. Thus, the bonding in the solid often retains its collective rather than localized nature. When the difference in electronegativity is decreased, the bonding may then lead to a semiconductor, a semimetal or eventually a metallic conductor with metallic bonding.

CHAPTER 4
THE COVALENT BOND

A covalent bond is a chemical bond that involves the sharing of electrons to form electron pairs between atoms. These electron pairs are known as shared pairs or bonding pairs. The stable balance of attractive and repulsive forces between atoms, when they share electrons, is known as covalent bonding. For many molecules, the sharing of electrons allows each atom to attain the equivalent of a full valence shell, corresponding to a stable electronic configuration. In organic chemistry, covalent bonding is much more common than ionic bonding.

Covalent bonding also includes many kinds of interactions, including σ-bonding, π-bonding, metal-to-metal bonding, agostic interactions, bent bonds, three-center two-electron bonds and three-center four-electron bonds. The term *covalent bond* dates from 1939. The prefix *co-* means *jointly, associated in action, partnered to a lesser degree,* etc.; thus a "co-valent bond", in essence, means that the atoms share "valence", such as is discussed in valence bond theory.

In the molecule H_2, the hydrogen atoms share the two electrons via covalent bonding. Covalency is greatest between atoms of similar electronegativities. Thus, covalent bonding does not necessarily require that the two atoms be of the same elements, only that they be of comparable electronegativity. Covalent bonding that entails the sharing of electrons over more than two atoms is said to be delocalized.

4.1 HISTORY

The term *covalence* in regard to bonding was first used in 1919 by Irving Langmuir in a *Journal of the American Chemical Society* article entitled "The Arrangement of Electrons in Atoms and Molecules". Langmuir wrote that "we shall denote by the term *covalence* the number of pairs of electrons that a given atom shares with its neighbors."

The idea of covalent bonding can be traced several years before 1919 to Gilbert N. Lewis, who in 1916 described the sharing of electron pairs between atoms (and in 1926 he also coined the term "photon" for the smallest unit of radiant energy). He introduced the *Lewis notation* or *electron dot notation* or *Lewis dot structure*, in which valence electrons (those in the outer shell) are represented as dots around the atomic symbols. Pairs of electrons located between atoms represent covalent bonds. Multiple pairs represent multiple bonds, such as double bonds and triple bonds. An alternative form of representation, not shown here, has bond-forming electron pairs represented as solid lines.

Lewis proposed that an atom forms enough covalent bonds to form a full (or closed) outer electron shell. In the diagram of methane shown here, the carbon atom has a valence of four and is, therefore, surrounded by eight electrons (the octet rule), four from the carbon itself and four from the hydrogens bonded to it. Each hydrogen has a valence of one and is surrounded by two electrons (a duet rule) – its own one electron plus one from the carbon. The numbers of electrons correspond to full shells in the quantum theory of the atom; the outer shell of a carbon atom is the $n = 2$ shell, which can hold eight electrons, whereas the outer (and only) shell of a hydrogen atom is the $n = 1$ shell, which can hold only two.

While the idea of shared electron pairs provides an effective qualitative picture of covalent bonding, quantum mechanics is needed to understand the nature of these bonds and predict the structures and properties of simple molecules. Walter Heitler and Fritz London are credited with the first successful quantum mechanical explanation of a chemical bond (molecular hydrogen) in 1927. Their work was based on the valence bond model, which assumes that a chemical bond is formed when there is good overlap between the atomic orbitals of participating atoms.

4.2 TYPES OF COVALENT BONDS

Atomic orbitals (except for s orbitals) have specific directional properties leading to different types of covalent bonds. Sigma (σ) bonds are the strongest covalent bonds and are due to head-on overlapping of orbitals on two different atoms. A single bond is usually a σ bond. Pi (π) bonds are weaker and are due to lateral overlap between p (or d) orbitals. A double bond between two given atoms consists of one σ and one π bond, and a triple bond is one σ and two π bonds.

Covalent bonds are also affected by the electronegativity of the connected atoms which determines the chemical polarity of the bond. Two atoms with equal electronegativity will make nonpolar covalent bonds such as H–H. An unequal relationship creates a polar covalent bond such as with H-Cl. However polarity also requires geometric asymmetry, or else dipoles may cancel out, resulting in a non-polar molecule.

4.3 COVALENT STRUCTURES

There are several types of structures for covalent substances, including individual molecules, molecular structures, macromolecular structures and giant covalent structures. Individual molecules have strong bonds that hold the atoms together, but generally, there are negligible forces of attraction between molecules. Such covalent substances are usually gases, for example, HCl, SO_2, CO_2, and CH_4. In molecular structures, there are weak forces of attraction. Such covalent substances are low-boiling-temperature liquids (such as ethanol), and low-melting-temperature solids (such as iodine and solid CO_2). Macromolecular structures have large numbers of atoms linked by covalent bonds in chains, including synthetic polymers such as polyethylene and nylon, and biopolymers such as proteins and starch. Network covalent structures (or giant covalent structures) contain large numbers of atoms linked in sheets (such as graphite), or 3-dimensional structures (such as diamond and quartz). These substances have high melting and boiling points, are frequently brittle, and tend to have high electrical resistivity. Elements that have high electronegativity, and the ability to form three or four electron pair bonds, often form such large macromolecular structures.

4.4 ONE-AND THREE-ELECTRON BONDS

Bonds with one or three electrons can be found in radical species, which have an odd number of electrons. The simplest example of a 1-electron bond is found in the dihydrogen cation, $H^+{}_2$. One-electron bonds often have about half the bond energy of a 2-electron bond, and are therefore called "half bonds". However, there are exceptions: in the case of dilithium, the bond is actually stronger for the 1-electron $Li^+{}_2$ than for the 2-electron Li_2. This exception can be explained in terms of hybridization and inner-shell effects.

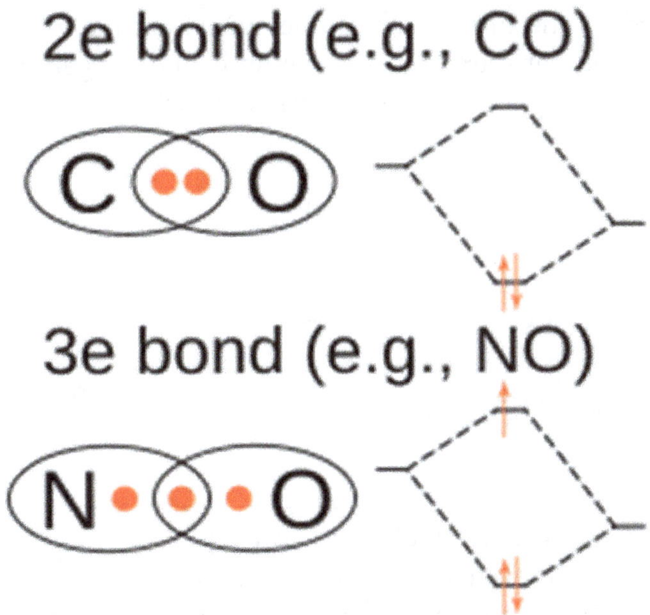

Fig 1: Lewis and MO diagrams of an individual 2e⁻ bond and 3e⁻ bond

$_2$. It is considered a "half bond" because it consists of only one shared electron (rather than two); in molecular orbital terms, the third electron is in an anti-bonding orbital which cancels out half of the bond formed by the other two electrons.

Another example of a molecule containing a 3-electron bond, in addition to two 2-electron bonds, is nitric oxide, NO. The oxygen molecule, O_2 can also be regarded as having two 3-electron bonds and one 2-electron bond, which accounts for its paramagnetism and its formal bond order of 2. Chlorine dioxide and its heavier analogues bromine dioxide and iodine dioxide also contain three-electron bonds.

Molecules with odd-electron bonds are usually highly reactive. These types of bond are only stable between atoms with similar electronegativities.

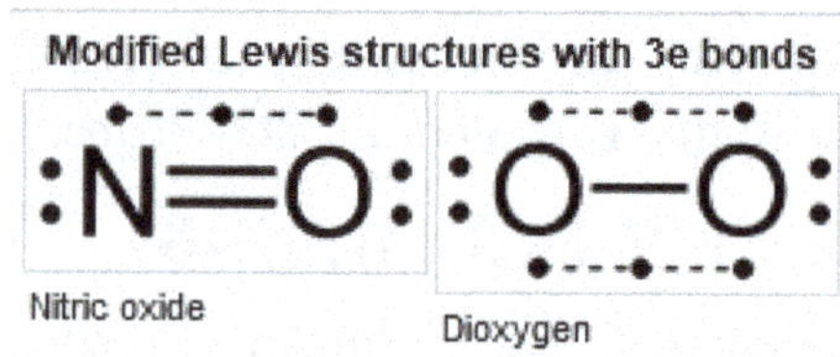

4.5 RESONANCE

There are situations whereby a single Lewis structure is insufficient to explain the electron configuration in a molecule and its resulting experimentally-determined properties, hence a superposition of structures is needed. The same two atoms in such molecules can be bonded differently in different Lewis structures (a single bond in one, a double bond in another, or even none at all), resulting in a non-integer bond order. The nitrate ion is one such example with three equivalent structures. The bond between the nitrogen and each oxygen is a double bond in one structure and a single bond in the other two, so that the average bond order for each N–O interaction is □

$$\frac{2+1+1}{3}=\frac{4}{3}.$$

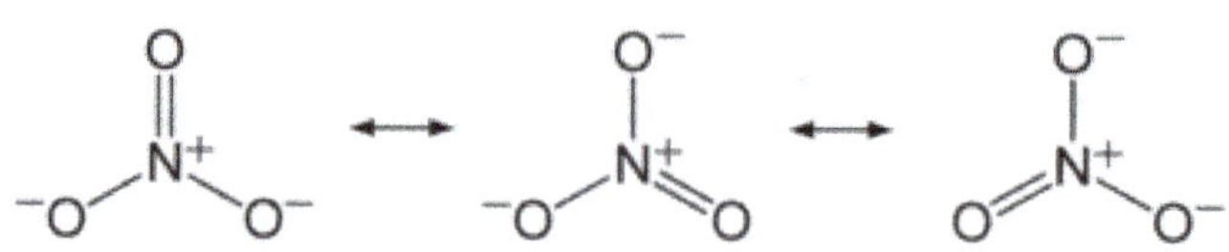

4.5.1 Aromaticity

In organic chemistry, when a molecule with a planar ring obeys Hückel's rule, where the number of π electrons fit the formula $4n + 2$ (where n is an integer), it attains extra stability and symmetry. In benzene, the prototypical aromatic compound, there are 6π bonding electrons ($n = 4 ,1n + 6 = 2$). These occupy three delocalized π molecular orbitals (molecular orbital theory) or form conjugate π bonds in two resonance structures that linearly combine (valence bond theory), creating a regular hexagon exhibiting a greater stabilization than the hypothetical 1,3,5-cyclohexatriene.

In the case of heterocyclic aromatics and substituted benzenes, the electronegativity differences between different parts of the ring may dominate the chemical behavior of aromatic ring bonds, which otherwise are equivalent.

4.5.2 Hypervalence

Certain molecules such as xenon difluoride and sulfur hexafluoride have higher co-ordination numbers than would be possible due to strictly covalent bonding according to the octet rule. This is explained by the three-center four-electron bond ("3c–4e") model which interprets the molecular wavefunction in terms of non-bonding highest occupied molecular orbitals in molecular orbital theory and resonance of sigma bonds in valence bond theory.

4.5.3 Electron Deficiency

In three-center two-electron bonds ("3c–2e") three atoms share two electrons in bonding. This type of bonding occurs in boron hydrides such as diborane (B_2H_6), which are often described as electron deficient because there are not enough valence electrons to form localized (2-centre 2-electron) bonds joining all the atoms. However the more modern description using 3c–2e bonds does provide enough bonding orbitals to connect all the atoms, so that the molecules can instead be classified as electron-precise.

Each such bond (2 per molecule in diborane) contains a pair of electrons which connect the boron atoms to each other in a banana shape, with a proton (the nucleus of a hydrogen atom) in the middle of the bond, sharing electrons with both boron atoms. In certain cluster compounds, so-called four-center two-electron bonds also have been postulated.

4.6 QUANTUM MECHANICAL DESCRIPTION

After the development of quantum mechanics, two basic theories were proposed to provide a quantum description of chemical bonding: valence bond (VB) theory and molecular orbital (MO) theory. A more recent quantum description is given in terms of atomic contributions to the electronic density of states.

4.6.1 Comparison of VB and MO Theories

The two theories represent two ways to build up the electron configuration of the molecule.

For valence bond theory, the atomic hybrid orbitals are filled with electrons first to produce a fully bonded valence configuration, followed by performing a linear combination of contributing structures (resonance) if there are several of them. In contrast, for molecular orbital theory a linear combination of atomic orbitals is performed first, followed by filling of the resulting molecular orbitals with electrons.

The two approaches are regarded as complementary, and each provides its own insights into the problem of chemical bonding. As valence bond theory builds the molecular wavefunction out of localized bonds, it is more suited for the calculation of bond energies and the understanding of reaction mechanisms. As molecular orbital theory builds the molecular wavefunction out of delocalized orbitals, it is more suited for the calculation of ionization energies and the understanding of spectral absorption bands.

At the qualitative level, both theories contain incorrect predictions. Simple (Heitler–London) valence bond theory correctly predicts the dissociation of homonuclear diatomic molecules into separate atoms, while simple (Hartree–Fock) molecular orbital theory incorrectly predicts dissociation into a mixture of atoms and ions. On the other hand, simple molecular orbital theory correctly predicts Hückel's rule of aromaticity, while simple valence bond theory incorrectly predicts that cyclobutadiene has larger resonance energy than benzene.

Although the wavefunctions generated by both theories at the qualitative level do not agree and do not match the stabilization energy by experiment, they can be corrected by configuration interaction. This is done by combining the valence bond covalent function with the functions describing all possible ionic structures or by combining the molecular orbital ground state function with the functions describing all possible excited states using unoccupied orbitals. It can then be seen that the simple molecular orbital approach overestimates the weight of the ionic structures while the simple valence bond approach neglects them. This can also be described as saying that the simple molecular orbital approach neglects electron correlation while the simple valence bond approach overestimates it.

Modern calculations in quantum chemistry usually start from (but ultimately go far beyond) a molecular orbital rather than a valence bond approach, not because of any intrinsic superiority in the former but rather because the MO approach is more readily adapted to numerical computations. Molecular orbitals are orthogonal, which significantly increases the feasibility and speed of computer calculations compared to nonorthogonal valence bond orbitals.

4.7 ANALOGOUS EFFECT IN NUCLEAR SYSTEMS

An analogous effect to covalent binding is believed to occur in some nuclear systems, with the difference that the shared fermions are quarks rather than electrons. High energy proton-proton scattering cross-section indicates that quark interchange of either u or d quarks is the dominant process of the nuclear force at short distance. In particular, it dominates over the Yukawa interaction where a meson is exchanged. Therefore, covalent binding by quark interchange is expected to be the dominating mechanism of nuclear binding at small distance when the bound hadrons have covalence quarks in common.

CHAPTER 5
THE METALLIC BOND

Metallic bonding is a type of chemical bonding that arises from the electrostatic attractive force between conduction electrons (in the form of an electron cloud of delocalized electrons) and positively charged metal ions. It may be described as the sharing of *free* electrons among a structure of positively charged ions (cations). Metallic bonding accounts for many physical properties of metals, such as strength, ductility, thermal and electrical resistivity and conductivity, opacity, and lustre.

Metallic bonding is not the only type of chemical bonding a metal can exhibit, even as a pure substance. For example, elemental gallium consists of covalently-bound pairs of atoms in both liquid and solid-state—these pairs form a crystal structure with metallic bonding between them. Another example of a metal–metal covalent bond is the mercurous ion (Hg^{2+}_2).

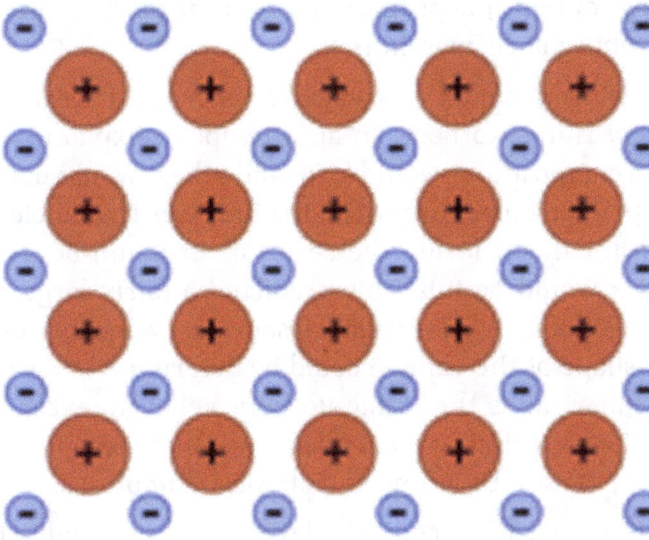

Fig 1: An Example Showing Metallic Bonding. + Represents Cations, - Represents the Free Floating Electrons

5.1 HISTORY

As chemistry developed into a science, it became clear that metals formed the majority of the periodic table of the elements, and great progress was made in the description of the salts that can be formed in reactions with acids. With the advent of electrochemistry, it became clear that metals generally go into solution as positively charged ions, and the oxidation reactions of the metals became well understood in their electrochemical series. A picture emerged of metals as positive ions held together by an ocean of negative electrons.

With the advent of quantum mechanics, this picture was given a more formal interpretation in the form of the free electron model and its further extension, the nearly free electron model. In both models, the electrons are seen as a gas traveling through the structure of the solid with an energy that is essentially isotropic, in that it depends on the square of the magnitude, *not* the direction of the momentum vector k. In three-dimensional k-space, the set of points of the highest filled levels (the Fermi surface) should therefore be a sphere. In the nearly-free model, box-like Brillouin zones are added to k-space by the periodic potential experienced from the (ionic) structure, thus mildly breaking the isotropy.

The advent of X-ray diffraction and thermal analysis made it possible to study the structure of crystalline solids, including metals and their alloys; and phase diagrams were developed. Despite all this progress, the nature of intermetallic compounds and alloys largely remained a mystery and their study was often merely empirical. Chemists generally steered away from anything that did not seem to follow Dalton's laws of multiple proportions; and the problem was considered the domain of a different science, metallurgy.

The nearly-free electron model was eagerly taken up by some researchers in metallurgy, notably Hume-Rothery, in an attempt to explain why intermetallic alloys with certain compositions would form and others would not. Initially Hume-Rothery's attempts were quite successful. His idea was to add electrons to inflate the spherical Fermi-balloon inside the series of Brillouin-boxes and determine when a certain box would be full. This predicted a fairly large number of alloy compositions that were later observed. As soon as cyclotron resonance became available and the shape of the balloon could be determined, it was found that the balloon was not spherical as the Hume-Rothery believed, except perhaps in the case of caesium. This revealed how a model can sometimes give a whole series of correct predictions, yet still be wrong in its basic assumptions.

The nearly-free electron debacle compelled researchers to modify the assumpition that ions flowed in a sea of free electrons. A number of quantum mechanical models were developed, such as band structure calculations based on molecular orbitals, and the density functional theory. These models either depart from the atomic orbitals of neutral atoms that share their electrons, or (in the case of density functional theory) departs from the total electron density. The free-electron picture has, nevertheless, remained a dominant one in introductory courses on metallurgy.

The electronic band structure model became a major focus for the study of metals and even more of semiconductors. Together with the electronic states, the vibrational states were also shown to form bands. Rudolf Peierls showed that, in the case of a one-dimensional row of metallic atoms—say, hydrogen—an inevitable instability would break such a chain into individual molecules. This sparked an interest in the general question: when is collective metallic bonding stable, and when will a localized bonding take its place? Much research went into the study of clustering of metal atoms.

As powerful as the band structure model proved to be in describing metallic bonding, it remains a one-electron approximation of a many-body problem: the energy states of an individual electron are described as if all the other electrons form a homogeneous background. Researchers such as Mott and Hubbard realized that the one-electron treatment was perhaps appropriate for strongly delocalized s- and p-electrons; but for d-electrons, and even more for f-electrons, the interaction with nearby individual electrons (and atomic displacements) may become stronger than the delocalized interaction that leads to broad bands. This gave a better explanation for the transition from localized unpaired electrons to itinerant ones partaking in metallic bonding.

5.2 THE NATURE OF METALLIC BONDING

The combination of two phenomena gives rise to metallic bonding: delocalization of electrons and the availability of a far larger number of delocalized energy states than of delocalized electrons. The latter could be called electron deficiency.

5.2.1 In 2D

Graphene is an example of two-dimensional metallic bonding. Its metallic bonds are similar to aromatic bonding in benzene, naphthalene, anthracene, ovalene, etc.

5.2.2 In 3D

Metal aromaticity in metal clusters is another example of delocalization, this time often in three-dimensional arrangements. Metals take the delocalization principle to its extreme, and one could say that a crystal of a metal represents a single molecule over which all conduction electrons are delocalized in all three dimensions. This means that inside the metal one can generally not distinguish molecules, so that the metallic bonding is neither intra- nor inter-molecular. 'Nonmolecular' would perhaps be a better term. Metallic bonding is mostly non-polar, because even in alloys there is little difference among the electronegativities of the atoms participating in the bonding interaction (and, in pure elemental metals, none at all). Thus, metallic bonding is an extremely delocalized communal form of covalent bonding. In a sense, metallic bonding is not a 'new' type of bonding at all. It describes the bonding only as present in a *chunk* of condensed matter: be it crystalline solid, liquid, or even glass. Metallic vapors, in contrast, are often atomic (Hg) or at times contain molecules, such as Na_2, held together by a more conventional covalent bond. This is why it is not correct to speak of a single 'metallic bond'.

Delocalization is most pronounced for s- and p-electrons. Delocalization in caesium is so strong that the electrons are virtually freed from the caesium atoms to form a gas constrained only by the surface of the metal. For caesium, therefore, the picture of Cs^+ ions held together by a negatively charged electron gas is very close to accurate (though not perfectly so). For other elements the electrons are less free, in that they still experience the potential of the metal atoms, sometimes quite strongly. They require a more intricate quantum mechanical treatment (e.g., tight binding) in which the atoms are viewed as neutral, much like the carbon atoms in benzene. For d- and especially f-electrons the delocalization is not strong at all and this explains why these electrons are able to continue behaving as unpaired electrons that retain their spin, adding interesting magnetic properties to these metals.

5.2.3 Electron Deficiency and Mobility

Metal atoms contain few electrons in their valence shells relative to their periods or energy levels. They are electron-deficient elements and the communal sharing does not change that. There remain far more available energy states than there are shared electrons. Both requirements for conductivity are therefore fulfilled: strong delocalization and partly filled energy bands. Such electrons can therefore easily change from one energy state to a slightly different one.

Thus, not only do they become delocalized, forming a sea of electrons permeating the structure, but they are also able to migrate through the structure when an external electrical field is applied, leading to electrical conductivity. Without the field, there are electrons moving equally in all directions. Within such a field, some electrons will adjust their state slightly, adopting a different wave vector. Consequently, there will be more moving one way than another and a net current will result.

The freedom of electrons to migrate also gives metal atoms, or layers of them, the capacity to slide past each other. Locally, bonds can easily be broken and replaced by new ones after a deformation. This process does not affect the communal metallic bonding very much, which gives rise to metals' characteristic malleability and ductility. This is particularly true for pure elements. In the presence of dissolved impurities, the normally easily formed cleavages may be blocked and the material become harder. Gold, for example, is very soft in pure form (24-karat), which is why alloys are preferred in jewelry.

Metals are typically also good conductors of heat, but the conduction electrons only contribute partly to this phenomenon. Collective (i.e., delocalized) vibrations of the atoms, known as phonons that travel through the solid as a wave, are bigger contributors.

However, a substance such as diamond, which conducts heat quite well, is not an electrical conductor. This is not a consequence of delocalization being absent in diamond, but simply that carbon is not electron deficient.

Electron deficiency is important in distinguishing metallic from more conventional covalent bonding. Thus, we should amend the expression given above to: *Metallic bonding is an extremely delocalized communal form of electron-deficient covalent bonding.*

5.3 METALLIC RADIUS

The metallic radius is defined as one-half of the distance between the two adjacent metal ions in the metallic structure. This radius depends on the nature of the atom as well as its environment—specifically, on the coordination number (CN), which in turn depends on the temperature and applied pressure.

When comparing periodic trends in the size of atoms it is often desirable to apply the so-called Goldschmidt correction, which converts atomic radii to the values the atoms would have if they were 12-coordinated. Since metallic radii are largest for the highest coordination number, correction for less dense coordinations involves multiplying by x, where $0 < x < 1$. Specifically, for CN = 4, $x = 0.88$; for CN = 6, $x = 0.96$, and for CN = 8, $x = 0.97$. The correction is named after Victor Goldschmidt who obtained the numerical values quoted above.

The radii follow general periodic trends: they decrease across the period due to the increase in the effective nuclear charge, which is not offset by the increased number of valence electrons; but the radii increase down the group due to an increase in the principal quantum number. Between the 4d and 5d elements, the lanthanide contraction is observed—there is very little increase of the radius down the group due to the presence of poorly shielding f orbitals.

5.4 STRENGTH OF THE BOND

The atoms in metals have a strong attractive force between them. Much energy is required to overcome it. Therefore, metals often have high boiling points, with tungsten (5828 K) being extremely high. A remarkable exception is the elements of the zinc group: Zn, Cd, and Hg. Their electron configurations end in ...ns^2, which resembles a noble gas configuration, like that of helium, more and more when going down the periodic table, because the energy differential to the empty np orbitals becomes larger. These metals are therefore relatively volatile, and are avoided in ultra-high vacuum systems.

Otherwise, metallic bonding can be very strong, even in molten metals, such as gallium. Even though gallium will melt from the heat of one's hand just above room temperature, its boiling point is not far from that of copper. Molten gallium is, therefore, a very nonvolatile liquid, thanks to its strong metallic bonding.

The strong bonding of metals in liquid form demonstrates that the energy of a metallic bond is not highly dependent on the direction of the bond; this lack of bond directionality is a direct consequence of electron delocalization, and is best understood in contrast to the directional bonding of covalent bonds. The energy of a metallic bond is thus mostly a function of the number of electrons which surround the metallic atom, as exemplified by the embedded atom model. This typically results in metals assuming relatively simple, close-packed crystal structures, such as FCC, BCC, and HCP.

Given high enough cooling rates and appropriate alloy composition, metallic bonding can occur even in glasses, which have amorphous structures.

Much biochemistry is mediated by the weak interaction of metal ions and biomolecules. Such interactions, and their associated conformational changes, have been measured using dual polarisation interferometry.

5.5 SOLUBILITY AND COMPOUND FORMATION

Metals are insoluble in water or organic solvents, unless they undergo a reaction with them. Typically, this is an oxidation reaction that robs the metal atoms of their itinerant electrons, destroying the metallic bonding. However metals are often readily soluble in each other while retaining the metallic character of their bonding. Gold, for example, dissolves easily in mercury, even at room temperature. Even in solid metals, the solubility can be extensive. If the structures of the two metals are the same, there can even be complete solid solubility, as in the case of electrum, an alloy of silver and gold. At times, however, two metals will form alloys with different structures than either of the two parents. One could call these materials metal compounds. But, because materials with metallic bonding are typically not molecular, Dalton's law of integral proportions is not valid; and often a range of stoichiometric ratios can be achieved. It is better to abandon such concepts as 'pure substance' or 'solute' in such cases and speak of phases instead. The study of such phases has traditionally been more the domain of metallurgy than of chemistry, although the two fields overlap considerably.

5.6 LOCALIZATION AND CLUSTERING: FROM BONDING TO BONDS

The metallic bonding in complex compounds does not necessarily involve all constituent elements equally. It is quite possible to have one or more elements that do not partake at all. One could picture the conduction electrons flowing around them like a river around an island or a big rock. It is possible to observe which elements do partake: e.g., by looking at the core levels in an X-ray photoelectron spectroscopy (XPS) spectrum. If an element partakes, its peaks tend to be skewed.

Some intermetallic materials, e.g., do exhibit metal clusters reminiscent of molecules; and these compounds are more a topic of chemistry than of metallurgy. The formation of the clusters could be seen as a way to 'condense out' (localize) the electron-deficient bonding into bonds of a more localized nature. Hydrogen is an extreme example of this form of condensation. At high pressures it is a metal. The core of the planet Jupiter could be said to be held together by a combination of metallic bonding and high pressure induced by gravity. At lower pressures, however, the bonding becomes entirely localized into a regular covalent bond. The localization is so complete that the (more familiar) H_2 gas results. A similar argument holds for an element such as boron. Though it is electron-deficient compared to carbon, it does not form a metal. Instead it has a number of complex structures in which icosahedral B_{12} clusters dominate. Charge density waves are a related phenomenon.

As these phenomena involve the movement of the atoms toward or away from each other, they can be interpreted as the coupling between the electronic and the vibrational states (i.e., the phonons) of the material. A different such electron-phonon interaction is thought to lead to a very different result at low temperatures, that of superconductivity. Rather than blocking the mobility of the charge carriers by forming electron pairs in localized bonds, Cooper pairs are formed that no longer experience any resistance to their mobility.

5.7 OPTICAL PROPERTIES

The presence of an ocean of mobile charge carriers has profound effects on the optical properties of metals, which can only be understood by considering the electrons as a *collective*, rather than considering the states of individual electrons involved in more conventional covalent bonds.

Light consists of a combination of an electrical and a magnetic field. The electrical field is usually able to excite an elastic response from the electrons involved in the metallic bonding.

The result is that photons cannot penetrate very far into the metal and are typically reflected, although some may also be absorbed. This holds equally for all photons in the visible spectrum, which is why metals are often silvery white or grayish with the characteristic specular reflection of metallic lustre. The balance between reflection and absorption determines how white or how gray a metal is, although surface tarnish can obscure the lustre. Silver, a metal with high conductivity, is one of the whitest.

Notable exceptions are reddish copper and yellowish gold. The reason for their color is that there is an upper limit to the frequency of the light that metallic electrons can readily respond to: the plasmon frequency. At the plasmon frequency, the frequency-dependent dielectric function of the free electron gas goes from negative (reflecting) to positive (transmitting); higher frequency photons are not reflected at the surface, and do not contribute to the color of the metal. There are some materials, such as indium tin oxide (ITO), that are metallic conductors (actually degenerate semiconductors) for which this threshold is in the infrared, which is why they are transparent in the visible, but good reflectors in the infrared.

For silver the limiting frequency is in the far ultraviolet, but for copper and gold it is closer to the visible. This explains the colors of these two metals. At the surface of a metal, resonance effects known as surface plasmons can result. They are collective oscillations of the conduction electrons, like a ripple in the electronic ocean. However, even if photons have enough energy, they usually do not have enough momentum to set the ripple in motion. Therefore, plasmons are hard to excite on a bulk metal. This is why gold and copper look like lustrous metals albeit with a dash of color. However, in colloidal gold the metallic bonding is confined to a tiny metallic particle, which prevents the oscillation wave of the plasmon from 'running away'. The momentum selection rule is therefore broken, and the plasmon resonance causes an extremely intense absorption in the green, with a resulting purple-red color. Such colors are orders of magnitude more intense than ordinary absorptions seen in dyes and the like, which involve individual electrons and their energy states.

CHAPTER 6

GENERAL PROPERTIES OF THE ELEMENTS

A chemical element is a chemical substance that cannot be broken down into other substances by chemical reactions. The basic particle that constitutes a chemical element is the atom. Elements are identified by the number of protons in their nucleus, known as the element's atomic number. For example, oxygen has an atomic number of 8, meaning each oxygen atom has 8 protons in its nucleus. Atoms of the same element can have different numbers of neutrons in their nuclei, known as isotopes of the element.

Two or more atoms can combine to form molecules. Chemical compounds are molecules made of atoms of different elements, while mixtures contain atoms of different elements not necessarily combined as molecules. Atoms can be transformed into different elements in nuclear reactions, which change an atom's atomic number.

Almost all baryonic matter in the universe is composed of elements (among rare exceptions are neutron stars). When different elements undergo chemical reactions, atoms are rearranged into new compounds held together by chemical bonds. Only a few elements, such as silver and gold, are found uncombined as relatively pure native element minerals. Nearly all other naturally occurring elements occur in the Earth as compounds or mixtures. Air is mostly a mixture of molecular nitrogen and oxygen, though it does contain compounds including carbon dioxide and water, as well as atomic argon, a noble gas which is chemically inert and therefore does not undergo chemical reactions.

The history of the discovery and use of elements began with early human societies that discovered native minerals like carbon, sulfur, copper and gold (though the modern concept of an element was not yet understood). Attempts to classify materials such as these resulted in the concepts of classical elements, alchemy, and similar theories throughout history.

Much of the modern understanding of elements developed from the work of Dmitri Mendeleev, a Russian chemist who published the first recognizable periodic table in 1869. This table organizes the elements by increasing atomic number into rows ("periods") in which the columns ("groups") share recurring ("periodic") physical and chemical properties. The periodic table summarizes various properties of the elements, allowing chemists to derive relationships between them and to make predictions about elements not yet discovered, and potential new compounds.

By November 2016, the International Union of Pure and Applied Chemistry (IUPAC) had recognized a total of 118 elements. The first 94 occur naturally on Earth, and the remaining 24 are synthetic elements produced in nuclear reactions. Save for unstable radioactive elements (radioelements) which decay quickly, nearly all elements are available industrially in varying amounts. The discovery and synthesis of further new elements is an ongoing area of scientific study.

6.1 DESCRIPTION

The lightest elements are hydrogen and helium, both created by Big Bang nucleosynthesis in the first 20 minutes of the universe in a ratio of around 3:1 by mass (or 12:1 by number of atoms), along with tiny traces of the next two elements, lithium and beryllium. Almost all other elements found in nature were made by various natural methods of nucleosynthesis. On Earth, small amounts of new atoms are naturally produced in nucleogenic reactions, or in cosmogenic processes, such as cosmic ray spallation. New atoms are also naturally produced on Earth as radiogenic daughter isotopes of ongoing radioactive decay processes such as alpha decay, beta decay, spontaneous fission, cluster decay, and other rarer modes of decay.

Of the 94 naturally occurring elements, those with atomic numbers 1 through 82 each have at least one stable isotope (except for technetium, element 43 and promethium, element 61, which have no stable isotopes). Isotopes considered stable are those for which no radioactive decay has yet been observed. Elements with atomic numbers 83 through 94 are unstable to the point that radioactive decay of all isotopes can be detected. Some of these elements, notably bismuth (atomic number 83), thorium (atomic number 90), and uranium (atomic number 92), have one or more isotopes with half-lives long enough to survive as remnants of the explosive stellar nucleosynthesis that produced the heavy metals before the formation of our Solar System. At over 1.9×10^{19} years, over a billion times longer than the estimated age of the universe, bismuth-209 has the longest known alpha decay half-life of any isotope, and is almost always considered on par with the 80 stable elements. The heaviest elements (those beyond plutonium, element 94) undergo radioactive decay with half-lives so short that they are not found in nature and must be synthesized.

There are now 118 known elements. In this context, "known" means observed well enough, even from just a few decay products, to have been differentiated from other elements. Most recently, the synthesis of element 118 (since named oganesson) was reported in October 2006, and the synthesis of element 117 (tennessine) was reported in April 2010. Of these 118 elements, 94 occur naturally on Earth. Six of these occur in extreme trace quantities: technetium, atomic number 43; promethium, number 61; astatine, number 85; francium, number 87; neptunium, number 93; and plutonium, number 94. These 94 elements have been detected in the universe at large, in the spectra of stars and also supernovae, where short-lived radioactive elements are newly being made. The first 94 elements have been detected directly on Earth as primordial nuclides present from the formation of the Solar System, or as naturally occurring fission or transmutation products of uranium and thorium.

The remaining 24 heavier elements, not found today either on Earth or in astronomical spectra, have been produced artificially: all are radioactive, with short half-lives; if any of these elements were present at the formation of Earth, they are certain to have completely decayed, and if present in novae, are in quantities too small to have been noted. Technetium was the first purportedly non-naturally occurring element synthesized, in 1937, though trace amounts of technetium have since been found in nature (and also the element may have been discovered naturally in 1925). This pattern of artificial production and later natural discovery has been repeated with several other radioactive naturally occurring rare elements.

List of the elements are available by name, atomic number, density, melting point, boiling point and chemical symbol, as well as ionization energy. The nuclides of stable and radioactive elements are also available as a list of nuclides, sorted by length of half-life for those that are unstable. One of the most convenient, and certainly the most traditional presentation of the elements, is in the form of the periodic table, which groups together elements with similar chemical properties (and usually also similar electronic structures).

6.1.1 Atomic Number

The atomic number of an element is equal to the number of protons in each atom, and defines the element.

For example, all carbon atoms contain 6 protons in their atomic nucleus; so the atomic number of carbon is 6. Carbon atoms may have different numbers of neutrons; atoms of the same element having different numbers of neutrons are known as isotopes of the element.

The number of protons in the nucleus also determines its electric charge, which in turn determines the number of electrons of the atom in its non-ionized state. The electrons are placed into atomic orbitals that determine the atom's chemical properties. The number of neutrons in a nucleus usually has very little effect on an element's chemical properties; except for hydrogen (for which the kinetic isotope effect is significant). Thus, all carbon isotopes have nearly identical chemical properties because they all have six electrons, even though they may have 6 to 8 neutrons. That is why atomic number, rather than mass number or atomic weight, is considered the identifying characteristic of an element. The symbol for atomic number is Z.

6.1.2 Isotopes

Isotopes are atoms of the same element (that is, with the same number of protons in their nucleus), but having *different* numbers of neutrons. Thus, for example, there are three main isotopes of carbon.

All carbon atoms have 6 protons, but they can have either 6, 7, or 8 neutrons. Since the mass numbers of these are 12, 13 and 14 respectively, said three isotopes are known as carbon-12, carbon-13, and carbon-14 (^{12}C, ^{13}C, and ^{14}C). Natural carbon is a mixture of ^{12}C (about 98.9%), ^{13}C (about 1.1%) and about 1 atom per trillion of ^{14}C.

Most (66 of 94) naturally occurring elements have more than one stable isotope. Except for the isotopes of hydrogen (which differ greatly from each other in relative mass—enough to cause chemical effects), the isotopes of a given element are chemically nearly indistinguishable.

All elements have radioactive isotopes (radioisotopes); most of these radioisotopes do not occur naturally. Radioisotopes typically decay into other elements via alpha decay, beta decay, or inverse beta decay; some isotopes of the heaviest elements also undergo spontaneous fission. Isotopes that are not radioactive, are termed "stable" isotopes.

All known stable isotopes occur naturally. The many radioisotopes that are not found in nature have been characterized after being artificially produced. Certain elements have no stable isotopes and are composed *only* of radioisotopes: specifically the elements without any stable isotopes are technetium (atomic number 43), promethium (atomic number 61), and all observed elements with atomic number greater than 82.

Of the 80 elements with at least one stable isotope, 26 have only one stable isotope. The mean number of stable isotopes for the 80 stable elements is 3.1 stable isotopes per element. The largest number of stable isotopes for a single element is 10 (for tin, element 50).

6.1.3 Isotopic Mass and Atomic Mass

The mass number of an element, *A*, is the number of nucleons (protons and neutrons) in the atomic nucleus. Different isotopes of a given element are distinguished by their mass number, which is written as a superscript on the left hand side of the chemical symbol (e.g., ^{238}U). The mass number is always an integer and has units of "nucleons". Thus, magnesium-24 (24 is the mass number) is an atom with 24 nucleons (12 protons and 12 neutrons).

Whereas the mass number simply counts the total number of neutrons and protons and is thus an integer, the atomic mass of a particular isotope (or "nuclide") of the element is the mass of a single atom of that isotope, and is typically expressed in daltons (symbol: Da), or universal atomic mass units (symbol: u). Its relative atomic mass is a dimensionless number equal to the atomic mass divided by the atomic mass constant, which equals 1 Da.

In general, the mass number of a given nuclide differs in value slightly from its relative atomic mass, since the mass of each proton and neutron is not exactly 1 Da; since the electrons contribute a lesser share to the atomic mass as neutron number exceeds proton number; and because of the nuclear binding energy and electron binding energy. For example, the atomic mass of chlorine-35 to five significant digits is 34.969 Da and that of chlorine-37 is 36.966 Da. However, the relative atomic mass of each isotope is quite close to its mass number (always within 1%). The only isotope whose atomic mass is exactly a natural number is ^{12}C, which has a mass of 12 Da; because the dalton is defined as 1/12 of the mass of a free neutral carbon-12 atom in the ground state.

The standard atomic weight (commonly called "atomic weight") of an element is the *average* of the atomic masses of all the chemical element's isotopes as found in a particular environment, weighted by isotopic abundance, relative to the atomic mass unit. This number may be a fraction that is *not* close to a whole number. For example, the relative atomic mass of chlorine is 35.453 u, which differs greatly from a whole number as it is an average of about 76% chlorine-35 and 24% chlorine-37. Whenever a relative atomic mass value differs by more than ~1% from a whole number, it is due to this averaging effect, as significant amounts of more than one isotope are naturally present in a sample of that element.

6.1.4 Chemically Pure and Isotopically Pure

Chemists and nuclear scientists have different definitions of a *pure element*. In chemistry, a pure element means a substance whose atoms all (or in practice almost all) have the same atomic number, or number of protons. Nuclear scientists, however, define a pure element as one that consists of only one isotope.

For example, a copper wire is 99.99% chemically pure if 99.99% of its atoms are copper, with 29 protons each. However it is not isotopically pure since ordinary copper consists of two stable isotopes, 69% ^{63}Cu and 31% ^{65}Cu, with different numbers of neutrons. However, pure gold would be both chemically and isotopically pure, since ordinary gold consists only of one isotope, ^{197}Au.

6.1.5 Allotropes

Atoms of chemically pure elements may bond to each other chemically in more than one way, allowing the pure element to exist in multiple chemical structures (spatial arrangements of atoms), known as allotropes, which differ in their properties. For example, carbon can be found as diamond, which has a tetrahedral structure around each carbon atom; graphite, which has layers of carbon atoms with a hexagonal structure stacked on top of each other; graphene, which is a single layer of graphite that is very strong; fullerenes, which have nearly spherical shapes; and carbon nanotubes, which are tubes with a hexagonal structure (even these may differ from each other in electrical properties). The ability of an element to exist in one of many structural forms is known as 'allotropy'.

The reference state of an element is defined by convention, usually as the thermodynamically most stable allotrope and physical state at a pressure of 1 bar and a given temperature (typically at 298.15K). However, for phosphorus, the reference state is white phosphorus even though it is not the most stable allotrope, and the reference state for carbon is graphite, because the structure of graphite is more stable than that of the other allotropes. In thermochemistry, an element is defined to have an enthalpy of formation of zero in its reference state.

6.1.6 Properties

Several kinds of descriptive categorizations can be applied broadly to the elements, including consideration of their general physical and chemical properties, their states of matter under familiar conditions, their melting and boiling points, their densities, their crystal structures as solids, and their origins.

General Properties

Several terms are commonly used to characterize the general physical and chemical properties of the chemical elements. A first distinction is between metals, which readily conduct electricity, nonmetals, which do not, and a small group, (the *metalloids*), having intermediate properties and often behaving as semiconductors.

A more refined classification is often shown in colored presentations of the periodic table. This system restricts the terms "metal" and "nonmetal" to only certain of the more broadly defined metals and nonmetals, adding additional terms for certain sets of the more broadly viewed metals and nonmetals. The version of this classification used in the periodic tables presented here includes: actinides, alkali metals, alkaline earth metals, halogens, lanthanides, transition metals, post-transition metals, metalloids, reactive nonmetals, and noble gases. In this system, the alkali metals, alkaline earth metals, and transition metals, as well as the lanthanides and the actinides, are special groups of the metals viewed in a broader sense. Similarly, the reactive nonmetals and the noble gases are nonmetals viewed in the broader sense. In some presentations, the halogens are not distinguished, with astatine identified as a metalloid and the others identified as nonmetals.

States of Matter

Another commonly used basic distinction among the elements is their state of matter (phase), whether solid, liquid, or gas, at standard temperature and pressure (STP). Most elements are solids at STP, while several are gases. Only bromine and mercury are liquid at 0 degrees Celsius (32 degrees Fahrenheit) and 1 atmosphere pressure; caesium and gallium are solid at that temperature, but melt at 28.4°C (83.2°F) and 29.8°C (85.6°F), respectively.

Melting and Boiling Points

Melting and boiling points, typically expressed in degrees Celsius at a pressure of one atmosphere, are commonly used in characterizing the various elements. While known for most elements, either or both of these measurements is still undetermined for some of the radioactive elements available in only tiny quantities. Since helium remains a liquid even at absolute zero at atmospheric pressure, it has only a boiling point, and not a melting point, in conventional presentations.

Densities

The density at selected standard temperature and pressure (STP) is often used in characterizing the elements. Density is often expressed in grams per cubic centimetre (g/cm^3). Since several elements are gases at commonly encountered temperatures, their densities are usually stated for their gaseous forms; when liquefied or solidified, the gaseous elements have densities similar to those of the other elements.

When an element has allotropes with different densities, one representative allotrope is typically selected in summary presentations, while densities for each allotrope can be stated where more detail is provided. For example, the three familiar allotropes of carbon (amorphous carbon, graphite, and diamond) have densities of 1.8–2.1, 2.267, and 3.515 g/cm^3, respectively.

Crystal Structures

The elements studied to date as solid samples have eight kinds of crystal structures: cubic, body-centered cubic, face-centered cubic, hexagonal, monoclinic, orthorhombic, rhombohedral, and tetragonal. For some of the synthetically produced transuranic elements, available samples have been too small to determine crystal structures.

Occurrence and Origin on Earth

Chemical elements may also be categorized by their origin on Earth, with the first

94 considered naturally occurring, while those with atomic numbers beyond 94 have only been produced artificially via human-made nuclear reactions.

Of the 94 naturally occurring elements, 83 are considered primordial and either stable or weakly radioactive. The longest-lived isotopes of the remaining 11 elements have half lives too short for them to have been present at the beginning of the Solar System, and are therefore considered transient elements. Of these 11 transient elements, five (polonium, radon, radium, actinium, and protactinium) are relatively common decay products of thorium and uranium. The remaining six transient elements (technetium, promethium, astatine, francium, neptunium, and plutonium) occur only rarely, as products of rare decay modes or nuclear reaction processes involving uranium or other heavy elements.

Elements with atomic numbers 1 through 82, except 43 (technetium) and 61 (promethium), each have at least one isotope for which no radioactive decay has been observed. Observationally stable isotopes of some elements (such as tungsten and lead), however, are predicted to be slightly radioactive with very long half-lives: for example, the half-lives predicted for the observationally stable lead isotopes range from 10^{35} to 10^{189} years.

Elements with atomic numbers 43, 61, and 83 through 94 are unstable enough that their radioactive decay can be detected. Three of these elements, bismuth (element 83), thorium (90), and uranium (92) have one or more isotopes with half-lives long enough to survive as remnants of the explosive stellar nucleosynthesis that produced the heavy elements before the formation of the Solar System. For example, at over 1.9×10^{19} years, over a billion times longer than the estimated age of the universe, bismuth-209 has the longest known alpha decay half-life of any isotope. The last 24 elements (those beyond plutonium, element 94) undergo radioactive decay with short half-lives and cannot be produced as daughters of longer-lived elements, and thus are not known to occur in nature at all.

6.1.7 Periodic table

The properties of the elements are often summarized using the periodic table, which powerfully and elegantly organizes the elements by increasing atomic number into rows ("periods") in which the columns ("groups") share recurring ("periodic") physical and chemical properties. The table contains 118 confirmed elements as of 2021.

Although earlier precursors to this presentation exist, its invention is generally credited to Russian chemist Dmitri Mendeleev in 1869, who intended the table to illustrate recurring trends in the properties of the elements. The layout of the table has been refined and extended over time as new elements have been discovered and new theoretical models have been developed to explain chemical behavior.

Use of the periodic table is now ubiquitous in chemistry, providing an extremely useful framework to classify, systematize and compare all the many different forms of chemical behavior. The table has also found wide application in physics, geology, biology, materials science, engineering, agriculture, medicine, nutrition, environmental health, and astronomy. Its principles are especially important in chemical engineering.

6.2 NOMENCLATURE AND SYMBOLS

The various chemical elements are formally identified by their unique atomic numbers, their accepted names, and their chemical symbols.

6.2.1 Atomic Numbers

The known elements have atomic numbers from 1 to 118, conventionally presented as Arabic numerals. Since the elements can be uniquely sequenced by atomic number, conventionally from lowest to highest (as in a periodic table), sets of elements are sometimes specified by such notation as "through", "beyond", or "from ... through", as in "through iron", "beyond uranium", or "from lanthanum through lutetium". The terms "light" and "heavy" are sometimes also used informally to indicate relative atomic numbers (not densities), as in "lighter than carbon" or "heavier than lead", though the atomic masses of the elements (their atomic weights or atomic masses) do not always increase monotonically with their atomic numbers.

6.2.2 Element Names

The naming of various substances now known as elements precedes the atomic theory of matter, as names were given locally by various cultures to various minerals, metals, compounds, alloys, mixtures, and other materials, though at the time it was not known which chemicals were elements and which compounds.

As they were identified as elements, the existing names for anciently known elements (e.g., gold, mercury, iron) were kept in most countries. National differences emerged over the element names either for convenience, linguistic niceties, or nationalism.

For purposes of international communication and trade, the official names of the chemical elements both ancient and more recently recognized are decided by the International Union of Pure and Applied Chemistry (IUPAC), which has decided on a sort of international English language, drawing on traditional English names even when an element's chemical symbol is based on a Latin or other traditional word, for example adopting "gold" rather than "aurum" as the name for the 79th element (Au). IUPAC prefers the British spellings "aluminium" and "caesium" over the U.S. spellings "aluminum" and "cesium", and the U.S. "sulfur" over British "sulphur". However, elements that are practical to sell in bulk in many countries often still have locally used national names, and countries whose national language does not use the Latin alphabet are likely to use the IUPAC element names.

According to IUPAC, element names are not proper nouns; therefore, the full name of an element is not capitalized in English, even if derived from a proper noun, as in californium and einsteinium. Isotope names are also uncapitalized if written out, *e.g.*, carbon-12 or uranium-235. Chemical element *symbols* are always capitalized.

In the second half of the 20th century, physics laboratories became able to produce elements with half-lives too short for an appreciable amount of them to exist at any time. These are also named by IUPAC, which generally adopts the name chosen by the discoverer. This practice can lead to the controversial question of which research group actually discovered an element, a question that delayed the naming of elements with atomic number of 104 and higher for a considerable amount of time.

Precursors of such controversies involved the nationalistic namings of elements in the late 19th century. For example, *lutetium* was named in reference to Paris, France. The Germans were reluctant to relinquish naming rights to the French, often calling it *cassiopeium*. Similarly, the British discoverer of *niobium* originally named it *columbium*, in reference to the New World. It was used extensively as such by American publications before the international standardization (in 1950).

6.2.3 Chemical Symbols

Specific Elements

Before chemistry became a science, alchemists designed arcane symbols for both metals and common compounds. These were however used as abbreviations in diagrams or procedures; there was no concept of atoms combining to form molecules. With his advances in the atomic theory of matter, John Dalton devised his own simpler symbols, based on circles, to depict molecules.

The current system of chemical notation was invented by Jöns Jacob Berzelius in 1814. In this system, chemical symbols are not mere abbreviations—though each consists of letters of the Latin alphabet. They are intended as universal symbols for people of all languages and alphabets.

Since Latin was the common language of science at Berzelius' time, his symbols were abbreviations based on the Latin names of elements (they may be Classical Latin names of elements known since antiquity or Neo-Latin coinages for later elements). The symbols are not followed by a period (full stop) as with abbreviations. In most cases, Latin names of elements as used by Berzelius have the same roots as the modern English name. For example, hydrogen has the symbol "H" from Neo-

Latin *hydrogenium*, which has the same Greek roots as English *hydrogen*. However, in eleven cases Latin (as used by Berzelius) and English names of elements have different roots. Eight of them are the seven metals of antiquity and a metalloid also known since antiquity: "Fe" (Latin *ferrum*) for iron, "Hg" (Latin *hydrargyrum*) for mercury, "Sn" (Latin *stannum*) for tin, "Au" (Latin *aurum*) for gold, "Ag" (Latin *argentum*) for silver, "Pb" (Latin *plumbum*) for lead, "Cu" (Latin *cuprum*) for copper, and "Sb" (Latin *stibium*) for antimony. The three other mismatches between Neo-Latin (as used by Berzelius) and English names are "Na" (Neo-Latin *natrium*) for sodium, "K" (Neo-Latin *kalium*) for potassium, and "W" (Neo-Latin *wolframium*) for tungsten. These mismatches came from different suggestings of naming the elements in the Modern era. Initially Berzelius had suggested "So" and "Po" for sodium and potassium, but he changed the symbols to "Na" and "K" later in the same year.

Elements discovered after 1814 were also assigned unique chemical symbols, based on the name of the element. The use of Latin as the universal language of science was fading, but chemical names of newly discovered elements came to be borrowed from language to language with little or no modifications.

Chemical symbols are understood internationally when element names might require translation. There have been some differences in the past. For example, Germans in the past have used "J" (for the name *Jod*) for iodine, but now use "I" and *Iod*.

The first letter of a chemical symbol is always capitalized, as in the preceding examples, and the subsequent letters, if any, are always lower case. Thus, the symbols for californium and einsteinium are Cf and Es.

General Chemical Symbols

There are also symbols in chemical equations for groups of elements, for example in comparative formulas. These are often a single capital letter, and the letters are reserved and not used for names of specific elements. For example, "X" indicates a variable group (usually a halogen) in a class of compounds, while "R" is a radical, meaning a compound structure such as a hydrocarbon chain. The letter "Q" is reserved for "heat" in a chemical reaction. "Y" is also often used as a general chemical symbol, though it is also the symbol of yttrium. "Z" is also often used as a general variable group. "E" is used in organic chemistry to denote an electron-withdrawing group or an electrophile; similarly "Nu" denotes a nucleophile. "L" is used to represent a general ligand in inorganic and organometallic chemistry. "M" is also often used in place of a general metal.

At least two other, two-letter generic chemical symbols are also in informal use, "Ln" for any lanthanide and "An" for any actinide. "Rg" was formerly used for any rare gas element, but the group of rare gases has now been renamed noble gases and "Rg" now refers to roentgenium.

Isotope Symbols

Isotopes of an element are distinguished by mass number (total protons and neutrons), with this number combined with the element's symbol. IUPAC prefers that isotope symbols be written in superscript notation when practical, for example ^{12}C and ^{235}U. However, other notations, such as carbon-12 and uranium-235, or C-12 and U-235, are also used.

As a special case, the three naturally occurring isotopes of hydrogen are often specified as H for ^1H (protium), D for ^2H (deuterium), and T for ^3H (tritium). This convention is easier to use in chemical equations, replacing the need to write out the mass number each time. Thus, the formula for heavy water may be written D_2O instead of ^2H_2O.

6.3 ORIGIN OF THE ELEMENTS

Only about 4% of the total mass of the universe is made of atoms or ions, and thus represented by elements. This fraction is about 15% of the total matter, with the remainder of the matter (85%) being dark matter. The nature of dark matter is unknown, but it is not composed of atoms of elements because it contains no protons, neutrons, or electrons. (The remaining non-matter part of the mass of the universe is composed of the even less well understood dark energy).

The 94 naturally occurring elements were produced by at least four classes of astrophysical process. Most of the hydrogen, helium and a very small quantity of lithium were produced in the first few minutes of the Big Bang. This Big Bang nucleosynthesis happened only once; the other processes are ongoing. Nuclear fusion inside stars produces elements through stellar nucleosynthesis, including all elements from carbon to iron in atomic number. Elements higher in atomic number than iron, including heavy elements like uranium and plutonium, are produced by various forms of explosive nucleosynthesis in supernovae and neutron star mergers. The light elements lithium, beryllium and boron are produced mostly through cosmic ray spallation (fragmentation induced by cosmic rays) of carbon, nitrogen, and oxygen.

In the early phases of the Big Bang, nucleosynthesis of hydrogen resulted in the production of hydrogen-1 (protium, ^1H) and helium-4 (^4He), as well as a smaller amount of deuterium (^2H) and tiny amounts (on the order of 10^{-10}) of lithium and beryllium. Even smaller amounts of boron may have been produced in the Big Bang, since it has been observed in some very old stars, while carbon has not. No elements heavier than boron were produced in the Big Bang. As a result, the primordial abundance of atoms (or ions) consisted of ~75% ^1H, 25% ^4He, and

0.01% deuterium, with only tiny traces of lithium, beryllium, and perhaps boron. Subsequent enrichment of galactic halos occurred due to stellar nucleosynthesis and supernova nucleosynthesis. However, the element abundance in intergalactic space can still closely resemble primordial conditions, unless it has been enriched by some means.

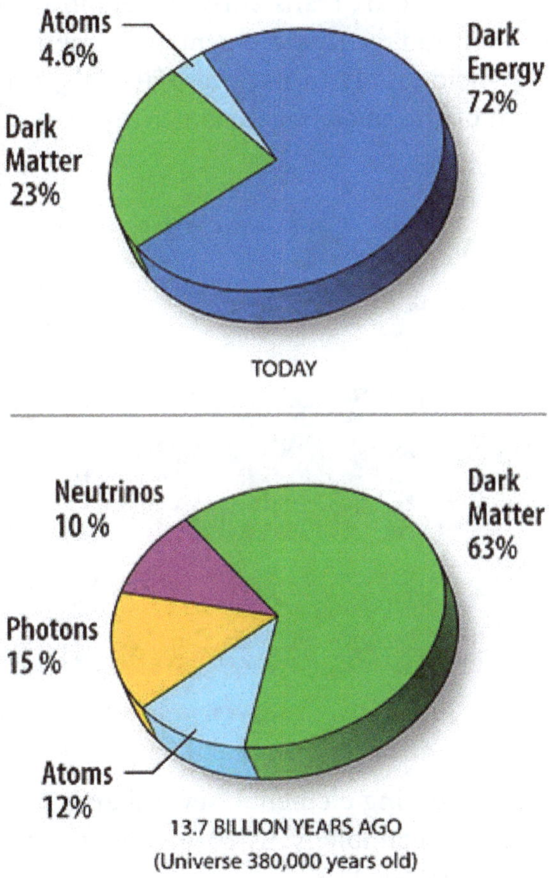

Fig 1: Estimated Distribution of Dark Matter and Dark Energy in the Universe. Only the Fraction of the Mass and Energy Labeled "Atoms" is Composed of Elements

Small stars can produce certain elements up to sulfur, by the alpha process. Supernovae are needed to produce "heavy" elements (those beyond iron and nickel) rapidly by neutron buildup, in the r-process. Certain large stars slowly produce other elements heavier than iron, in the s-process; these may then be blown into space in the off-gassing of planetary nebulae

On Earth (and elsewhere), trace amounts of various elements continue to be produced from other elements as products of nuclear transmutation processes. These include some produced by cosmic rays or other nuclear reactions and others produced as decay products of long-lived primordial nuclides. For example, trace (but detectable) amounts of carbon-14 (^{14}C) are continually produced in the air by cosmic rays impacting nitrogen atoms, and argon-40 (^{40}Ar) is continually produced by the decay of primordially occurring but unstable potassium-40 (^{40}K). Also, three primordially occurring but radioactive actinides, thorium, uranium, and plutonium, decay through a series of recurrently produced but unstable elements such as radium and radon, which are transiently present in any sample of containing these metals. Three other radioactive elements, technetium, promethium, and neptunium, occur only incidentally in natural materials, produced as individual atoms by nuclear fission of the nuclei of various heavy elements or in other rare nuclear processes.

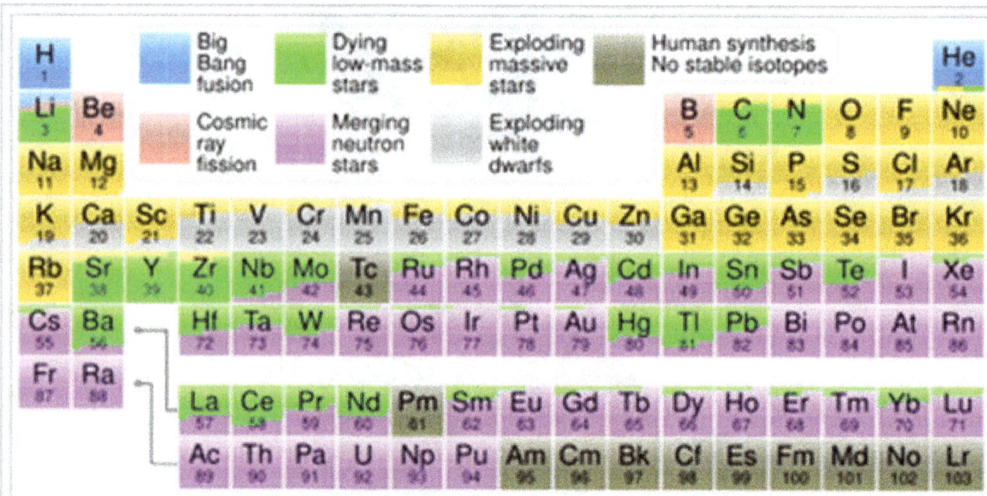

Fig 2: Periodic Table Showing the Cosmogenic Origin of Each Element in the Big Bang, or in Large or Small Stars

Besides the 94 naturally occurring elements, several artificial elements have been produced by nuclear physics technology. By 2016, these experiments had produced all elements up to atomic number 118.

6.4 ABUNDANCE

The following graph (note log scale) shows the abundance of elements in our Solar System. The table shows the 12 most common elements in our galaxy (estimated spectroscopically), as measured in parts per million by mass. Nearby galaxies that have evolved along similar lines have a corresponding enrichment of elements heavier than hydrogen and helium. The more distant galaxies are being viewed as they appeared in the past, so their abundances of elements appear closer to the primordial mixture. As physical laws and processes appear common throughout the visible universe, however, scientists expect that these galaxies evolved elements in similar abundance.

The abundance of elements in the Solar System is in keeping with their origin from nucleosynthesis in the Big Bang and a number of progenitor supernova stars. Very abundant hydrogen and helium are products of the Big Bang, but the next three elements are rare since they had little time to form in the Big Bang and are not made in stars (they are, however, produced in small quantities by the breakup of heavier elements in interstellar dust, as a result of impact by cosmic rays). Beginning with carbon, elements are produced in stars by buildup from alpha particles (helium nuclei), resulting in an alternatingly larger abundance of elements with even atomic numbers (these are also more stable). In general, such elements up to iron are made in large stars in the process of becoming supernovas. Iron-56 is particularly common, since it is the most stable nuclide that can easily be made from alpha particles (being a product of decay of radioactive nickel-56, ultimately made from 14 helium nuclei). Elements heavier than iron are made in energy-absorbing processes in large stars, and their abundance in the universe (and on Earth) generally decreases with their atomic number.

The abundance of the chemical elements on Earth varies from air to crust to ocean, and in various types of life. The abundance of elements in Earth's crust differs from that in the Solar System mainly in selective loss of the very lightest elements (hydrogen and helium) and also volatile neon, carbon (as hydrocarbons), nitrogen and sulfur, as a result of solar heating in the early formation of the Solar System. Oxygen, the most abundant Earth element by mass, is retained on Earth by combination with silicon.

Aluminium at 8% by mass is more common in the Earth's crust than in the universe and solar system, but the composition of the far more bulky mantle, which has magnesium and iron in place of aluminium (which occurs there only at 2% of mass) more closely mirrors the elemental composition of the solar system, save for the noted loss of volatile elements to space, and loss of iron which has migrated to the Earth's core.

The composition of the human body, by contrast, more closely follows the composition of seawater—save that the human body has additional stores of carbon and nitrogen necessary to form the proteins and nucleic acids, together with phosphorus in the nucleic acids and energy transfer molecule adenosine triphosphate (ATP) that occurs in the cells of all living organisms. Certain kinds of organisms require particular additional elements, for example the magnesium in chlorophyll in green plants, the calcium in mollusc shells, or the iron in the hemoglobin in vertebrates' red blood cells.

CHAPTER 7
COORDINATION COMPOUNDS

Coordination compounds, also known as coordination complexes, are molecules that consist of a central metal atom or ion bonded to surrounding molecules or ions called ligands.

These ligands can be atoms, ions, or molecules that donate pairs of electrons to the metal center, forming coordinate covalent bonds. Coordination compounds are essential in many fields, including chemistry, biology, and medicine, due to their unique chemical properties and diverse applications.

7.1 HISTORICAL BACKGROUND OF COORDINATION COMPOUNDS

The concept of coordination compounds was first introduced by Swiss chemist Alfred Werner in the late 19th and early 20th centuries. Before Werner's work, scientists struggled to understand the peculiar bonding and properties exhibited by certain metal compounds, particularly those involving transition metals. Werner proposed the idea of a coordination number, which describes the number of ligands attached to a metal center, and developed the coordination theory to explain the structure and bonding in these compounds. His pioneering work earned him the Nobel Prize in Chemistry in 1913 and laid the foundation for modern coordination chemistry.

7.2 STRUCTURE OF COORDINATION COMPOUNDS

Coordination compounds consist of a central metal atom or ion, typically a transition metal, surrounded by ligands. The arrangement of ligands around the metal center determines the geometry of the complex, which can vary based on the metal's coordination number. Common geometries include octahedral, tetrahedral, and square planar structures.

- **Octahedral Geometry**: This is the most common geometry for coordination compounds with a coordination number of six. In octahedral complexes, six ligands are symmetrically arranged around the metal center, forming an octahedron.
- **Tetrahedral Geometry**: In complexes with a coordination number of four, the ligands are arranged in a tetrahedral shape, with the metal at the center and ligands at the corners.

- **Square Planar Geometry**: Another common structure for coordination number four, square planar geometry involves four ligands arranged in a plane around the metal center. This geometry is common in compounds with metals like platinum.

The structure of a coordination compound influences its physical and chemical properties, including color, magnetic behavior, and reactivity.

7.3 TYPES OF COORDINATION COMPOUNDS

Coordination compounds can be classified based on the nature of the metal center, the type of ligands, and the bonding characteristics. The major types of coordination compounds include:

- **Homoleptic Complexes**: These complexes have only one type of ligand bonded to the metal center. For example, $[Co(NH_3)_6]^{3+}$ is a homoleptic complex with only ammonia ligands.
- **Heteroleptic Complexes**: Heteroleptic complexes contain different types of ligands attached to the metal. An example is $[Cr(NH_3)_4Cl_2]^+$, where both ammonia and chloride ligands are present.
- **Monodentate Ligands**: These ligands bind to the metal center through a single donor atom. Examples include water (H_2O), ammonia (NH_3), and chloride ions (Cl^-).
- **Polydentate Ligands**: Polydentate ligands, also known as chelating agents, can bind to the metal center through multiple donor atoms. Ethylenediamine (en) is a bidentate ligand, meaning it can attach to the metal at two points.
- **Bridging Ligands**: Some ligands can bridge between two or more metal centers, linking them together. An example is the oxide ion (O^{2-}), which can act as a bridge in metal oxo complexes.

7.4 NOMENCLATURE OF COORDINATION COMPOUNDS

Naming coordination compounds follows specific rules established by the International Union of Pure and Applied Chemistry (IUPAC). These rules ensure that each compound has a unique and systematic name:

- **Name the Ligands**: Ligands are named before the metal in the complex. Neutral ligands retain their names, such as water and ammonia, while anionic ligands end in -o, such as chloride (chloro) and cyanide (cyano).
- **List Ligands in Alphabetical Order**: When multiple ligands are present, they are listed in alphabetical order, regardless of their charge.

- **Specify the Number of Ligands:** Prefixes like di-, tri-, and tetra- are used to indicate the number of each ligand type. For example, $[Co(NH_3)_6]^{3+}$ is named hexaamminecobalt(III).
- **State the Oxidation State of the Metal:** The oxidation state of the metal is indicated in Roman numerals in parentheses after the metal's name.
- **Name the Complex Ion:** If the complex ion is an anion, the metal's name ends in -ate. For example, $[PtCl_4]^{2-}$ is named tetrachloroplatinate(II).

Using these rules, chemists can accurately describe and communicate the structures of various coordination compounds.

7.5 PROPERTIES OF COORDINATION COMPOUNDS

Coordination compounds exhibit unique properties due to the interactions between the metal center and the ligands.

These properties are influenced by the metal's electronic structure, the nature of the ligands, and the geometry of the complex.

- **Color:** Many coordination compounds are colored due to d-d transitions, where electrons in the d orbitals of the metal are excited to higher energy levels. The color observed is the result of light absorption at specific wavelengths.
- **Magnetism:** The magnetic properties of coordination compounds depend on the metal center and its electronic configuration. Compounds with unpaired electrons are paramagnetic, while those without unpaired electrons are diamagnetic.
- **Reactivity:** The reactivity of coordination compounds varies widely and is influenced by factors such as the metal's oxidation state, the type of ligands, and the complex's overall stability.

7.6 APPLICATIONS OF COORDINATION COMPOUNDS

Coordination compounds have a wide range of applications in various fields due to their diverse chemical properties:

- **Medicine:** Some coordination compounds are used in medical treatments, such as cisplatin, a platinum-based drug used in chemotherapy. Metal complexes can also serve as contrast agents in MRI scans.
- **Catalysis:** Coordination compounds are widely used as catalysts in chemical reactions, particularly in industrial processes. For example, the Wilkinson's catalyst is a rhodium-based complex used in hydrogenation reactions.

- **Biological Systems:** Many biological processes involve coordination compounds. Hemoglobin, the oxygen-transporting protein in blood, is a coordination complex with iron at its center.
- **Environmental Science:** Coordination compounds can be used in environmental applications, such as pollution control and waste treatment. Chelating agents, for example, can bind heavy metals, making them easier to remove from contaminated environments.

7.7 CONCLUSION

Coordination compounds are an integral part of both natural and synthetic chemistry. Their diverse structures, properties, and applications make them invaluable in fields ranging from medicine and industry to environmental science. Understanding coordination compounds requires knowledge of chemical bonding, electronic configuration, and molecular geometry. With ongoing research, the potential uses for these compounds continue to expand, promising new advancements and applications in science and technology.

CHAPTER 8
HYDROGEN AND THE HYDRIDES

8.1 HYDROGEN

Hydrogen is a chemical element; it has symbol H and atomic number 1. It is the lightest element and, at standard conditions, is a gas of diatomic molecules with the formula H_2, sometimes called dihydrogen, but more commonly called hydrogen gas, molecular hydrogen or simply hydrogen. It is colorless, odorless, non-toxic, and highly combustible. Constituting about 75% of all normal matter, hydrogen is the most abundant chemical element in the universe. Stars, including the Sun, mainly consist of hydrogen in a plasma state, while on Earth, hydrogen is found in water, organic compounds, as dihydrogen, and in other molecular forms. The most common isotope of hydrogen (protium, ^1H) consists of one proton, one electron, and no neutrons.

In the early universe, the formation of hydrogen's protons occurred in the first second after the Big Bang; neutral hydrogen atoms only formed about 370,000 years later during the recombination epoch as the universe cooled and plasma had cooled enough for electrons to remain bound to protons. Hydrogen, typically nonmetallic except under extreme pressure, readily forms covalent bonds with most nonmetals, contributing to the formation of compounds like water and various organic substances. Its role is crucial in acid-base reactions, which mainly involve proton exchange among soluble molecules. In ionic compounds, hydrogen can take the form of either a negatively charged anion, where it is known as hydride, or as a positively charged cation, H^+.

The cation, usually just a proton (symbol p), exhibits specific behavior in aqueous solutions and in ionic compounds involves screening of its electric charge by surrounding polar molecules or anions. Hydrogen's unique position as the only neutral atom for which the Schrödinger equation can be directly solved, has significantly contributed to the foundational principles of quantum mechanics through the exploration of its energetics and chemical bonding.

Hydrogen gas was first produced artificially in the early 16th century by reacting acids with metals. Henry Cavendish, in 1766–81, identified hydrogen gas as a distinct substance and discovered its property of producing water when burned; hence its name means "water-former" in Greek.

Most hydrogen production occurs through steam reforming of natural gas; a smaller portion comes from energy-intensive methods such as the electrolysis of water. Its main industrial uses include fossil fuel processing, such as hydrocracking, and ammonia production, with emerging uses in fuel cells for electricity generation and as a heat source. When used in fuel cells, hydrogen's only emission at point of use is water vapor, though combustion can produce nitrogen oxides. Hydrogen's interaction with metals may cause embrittlement.

8.1.1 Properties

Combustion

Hydrogen gas is highly flammable:

$$2H_2(g)+O_2(g)\rightarrow 2H_2O(l)(572\,kJ/2mol=286\,kJ/mol=141.865\ MJ/kg)$$

Enthalpy of combustion: -286 kJ/mol. Hydrogen gas forms explosive mixtures with air in concentrations from 4–74% and with chlorine at 5–95%. The hydrogen autoignition temperature, the temperature of spontaneous ignition in air, is 500°C (932°F).

Flame

Pure hydrogen-oxygen flames emit ultraviolet light and with high oxygen mix are nearly invisible to the naked eye, as illustrated by the faint plume of the Space Shuttle Main Engine, compared to the highly visible plume of a Space Shuttle Solid Rocket Booster, which uses an ammonium perchlorate composite.

The detection of a burning hydrogen leak, may require a flame detector; such leaks can be very dangerous. Hydrogen flames in other conditions are blue, resembling blue natural gas flames.

The destruction of the *Hindenburg* airship was a notorious example of hydrogen combustion and the cause is still debated. The visible flames in the photographs were the result of carbon compounds in the airship skin burning.

Reactants

H_2 is unreactive compared to diatomic elements such as halogens or oxygen. The thermodynamic basis of this low reactivity is the very strong H–H bond, with a bond dissociation energy of 435.7 kJ/mol. The kinetic basis of the low reactivity is the nonpolar nature of H_2 and its weak polarizability. It spontaneously reacts with chlorine and fluorine to form hydrogen chloride and hydrogen fluoride, respectively. The reactivity of H_2 is strongly affected by the presence of metal catalysts. Thus, while mixtures of H_2 with O_2 or air combust readily when heated to at least 500°C by a spark or flame, they do not react at room temperature in the absence of a catalyst.

Electron Energy Levels

The ground state energy level of the electron in a hydrogen atom is -13.6 eV, equivalent to an ultraviolet photon of roughly 91 nm wavelength. The energy levels of hydrogen can be calculated fairly accurately using the Bohr model of the atom, in which the electron "orbits" the proton, like how Earth orbits the Sun. However, the electron and proton are held together by electrostatic attraction, while planets and celestial objects are held by gravity.

Due to the discretization of angular momentum postulated in early quantum mechanics by Bohr, the electron in the Bohr model can only occupy certain allowed distances from the proton, and therefore only certain allowed energies.

A more accurate description of the hydrogen atom comes from a quantum analysis that uses the Schrödinger equation, Dirac equation or Feynman path integral formulation to calculate the probability density of the electron around the proton.

The most complex formulas include the small effects of special relativity and vacuum polarization. In the quantum mechanical treatment, the electron in a ground state hydrogen atom has no angular momentum—illustrating how the "planetary orbit" differs from electron motion.

Spin Isomers

Molecular H_2 exists as two spin isomers, i.e. compounds that differ only in the spin states of their nuclei. In the orthohydrogen form, the spins of the two nuclei are parallel, forming a spin triplet state having a total molecular spin $S = 1$; in the parahydrogen form the spins are antiparallel and form a spin singlet state having spin $S = 0$.

The equilibrium ratio of ortho- to para-hydrogen depends on temperature. At room temperature or warmer, equilibrium hydrogen gas contains about 25% of the para form and 75% of the ortho form. The ortho form is an excited state, having higher energy than the para form by 1.455 kJ/mol, and it converts to the para form over the course of several minutes when cooled to low temperature.

The thermal properties of the forms differ because they differ in their allowed rotational quantum states, resulting in different thermal properties such as the heat capacity.

The ortho-to-para ratio in H_2 is an important consideration in the liquefaction and storage of liquid hydrogen: the conversion from ortho to para is exothermic and produces enough heat to evaporate most of the liquid if not converted first to parahydrogen during the cooling process. Catalysts for the ortho-para interconversion, such as ferric oxide and activated carbon compounds, are used during hydrogen cooling to avoid this loss of liquid.

Phases

- Gaseous hydrogen.
- Liquid hydrogen.
- Slush hydrogen.
- Solid hydrogen.
- Metallic hydrogen.
- Plasma hydrogen.

Compounds

Covalent and Organic Compounds

While H_2 is not very reactive under standard conditions, it does form compounds with most elements. Hydrogen can form compounds with elements that are more electronegative, such as halogens (F, Cl, Br, I), or oxygen; in these compounds hydrogen takes on a partial positive charge. When bonded to a more electronegative element, particularly fluorine, oxygen, or nitrogen, hydrogen can participate in a form of medium-strength noncovalent bonding with another electronegative element with a lone pair, a phenomenon called hydrogen bonding that is critical to the stability of many biological molecules. Hydrogen also forms compounds with less electronegative elements, such as metals and metalloids, where it takes on a partial negative charge. These compounds are often known as hydrides.

Hydrogen forms many compounds with carbon called the hydrocarbons, and even more with heteroatoms that, due to their association with living things, are called organic compounds. The study of their properties is known as organic chemistry and their study in the context of living organisms is called biochemistry. By some definitions, "organic" compounds are only required to contain carbon.

However, most of them also contain hydrogen, and because it is the carbon-hydrogen bond that gives this class of compounds most of its particular chemical characteristics, carbon-hydrogen bonds are required in some definitions of the word "organic" in chemistry. Millions of hydrocarbons are known, and they are usually formed by complicated pathways that seldom involve elemental hydrogen.

Hydrogen is highly soluble in many rare earth and transition metals and is soluble in both nanocrystalline and amorphous metals. Hydrogen solubility in metals is influenced by local distortions or impurities in the crystal lattice. These properties may be useful when hydrogen is purified by passage through hot palladium disks, but the gas's high solubility is a metallurgical problem, contributing to the embrittlement of many metals, complicating the design of pipelines and storage tanks.

Hydrides

Hydrogen compounds are often called hydrides, a term that is used fairly loosely. The term "hydride" suggests that the H atom has acquired a negative or anionic character, denoted H^-; and is used when hydrogen forms a compound with a more electropositive element. The existence of the hydride anion, suggested by Gilbert N. Lewis in 1916 for group 1 and 2 salt-like hydrides, was demonstrated by Moers in 1920 by the electrolysis of molten lithium hydride (LiH), producing a stoichiometric quantity of hydrogen at the anode. For hydrides other than group 1 and 2 metals, the term is quite misleading, considering the low electronegativity of hydrogen. An exception in group 2 hydrides is BeH_2, which is polymeric. In lithium aluminium hydride, the $[AlH_4]^-$ anion carries hydridic centers firmly attached to the Al(III).

Although hydrides can be formed with almost all main-group elements, the number and combination of possible compounds varies widely; for example, more than 100 binary borane hydrides are known, but only one binary aluminium hydride. Binary indium hydride has not yet been identified, although larger complexes exist.

In inorganic chemistry, hydrides can also serve as bridging ligands that link two metal centers in a coordination complex. This function is particularly common in group 13 elements, especially in boranes (boron hydrides) and aluminium complexes, as well as in clustered carboranes.

Protons and Acids

Oxidation of hydrogen removes its electron and gives H^+, which contains no electrons and a nucleus which is usually composed of one proton. That is why H^+ is often called a proton. This species is central to discussion of acids. Under the Brønsted–Lowry acid–base theory, acids are proton donors, while bases are proton acceptors.

A bare proton, H^+, cannot exist in solution or in ionic crystals because of its strong attraction to other atoms or molecules with electrons. Except at the high temperatures associated with plasmas, such protons cannot be removed from the electron clouds of atoms and molecules, and will remain attached to them. However, the term 'proton' is sometimes used loosely and metaphorically to refer to positively charged or cationic hydrogen attached to other species in this fashion, and as such is denoted "H^+" without any implication that any single protons exist freely as a species.

To avoid the implication of the naked "solvated proton" in solution, acidic aqueous solutions are sometimes considered to contain a less unlikely fictitious species, termed the "hydronium ion" ($[H_3O]^+$). However, even in this case, such solvated hydrogen cations are more realistically conceived as being organized into clusters that form species closer to $[H_9O_4]^+$. Other oxonium ions are found when water is in acidic solution with other solvents. Although exotic on Earth, one of the most common ions in the universe is the H+3 ion, known as protonated molecular hydrogen or the trihydrogen cation.

Isotopes

Hydrogen has three naturally occurring isotopes, denoted ^1H, ^2H and ^3H. Other, highly unstable nuclei (^4H to ^7H) have been synthesized in the laboratory but not observed in nature.

- ^1H is the most common hydrogen isotope, with an abundance of >99.98%. Because the nucleus of this isotope consists of only a single proton, it is given the descriptive but rarely used formal name *protium*. It is the only stable isotope with no neutrons.
- ^2H, the other stable hydrogen isotope, is known as deuterium and contains one proton and one neutron in the nucleus. Nearly all deuterium in the universe is thought to have been produced at the time of the Big Bang, and has endured since then. Deuterium is not radioactive, and is not a significant toxicity hazard. Water enriched in molecules that include deuterium instead of normal hydrogen is called heavy water. Deuterium and its compounds are used as a non-radioactive label in chemical experiments and in solvents for ^{1} H-NMR spectroscopy. Heavy water is used as a neutron moderator and coolant for nuclear reactors. Deuterium is also a potential fuel for commercial nuclear fusion.
- ^3H is known as tritium and contains one proton and two neutrons in its nucleus. It is radioactive, decaying into helium-3 through beta decay with a half-life of 12.32 years. It is radioactive enough to be used in luminous paint to enhance the visibility of data displays, such as for painting the hands and dial-markers of watches. The watch glass prevents the small amount of radiation from escaping the case. Small amounts of tritium are produced naturally by cosmic rays striking atmospheric gases; tritium has also been released in nuclear weapons tests. It is used in nuclear fusion, as a tracer in isotope geochemistry, and in specialized self-powered lighting devices. Tritium has also been used in chemical and biological labeling experiments as a radiolabel.

Unique among the elements, distinct names are assigned to its isotopes in common use. During the early study of radioactivity, heavy radioisotopes were given their own names, but these are mostly no longer used.

The symbols D and T (instead of ^2H and ^3H) are sometimes used for deuterium and tritium, but the symbol P was already used for phosphorus and thus was not available for protium. In its nomenclatural guidelines, the International Union of Pure and Applied Chemistry (IUPAC) allows any of D, T, ^{2}H, and ^{3}H to be used, though ^{2}H and ^{3}H are preferred.

The exotic atom muonium (symbol Mu), composed of an antimuon and an electron, can also be considered a light radioisotope of hydrogen. Because muons decay with lifetime 2.2μs, muonium is too unstable for observable chemistry. Nevertheless, muonium compounds are important test cases for quantum simulation, due to the mass difference between the antimuon and the proton, and IUPAC nomenclature incorporates such hypothetical compounds as muonium chloride (MuCl) and sodium muonide (NaMu), analogous to hydrogen chloride and sodium hydride respectively.

8.1.2 History

Discovery and Use

Robert Boyle

In 1671, Irish scientist Robert Boyle discovered and described the reaction between iron filings and dilute acids, which results in the production of hydrogen gas.

- Having provided a saline spirit [hydrochloric acid], which by an uncommon way of preparation was made exceeding sharp and piercing, we put into a vial, capable of containing three or four ounces of water, a convenient quantity of filings of steel, which were not such as are commonly sold in shops to Chymists and Apothecaries, (those being usually not free enough from rust) but such as I had a while before caus'd to be purposely fil'd off from a piece of good steel. This metalline powder being moistn'd in the viol with a little of the menstruum, was afterwards drench'd with more; whereupon the mixture grew very hot, and belch'd up copious and stinking fumes; which whether they consisted altogether of the volatile sulfur of the Mars [iron], or of metalline steams participating of a sulfureous nature, and join'd with the saline exhalations of the menstruum, is not necessary to be here discuss'd. But whencesoever this stinking smoak proceeded, so inflammable it was, that upon the approach of a lighted candle to it, it would readily enough take fire, and burn with a blewish and somewhat greenish flame at the mouth of the viol for a good while together; and that, though with little light, yet with more strength than one would easily suspect.—*Robert Boyle, Tracts written by the Honourable Robert Boyle containing new experiments, touching the relation betwixt flame and air...*

The word "sulfureous" may be somewhat confusing, especially since Boyle did a similar experiment with iron and sulfuric acid. However, in all likelihood, "sulfureous" should here be understood to mean "combustible".

Henry Cavendish

In 1766, Henry Cavendish was the first to recognize hydrogen gas as a discrete substance, by naming the gas from a metal-acid reaction "inflammable air". He speculated that "inflammable air" was in fact identical to the hypothetical substance "phlogiston" and further finding in 1781 that the gas produces water when burned. He is usually given credit for the discovery of hydrogen as an element.

Antoine Lavoisier

In 1783, Antoine Lavoisier identified the element that came to be known as hydrogen when he and Laplace reproduced Cavendish's finding that water is produced when hydrogen is burned. Lavoisier produced hydrogen for his experiments on mass conservation by reacting a flux of steam with metallic iron through an incandescent iron tube heated in a fire. Anaerobic oxidation of iron by the protons of water at high temperature can be schematically represented by the set of following reactions:

- $Fe + H_2O \rightarrow FeO + H_2$
- $Fe + 3\ H_2O \rightarrow Fe_2O_3 + 3\ H_2$
- $Fe + 4\ H_2O \rightarrow Fe_3O_4 + 4\ H_2$

Many metals such as zirconium undergo a similar reaction with water leading to the production of hydrogen.

19th Century

François Isaac de Rivaz built the first de Rivaz engine, an internal combustion engine powered by a mixture of hydrogen and oxygen in 1806. Edward Daniel Clarke invented the hydrogen gas blowpipe in 1819. The Döbereiner's lamp and limelight were invented in 1823.

Hydrogen was liquefied for the first time by James Dewar in 1898 by using regenerative cooling and his invention, the vacuum flask. He produced solid hydrogen the next year.

Hydrogen-Lifted Airship

The first hydrogen-filled balloon was invented by Jacques Charles in 1783. Hydrogen provided the lift for the first reliable form of air-travel following the 1852 invention of the first hydrogen-lifted airship by Henri Giffard. German count Ferdinand von Zeppelin promoted the idea of rigid airships lifted by hydrogen that later were called Zeppelins; the first of which had its maiden flight in 1900. Regularly scheduled flights started in 1910 and by the outbreak of World War I in August 1914, they had carried 35,000 passengers without a serious incident. Hydrogen-lifted airships were used as observation platforms and bombers during the war.

The first non-stop transatlantic crossing was made by the British airship *R34* in 1919. Regular passenger service resumed in the 1920s and the discovery of helium reserves in the United States promised increased safety, but the U.S. government refused to sell the gas for this purpose.

Therefore, H_2 was used in the *Hindenburg* airship, which was destroyed in a midair fire over New Jersey on 6 May 1937. The incident was broadcast live on radio and filmed. Ignition of leaking hydrogen is widely assumed to be the cause, but later investigations pointed to the ignition of the aluminized fabric coating by static electricity. But the damage to hydrogen's reputation as a lifting gas was already done and commercial hydrogen airship travel ceased. Hydrogen is still used, in preference to non-flammable but more expensive helium, as a lifting gas for weather balloons.

Deuterium and Tritium

Deuterium was discovered in December 1931 by Harold Urey, and tritium was prepared in 1934 by Ernest Rutherford, Mark Oliphant, and Paul Harteck. Heavy water, which consists of deuterium in the place of regular hydrogen, was discovered by Urey's group in 1932.

Hydrogen-Cooled Turbogenerator

The first hydrogen-cooled turbogenerator went into service using gaseous hydrogen as a coolant in the rotor and the stator in 1937 at Dayton, Ohio, owned by the Dayton Power & Light Co. This was justified by the high thermal conductivity and very low viscosity of hydrogen gas, thus lower drag than air. This is the most common coolant used for generators 60 MW and larger; smaller generators are usually air-cooled.

Nickel–Hydrogen Battery

The nickel–hydrogen battery was used for the first time in 1977 aboard the U.S. Navy's Navigation technology satellite-2 (NTS-2). The International Space Station, Mars Odyssey and the Mars Global Surveyor are equipped with nickel-hydrogen batteries. In the dark part of its orbit, the Hubble Space Telescope is also powered by nickel-hydrogen batteries, which were finally replaced in May 2009, more than 19 years after launch and 13 years beyond their design life.

Role in Quantum Theory

Because of its simple atomic structure, consisting only of a proton and an electron, the hydrogen atom, together with the spectrum of light produced from it or absorbed by it, has been central to the development of the theory of atomic structure. Furthermore, study of the corresponding simplicity of the hydrogen molecule and the corresponding cation H+2 brought understanding of the nature of the chemical bond, which followed shortly after the quantum mechanical treatment of the hydrogen atom had been developed in the mid-1920s.

One of the first quantum effects to be explicitly noticed (but not understood at the time) was a Maxwell observation involving hydrogen, half a century before full quantum mechanical theory arrived. Maxwell observed that the specific heat capacity of H_2 unaccountably departs from that of a diatomic gas below room temperature and begins to increasingly resemble that of a monatomic gas at cryogenic temperatures. According to quantum theory, this behavior arises from the spacing of the (quantized) rotational energy levels, which are particularly wide-spaced in H_2 because of its low mass. These widely spaced levels inhibit equal partition of heat energy into rotational motion in hydrogen at low temperatures. Diatomic gases composed of heavier atoms do not have such widely spaced levels and do not exhibit the same effect.

Antihydrogen (H) is the antimatter counterpart to hydrogen. It consists of an antiproton with a positron. Antihydrogen is the only type of antimatter atom to have been produced as of 2015.

8.1.3 Cosmic Prevalence and Distribution

Hydrogen, as atomic H, is the most abundant chemical element in the universe, making up 75% of normal matter by mass and >90% by number of atoms. Most of the mass of the universe, however, is not in the form of chemical-element type matter, but rather is postulated to occur as yet-undetected forms of mass such as dark matter and dark energy.

Hydrogen is found in great abundance in stars and gas giant planets. Molecular clouds of H_2 are associated with star formation. Hydrogen plays a vital role in powering stars through the proton-proton reaction in case of stars with very low to approximately 1 mass of the Sun and the CNO cycle of nuclear fusion in case of stars more massive than the Sun.

States

Throughout the universe, hydrogen is mostly found in the atomic and plasma states, with properties quite distinct from those of molecular hydrogen. As a plasma, hydrogen's electron and proton are not bound together, resulting in very high electrical conductivity and high emissivity (producing the light from the Sun and other stars). The charged particles are highly influenced by magnetic and electric fields. For example, in the solar wind they interact with the Earth's magnetosphere giving rise to Birkeland currents and the aurora.

Hydrogen is found in the neutral atomic state in the interstellar medium because the atoms seldom collide and combine. They are the source of the 21-cm hydrogen line at 1420 MHz that is detected in order to probe primordial hydrogen. The large amount of neutral hydrogen found in the damped Lyman-alpha systems is thought to dominate the cosmological baryonic density of the universe up to a redshift of z = 4.

Under ordinary conditions on Earth, elemental hydrogen exists as the diatomic gas, H_2. Hydrogen gas is very rare in Earth's atmosphere (around 0.53 ppm on a molar basis) because of its light weight, which enables it to escape the atmosphere more rapidly than heavier gases. However, hydrogen is the third most abundant element on the Earth's surface, mostly in the form of chemical compounds such as hydrocarbons and water.

A molecular form called protonated molecular hydrogen (H+3) is found in the interstellar medium, where it is generated by ionization of molecular hydrogen from cosmic rays. This ion has also been observed in the upper atmosphere of Jupiter. The ion is relatively stable in outer space due to the low temperature and density. H+3 is one of the most abundant ions in the universe, and it plays a notable role in the chemistry of the interstellar medium. Neutral triatomic hydrogen H_3 can exist only in an excited form and is unstable. By contrast, the positive hydrogen molecular ion (H+2) is a rare molecule in the universe.

8.1.4 Production

Many methods exist for producing H_2, but three dominate commercially: steam reforming often coupled to water-gas shift, partial oxidation of hydrocarbons, and water electrolysis.

Steam Reforming

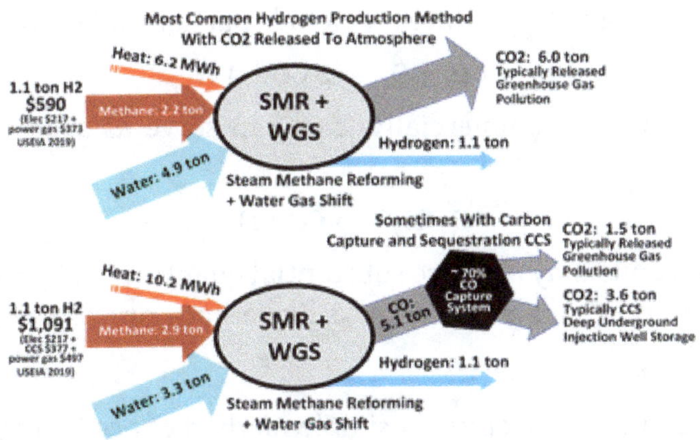

Fig 1: Inputs and Outputs of Steam Reforming (SMR) and Water Gas Shift (WGS) Reaction of Natural Gas, A Process Used in Hydrogen Production

Hydrogen is mainly produced by steam methane reforming (SMR), the reaction of water and methane. Thus, at high temperature (1000–1400 K, 700–1100°C or

1300–2000°F), steam (water vapor) reacts with methane to yield carbon monoxide and H_2.

$$CH_4 + H_2O \rightarrow CO + 3\,H_2$$

Steam reforming is also used for the industrial preparation of ammonia.

This reaction is favored at low pressures, Nonetheless, conducted at high pressures (2.0MPa, 20atm or 600inHg) because high-pressure H_2 is the most marketable product, and pressure swing adsorption (PSA) purification systems work better at higher pressures.

The product mixture is known as "synthesis gas" because it is often used directly for the production of methanol and many other compounds. Hydrocarbons other than methane can be used to produce synthesis gas with varying product ratios. One of the many complications to this highly optimized technology is the formation of coke or carbon:

$$CH_4 \rightarrow C + 2\,H_2$$

Therefore, steam reforming typically employs an excess of H_2O. Additional hydrogen can be recovered from the steam by using carbon monoxide through the water gas shift reaction (WGS). This process requires an iron oxide catalyst:

$$CO + H_2O \rightarrow CO_2 + H_2$$

Hydrogen is sometimes produced and consumed in the same industrial process, without being separated. In the Haber process for ammonia production, hydrogen is generated from natural gas.

Partial Oxidation of Hydrocarbons

Other methods for CO and H_2 production include partial oxidation of hydrocarbons:

$$2\,CH_4 + O_2 \rightarrow 2\,CO + 4\,H_2$$

Although less important commercially, coal can serve as a prelude to the shift reaction above:

$$C + H_2O \rightarrow CO + H_2$$

Olefin production units may produce substantial quantities of byproduct hydrogen particularly from cracking light feedstocks like ethane or propane.

Water Electrolysis

Electrolysis of water is a conceptually simple method of producing hydrogen.

$$2\,H_2O(l) \rightarrow 2\,H_2(g) + O_2(g)$$

Commercial electrolyzers use nickel-based catalysts in strongly alkaline solution. Platinum is a better catalyst but is expensive. Electrolysis of brine to yield chlorine also produces hydrogen as a co-product.

Methane Pyrolysis

Hydrogen can be produced by pyrolysis of natural gas (methane). This route has a lower carbon footprint than commercial hydrogen production processes. Developing a commercial methane pyrolysis process could expedite the expanded use of hydrogen in industrial and transportation applications. Methane pyrolysis is accomplished by passing methane through a molten metal catalyst containing dissolved nickel. Methane is converted to hydrogen gas and solid carbon.

$$CH_4(g) \rightarrow C(s) + 2\ H_2(g)\ (\Delta H° = 74\ kJ/mol)$$

The carbon may be sold as a manufacturing feedstock or fuel, or landfilled. Further research continues in several laboratories, including at Karlsruhe Liquid-metal Laboratory and at University of California – Santa Barbara. BASF built a methane pyrolysis pilot plant.

Thermochemical

More than 200 thermochemical cycles can be used for water splitting. Many of these cycles such as the iron oxide cycle, cerium(IV) oxide–cerium(III) oxide cycle, zinc zinc-oxide cycle, sulfur-iodine cycle, copper-chlorine cycle and hybrid sulfur cycle have been evaluated for their commercial potential to produce hydrogen and oxygen from water and heat without using electricity.

A number of labs (including in France, Germany, Greece, Japan, and the United States) are developing thermochemical methods to produce hydrogen from solar energy and water.

Laboratory Methods

H_2 is produced in labs, often as a by-product of other reactions. Many metals react with water to produce H_2, but the rate of hydrogen evolution depends on the metal, the pH, and the presence of alloying agents. Most often, hydrogen evolution is induced by acids. The alkali and alkaline earth metals, aluminium, zinc, manganese, and iron react readily with aqueous acids. This reaction is the basis of the Kipp's apparatus, which once was used as a laboratory gas source:

$$Zn + 2\ H^+ \rightarrow Zn^{2+} + H_2$$

In the absence of acid, the evolution of H_2 is slower. Because iron is widely used structural material, its anaerobic corrosion is of technological significance:

$$Fe + 2\ H_2O \rightarrow Fe(OH)_2 + H_2$$

Many metals, such as aluminium, are slow to react with water because they form passivated oxide coatings of oxides. An alloy of aluminium and gallium, however, does react with water. At high pH, aluminium can produce H_2:

$$2\ Al + 6\ H_2O + 2\ OH^- \rightarrow 2\ [Al(OH)_4]^- + 3\ H_2$$

Some metal-containing compounds react with acids to evolve H_2. Under anaerobic conditions, ferrous hydroxide ($Fe(OH)_2$) can be oxidized by the protons of water to form magnetite and H_2. This process is described by the Schikorr reaction:

$$3\ Fe(OH)_2 \rightarrow Fe_3O_4 + 2\ H_2O + H_2$$

This process occurs during the anaerobic corrosion of iron and steel in oxygen-free groundwater and in reducing soils below the water table.

Biohydrogen

H_2 is produced by hydrogenase enzymes in some fermentation.

Wells

There is a well in Mali and deposits in several other countries, such as France.

8.1.5 Applications

Petrochemical Industry

Large quantities of H_2 are used in the "upgrading" of fossil fuels. Key consumers of H_2 include hydrodesulfurization, and hydrocracking. Many of these reactions can be classified as hydrogenolysis, i.e., the cleavage of bonds by hydrogen. Illustrative is the separation of sulfur from liquid fossil fuels:

$$R_2S + 2\ H_2 \rightarrow H_2S + 2\ RH$$

Hydrogenation

Hydrogenation, the addition of H_2 to various substrates, is done on a large scale. Hydrogenation of N_2 to produce ammonia by the Haber process, consumes a few percent of the energy budget in the entire industry. The resulting ammonia is used to supply most of the protein consumed by humans. Hydrogenation is used to convert unsaturated fats and oils to saturated (trans) fats and oils. The major application is the production of margarine. Methanol is produced by hydrogenation of carbon dioxide. It is similarly the source of hydrogen in the manufacture of hydrochloric acid. H_2 is also used as a reducing agent for the conversion of some ores to the metals.

Coolant

Hydrogen is commonly used in power stations as a coolant in generators due to a number of favorable properties that are a direct result of its light diatomic molecules. These include low density, low viscosity, and the highest specific heat and thermal conductivity of all gases.

Energy Carrier

Elemental hydrogen is widely discussed in the context of energy as an energy carrier with potential to help to decarbonize economies and mitigate greenhouse gas emissions. This therefore requires hydrogen to be produced cleanly in quantities to be supplied in sectors and applications where cheaper and more energy-efficient mitigation alternatives are limited. These include heavy industry and long-distance transport. Hydrogen is a *carrier* of energy rather than an energy resource, because there is no naturally occurring source of hydrogen in useful quantities.

Hydrogen can be deployed as an energy source in fuel cells to produce electricity or via combustion to generate heat. When hydrogen is consumed in fuel cells, the only emission at the point of use is water vapor. Combustion of hydrogen can lead to the thermal formation of harmful nitrogen oxides. The overall lifecycle emissions of hydrogen depend on how it is produced. Nearly all the world's current supply of hydrogen is created from fossil fuels. The main method is steam methane reforming, in which hydrogen is produced from a chemical reaction between steam and methane, the main component of natural gas. Producing one tonne of hydrogen through this process emits 6.6–9.3 tonnes of carbon dioxide. While carbon capture and storage (CCS) could remove a large fraction of these emissions, the overall carbon footprint of hydrogen from natural gas is difficult to assess as of 2021, in part because of emissions (including vented and fugitive methane) created in the production of the natural gas itself.

Electricity can be used to split water molecules, producing sustainable hydrogen, provided the electricity was generated sustainably. However, this electrolysis process is currently more expensive than creating hydrogen from methane without CCS and the efficiency of energy conversion is inherently low. Hydrogen can be produced when there is a surplus of variable renewable electricity, then stored and used to generate heat or to re-generate electricity. Hydrogen created through electrolysis using renewable energy is commonly referred to as "green hydrogen". It can be further transformed into synthetic fuels such as ammonia and methanol.

Innovation in hydrogen electrolyzers could make large-scale production of hydrogen from electricity more cost-competitive. There is potential for hydrogen produced this way to play a significant role in decarbonizing energy systems where there are challenges and limitations to replacing fossil fuels with direct use of electricity.

Hydrogen fuel can produce the intense heat required for industrial production of steel, cement, glass, and chemicals, thus contributing to the decarbonisation of industry alongside other technologies, such as electric arc furnaces for steelmaking. However, it is likely to play a larger role in providing industrial feedstock for cleaner production of ammonia and organic chemicals. For example, in steelmaking, hydrogen could function as a clean energy carrier and also as a low-carbon catalyst, replacing coal-derived coke. Hydrogen used to decarbonise transportation is likely to find its largest applications in shipping, aviation and, to a lesser extent, heavy goods vehicles, through the use of hydrogen-derived synthetic fuels such as ammonia and methanol and fuel cell technology. For light-duty vehicles including cars, hydrogen is far behind other alternative fuel vehicles, especially compared with the rate of adoption of battery electric vehicles, and may not play a significant role in future.

Disadvantages of hydrogen as an energy carrier include high costs of storage and distribution due to hydrogen's explosivity, its large volume compared to other fuels, and its tendency to make pipes brittle.

Semiconductor Industry

Hydrogen is employed to saturate broken ("dangling") bonds of amorphous silicon and amorphous carbon that helps stabilizing material properties. It is also a potential electron donor in various oxide materials, including ZnO, SnO_2, CdO, MgO, ZrO_2, HfO_2, La_2O_3, Y_2O_3, TiO_2, $SrTiO_3$, $LaAlO_3$, SiO_2, Al_2O_3, $ZrSiO_4$, $HfSiO_4$, and $SrZrO_3$.

Niche and Evolving Uses

- **Shielding Gas:** Hydrogen is used as a shielding gas in welding methods such as atomic hydrogen welding.
- **Cryogenic Research:** Liquid H_2 is used in cryogenic research, including superconductivity studies.
- **Buoyant Lifting:** Because H_2 is only 7% the density of air, it was once widely used as a lifting gas in balloons and airships.
- **Leak Detection:** Pure or mixed with nitrogen (sometimes called forming gas), hydrogen is a tracer gas for detection of minute leaks. Applications can be found in the automotive, chemical, power generation, aerospace, and telecommunications industries. Hydrogen is an authorized food additive (E 949) that allows food package leak testing, as well as having anti-oxidizing properties.
- **Neutron Moderation:** Deuterium (hydrogen-2) is used in nuclear fission applications as a moderator to slow neutrons.
- **Nuclear Fusion Fuel:** Deuterium is used in nuclear fusion reactions.
- **Isotopic Labeling:** Deuterium compounds have applications in chemistry and biology in studies of isotope effects on reaction rates.
- **Rocket Fuel:** Liquid hydrogen and liquid oxygen together serve as cryogenic propellants in liquid-propellant rockets, as in the Space Shuttle main engines. NASA has investigated the use of rocket propellant made from atomic hydrogen, boron or carbon that is frozen

into solid molecular hydrogen particles suspended in liquid helium. Upon warming, the mixture vaporizes to allow the atomic species to recombine, heating the mixture to high temperature.

- **Tritium Uses:** Tritium (hydrogen-3), produced in nuclear reactors, is used in the production of hydrogen bombs, as an isotopic label in the biosciences, and as a source of beta radiation in radioluminescent paint for instrument dials and emergency signage.

8.1.6 Biological Reactions

H_2 is a product of some types of anaerobic metabolism and is produced by several microorganisms, usually via reactions catalyzed by iron- or nickel-containing enzymes called hydrogenases. These enzymes catalyze the reversible redox reaction between H_2 and its component two protons and two electrons. Creation of hydrogen gas occurs in the transfer of reducing equivalents, produced during pyruvate fermentation, to water. The natural cycle of hydrogen production and consumption by organisms is called the hydrogen cycle. Bacteria such as *Mycobacterium smegmatis* can use the small amount of hydrogen in the atmosphere as a source of energy when other sources are lacking, using a hydrogenase with small channels that exclude oxygen and so permits the reaction to occur even though the hydrogen concentration is very low and the oxygen concentration is as in normal air.

Hydrogen is the most abundant element in the human body by numbers of atoms but the third most abundant by mass. H_2 occurs in human breath due to the metabolic activity of hydrogenase-containing microorganisms in the large intestine and is a natural component of flatus. The concentration in the breath of fasting people at rest is typically less than 5 parts per million (ppm) but can be 50 ppm when people with intestinal disorders consume molecules they cannot absorb during diagnostic hydrogen breath tests.

Water splitting, in which water is decomposed into its component protons, electrons, and oxygen, occurs in the light reactions in all photosynthetic organisms. Some such organisms, including the alga *Chlamydomonas reinhardtii* and cyanobacteria, have evolved a second step in the dark reactions in which protons and electrons are reduced to form H_2 gas by specialized hydrogenases in the chloroplast. Efforts have been undertaken to genetically modify cyanobacterial hydrogenases to efficiently synthesize H_2 gas even in the presence of oxygen. Efforts have also been undertaken with genetically modified alga in a bioreactor.

8.1.7 Safety and Precautions

Hydrogen poses a number of hazards to human safety, from potential detonations and fires when mixed with air to being an asphyxiant in its pure, oxygen-free form.

Also, liquid hydrogen is a cryogen and presents dangers (such as frostbite) associated with very cold liquids. Hydrogen dissolves in many metals and in addition to leaking out, may have adverse effects on them, such as hydrogen embrittlement, leading to cracks and explosions. Hydrogen gas leaking into external air may spontaneously ignite. Moreover, hydrogen fire, while being extremely hot, is almost invisible, and thus can lead to accidental burns.

Even interpreting the hydrogen data (including safety data) is confounded by a number of phenomena. Many physical and chemical properties of hydrogen depend on the parahydrogen/orthohydrogen ratio (it often takes days or weeks at a given temperature to reach the equilibrium ratio, for which the data is usually given). Hydrogen detonation parameters, such as critical detonation pressure and temperature, strongly depend on the container geometry.

8.2 THE HYDRIDES

In chemistry, a hydride is formally the anion of hydrogen (H^-), a hydrogen atom with two electrons. In modern usage, this is typically only used for ionic bonds, but it is sometimes (and more frequently in the past) been applied to all compounds containing covalently bound H atoms. In this broad and potentially archaic sense, water (H_2O) is a hydride of oxygen, ammonia is a hydride of nitrogen, etc. In covalent compounds, it implies hydrogen is attached to a less electronegative element. In such cases, the H centre has nucleophilic character, which contrasts with the protic character of acids. The hydride anion is very rarely observed.

Almost all of the elements form binary compounds with hydrogen, the exceptions being He, Ne, Ar, Kr, Pm, Os, Ir, Rn, Fr, and Ra. Exotic molecules such as positronium hydride have also been made.

8.2.1 Bonds

Bonds between hydrogen and the other elements range from being highly ionic to somewhat covalent. Some hydrides, e.g. boron hydrides, do not conform to classical electron counting rules and the bonding is described in terms of multi-centered bonds, whereas the interstitial hydrides often involve metallic bonding. Hydrides can be discrete molecules, oligomers or polymers, ionic solids, chemisorbed monolayers, bulk metals (interstitial), or other materials. While hydrides traditionally react as Lewis bases or reducing agents, some metal hydrides behave as hydrogen-atom donors and act as acids.

8.2.2 Applications

- Hydrides such as sodium borohydride, lithium aluminium hydride, diisobutylaluminium hydride (DIBAL) and super hydride, are commonly used as reducing agents in chemical synthesis. The hydride adds to an electrophilic center, typically unsaturated carbon.

- Hydrides such as sodium hydride and potassium hydride are used as strong bases in organic synthesis. The hydride reacts with the weak Bronsted acid releasing H_2.
- Hydrides such as calcium hydride are used as desiccants, i.e. drying agents, to remove trace water from organic solvents. The hydride reacts with water forming hydrogen and hydroxide salt. The dry solvent can then be distilled or vacuum transferred from the "solvent pot".
- Hydrides are important in storage battery technologies such as nickel-metal hydride battery. Various metal hydrides have been examined for use as a means of hydrogen storage for fuel cell-powered electric cars and other purposed aspects of a hydrogen economy.
- Hydride complexes are catalysts and catalytic intermediates in a variety of homogeneous and heterogeneous catalytic cycles. Important examples include hydrogenation, hydroformylation, hydrosilylation, hydrodesulfurization catalysts. Even certain enzymes, the hydrogenase, operate via hydride intermediates. The energy carrier nicotinamide adenine dinucleotide reacts as a hydride donor or hydride equivalent.

8.2.3 Hydride Ion

Free hydride anions exist only under extreme conditions and are not invoked for homogeneous solution. Instead, many compounds have hydrogen centres with hydridic character. Aside from electride, the hydride ion is the simplest possible anion, consisting of two electrons and a proton. Hydrogen has a relatively low electron affinity, 72.77kJ/mol and reacts exothermically with protons as a powerful Lewis base.

$$H^- + H^+ \rightarrow H2 \qquad \Delta H = \; -1676\, kJ\, /\, mol$$

The low electron affinity of hydrogen and the strength of the H–H bond (ΔH_{BE} = 436 kJ/mol) means that the hydride ion would also be a strong reducing agent

$$H_2 + 2e^- \square \;\; 2H\_ \qquad E^{\square} = -2.25V$$

8.2.4 Types of Hydrides

According to the general definition, every element of the periodic table (except some noble gases) forms one or more hydrides. These substances have been classified into three main types according to the nature of their bonding:

- **Ionic Hydrides**, which have significant ionic bonding character.
- **Covalent Hydrides,** which include the hydrocarbons and many other compounds which covalently bond to hydrogen atoms.
- **Interstitial Hydrides**, which may be described as having metallic bonding.

While these divisions have not been used universally, they are still useful to understand differences in hydrides.

Ionic Hydrides

These are stoichiometric compounds of hydrogen. Ionic or saline hydrides are composed of hydride bound to an electropositive metal, generally an alkali metal or alkaline earth metal. The divalent lanthanides such as europium and ytterbium form compounds similar to those of heavier alkaline earth metals. In these materials the hydride is viewed as a pseudohalide. Saline hydrides are insoluble in conventional solvents, reflecting their non-molecular structures. Ionic hydrides are used as bases and, occasionally, as reducing reagents in organic synthesis.

$$C_6H_5C(O)CH_3 + KH \rightarrow C_6H_5C(O)CH_2K + H_2$$

Typical solvents for such reactions are ethers. Water and other protic solvents cannot serve as a medium for ionic hydrides because the hydride ion is a stronger base than hydroxide and most hydroxyl anions. Hydrogen gas is liberated in a typical acid-base reaction.

$$NaH + H2O \rightarrow H2_{(g)} + NaOH$$

$$\Delta H = -83.6\,kJ/mol,\ \Delta G = -109.0\,kJ/mol$$

Often alkali metal hydrides react with metal halides. Lithium aluminium hydride (often abbreviated as LAH) arises from reactions of lithium hydride with aluminium chloride.

$$4\ LiH + AlCl_3 \rightarrow LiAlH_4 + 3\ LiCl$$

Covalent Hydrides

According to some definitions, covalent hydrides cover all other compounds containing hydrogen. Some definitions limit hydrides to hydrogen centres that formally react as hydrides, i.e. are nucleophilic, and hydrogen atoms bound to metal centers. These hydrides are formed by all the true non-metals (except zero group elements) and the elements like Al, Ga, Sn, Pb, Bi, Po, etc., which are normally metallic in nature, i.e., this class includes the hydrides of p-block elements. In these substances the hydride bond is formally a covalent bond much like the bond made by a proton in a weak acid. This category includes hydrides that exist as discrete molecules, polymers or oligomers, and hydrogen that has been chem-adsorbed to a surface. A particularly important segment of covalent hydrides are complex metal hydrides, powerful soluble hydrides commonly used in synthetic procedures.

Molecular hydrides often involve additional ligands; for example, diisobutylaluminium hydride (DIBAL) consists of two aluminum centers bridged by hydride ligands. Hydrides that are soluble in common solvents are widely used in organic synthesis. Particularly common are sodium borohydride ($NaBH_4$) and lithium aluminium hydride and hindered reagents such as DIBAL.

Interstitial Hydrides or Metallic Hydrides

Interstitial hydrides most commonly exist within metals or alloys. They are traditionally termed "compounds" even though they do not strictly conform to the definition of a compound, more closely resembling common alloys such as steel. In such hydrides, hydrogen can exist as either atomic or diatomic entities. Mechanical or thermal processing, such as bending, striking, or annealing, may cause the hydrogen to precipitate out of solution by degassing. Their bonding is generally considered metallic. Such bulk transition metals form interstitial binary hydrides when exposed to hydrogen.

These systems are usually non-stoichiometric, with variable amounts of hydrogen atoms in the lattice. In materials engineering, the phenomenon of hydrogen embrittlement results from the formation of interstitial hydrides. Hydrides of this type form according to either one of two main mechanisms. The first mechanism involves the adsorption of dihydrogen, succeeded by the cleaving of the H-H bond, the delocalisation of the hydrogen's electrons, and finally the diffusion of the protons into the metal lattice.

The other main mechanism involves the electrolytic reduction of ionised hydrogen on the surface of the metal lattice, also followed by the diffusion of the protons into the lattice.

The second mechanism is responsible for the observed temporary volume expansion of certain electrodes used in electrolytic experiments.

Palladium absorbs up to 900 times its own volume of hydrogen at room temperatures, forming palladium hydride.

This material has been discussed as a means to carry hydrogen for vehicular fuel cells. Interstitial hydrides show certain promise as a way for safe hydrogen storage. Neutron diffraction studies have shown that hydrogen atoms randomly occupy the octahedral interstices in the metal lattice (in an fcc lattice there is one octahedral hole per metal atom).

The limit of absorption at normal pressures is PdH0.7, indicating that approximately 70% of the octahedral holes are occupied.

Many interstitial hydrides have been developed that readily absorb and discharge hydrogen at room temperature and atmospheric pressure. They are usually based on intermetallic compounds and solid-solution alloys. However, their application is still limited, as they are capable of storing only about 2 weight percent of hydrogen, insufficient for automotive applications.

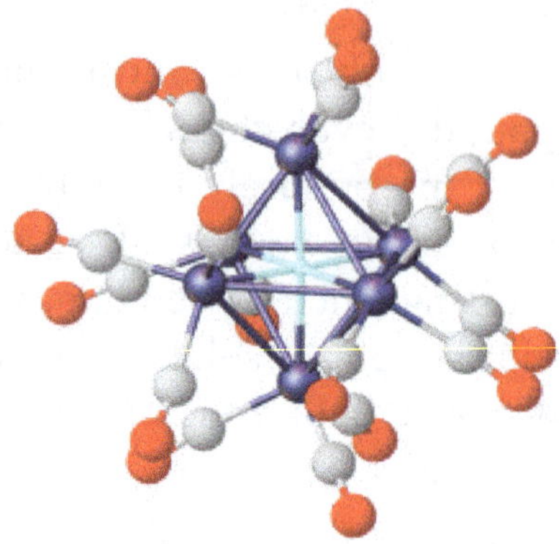

Fig 2: Structure of $[HRu_6(CO)_{18}]^-$, A Metal Cluster with an Interstitial Hydride Ligand (Small Turquoise Sphere at Center)

Transition Metal Hydride Complexes

Transition metal hydrides include compounds that can be classified as *covalent hydrides.* Some are even classified as interstitial hydrides and other bridging hydrides. Classical transition metal hydride feature a single bond between the hydrogen centre and the transition metal. Some transition metal hydrides are acidic, e.g., $HCo(CO)_4$ and $H_2Fe(CO)_4$. The anions potassium nonahydridorhenate $[ReH_9]^{2-}$ and $[FeH_6]^{4-}$ are examples from the growing collection of known molecular homoleptic metal hydrides. As pseudohalides, hydride ligands are capable of bonding with positively polarized hydrogen centres. This interaction, called dihydrogen bonding, is similar to hydrogen bonding, which exists between positively polarized protons and electronegative atoms with open lone pairs.

Protides

Hydrides containing protium are known as *protides.*

Deuterides

Hydrides containing deuterium are known as *deuterides.* Some deuterides, such as LiD, are important fusion fuels in thermonuclear weapons and useful moderators in nuclear reactors.

Tritides

Hydrides containing tritium are known as *tritides.*

Mixed Anion Compounds

Mixed anion compounds exist that contain hydride with other anions. These include boride hydrides, carbohydrides, hydridonitrides, oxyhydrides and others.

8.2.5 Appendix on Nomenclature

Protide, deuteride and *tritide* are used to describe ions or compounds that contain enriched hydrogen-1, deuterium or tritium, respectively. In the classic meaning, hydride refers to any compound hydrogen forms with other elements, ranging over groups 1–16 (the binary compounds of hydrogen). The following is a list of the nomenclature for the hydride derivatives of main group compounds according to this definition:

- **Alkali and Alkaline Earth Metals:** Metal hydride.
- **Boron:** Borane, BH_3.
- **Aluminium:** Alumane, AlH_3.
- **Gallium:** Gallane, GaH_3.
- **Indium:** Indigane, InH_3.
- **Thallium:** Thallane, TlH_3.
- **Carbon:** Alkanes, alkenes, alkynes, and all hydrocarbons.
- **Silicon:** Silane.
- **Germanium:** Germane.
- **Tin:** Stannane.
- **Lead:** Plumbane.
- **Nitrogen:** Ammonia ("azane" when substituted), hydrazine.
- **Phosphorus:** Phosphine (note "phosphane" is the IUPAC recommended name).
- **Arsenic:** Arsine (note "arsane" is the IUPAC recommended name).
- **Antimony:** Stibine (note "stibane" is the IUPAC recommended name).
- **Bismuth:** Bismuthine (note "bismuthane" is the IUPAC recommended name).
- **Helium:** Helium hydride (only exists as an ion).

According to the convention above, the following are "hydrogen compounds" and not "hydrides":

- **Oxygen:** Water ("oxidane" when substituted; synonym: hydrogen oxide), hydrogen peroxide.
- **Sulfur:** Hydrogen sulfide ("sulfane" when substituted).
- **Selenium:** Hydrogen selenide ("selane" when substituted).
- **Tellurium:** Hydrogen telluride ("tellane" when substituted).
- **Polonium:** Hydrogen polonide ("polane" when substituted).
- **Halogens:** Hydrogen halides.

Examples:

- **Nickel Hydride:** Used in NiMH batteries.
- **Palladium Hydride:** Electrodes in cold fusion experiments.
- **Lithium Aluminium Hydride:** A powerful reducing agent used in organic chemistry.
- **Sodium Borohydride:** Selective specialty reducing agent, hydrogen storage in fuel cells.
- **Sodium Hydride:** A powerful base used in organic chemistry.
- **Diborane:** Reducing agent, rocket fuel, semiconductor dopant, catalyst, used in organic synthesis; also borane, pentaborane and decaborane.
- **Arsine:** Used for doping semiconductors.
- **Stibine:** Used in semiconductor industry.
- **Phosphine:** Used for fumigation.
- **Silane:** Many industrial uses, e.g., manufacture of composite materials and water repellents.
- **Ammonia:** Coolant, fuel, fertilizer, many other industrial uses.
- **Hydrogen Sulfide:** Component of natural gas, important source of sulfur.
- **Chemically**, even water and hydrocarbons could be considered hydrides.

All metalloid hydrides are highly flammable. All solid non-metallic hydrides except ice are highly flammable. But when hydrogen combines with halogens it produces acids rather than hydrides, and they are not flammable.

Precedence Convention

According to IUPAC convention, by precedence (stylized electronegativity), hydrogen falls between group 15 and group 16 elements. Therefore, we have NH_3, "nitrogen hydride" (ammonia), versus H_2O, "hydrogen oxide" (water). This convention is sometimes broken for polonium, which on the grounds of polonium's metallicity is often referred to as "polonium hydride" instead of the expected "hydrogen polonide".

PART TWO
THE S-BLOCK ELEMENTS

CHAPTER 9
GROUP 1—THE ALKALI METALS

The alkali metals consist of the chemical elements lithium (Li), sodium (Na), potassium (K), rubidium (Rb), caesium (Cs), and francium (Fr). Together with hydrogen they constitute group 1, which lies in the s-block of the periodic table. All alkali metals have their outermost electron in an s-orbital: this shared electron configuration results in their having very similar characteristic properties. Indeed, the alkali metals provide the best example of group trends in properties in the periodic table, with elements exhibiting well-characterised homologous behaviour. This family of elements is also known as the lithium family after its leading element.

The alkali metals are all shiny, soft, highly reactive metals at standard temperature and pressure and readily lose their outermost electron to form cations with charge +1. They can all be cut easily with a knife due to their softness, exposing a shiny surface that tarnishes rapidly in air due to oxidation by atmospheric moisture and oxygen (and in the case of lithium, nitrogen). Because of their high reactivity, they must be stored under oil to prevent reaction with air, and are found naturally only in salts and never as the free elements. Caesium, the fifth alkali metal, is the most reactive of all the metals. All the alkali metals react with water, with the heavier alkali metals reacting more vigorously than the lighter ones.

All of the discovered alkali metals occur in nature as their compounds: in order of abundance, sodium is the most abundant, followed by potassium, lithium, rubidium, caesium, and finally francium, which is very rare due to its extremely high radioactivity; francium occurs only in minute traces in nature as an intermediate step in some obscure side branches of the natural decay chains. Experiments have been conducted to attempt the synthesis of element 119, which is likely to be the next member of the group; none were successful. However, ununennium may not be an alkali metal due to relativistic effects, which are predicted to have a large influence on the chemical properties of superheavy elements; even if it does turn out to be an alkali metal, it is predicted to have some differences in physical and chemical properties from its lighter homologues.

Most alkali metals have many different applications. One of the best-known applications of the pure elements is the use of rubidium and caesium in atomic clocks, of which caesium atomic clocks form the basis of the second. A common application of the compounds of sodium is the sodium-vapour lamp, which emits light very efficiently. Table salt, or sodium chloride, has been used since antiquity. Lithium finds use as a psychiatric medication and as an anode in lithium batteries. Sodium, potassium and possibly lithium are essential elements, having major biological roles as electrolytes, and although the other alkali metals are not essential, they also have various effects on the body, both beneficial and harmful.

9.1 HISTORY

Sodium compounds have been known since ancient times; salt (sodium chloride) has been an important commodity in human activities. While potash has been used since ancient times, it was not understood for most of its history to be a fundamentally different substance from sodium mineral salts. Georg Ernst Stahl obtained experimental evidence which led him to suggest the fundamental difference of sodium and potassium salts in 1702, and Henri-Louis Duhamel du Monceau was able to prove this difference in 1736. The exact chemical composition of potassium and sodium compounds, and the status as chemical element of potassium and sodium, was not known then, and thus Antoine Lavoisier did not include either alkali in his list of chemical elements in 1789.

Pure potassium was first isolated in 1807 in England by Humphry Davy, who derived it from caustic potash (KOH, potassium hydroxide) by the use of electrolysis of the molten salt with the newly invented voltaic pile. Previous attempts at electrolysis of the aqueous salt were unsuccessful due to potassium's extreme reactivity. Potassium was the first metal that was isolated by electrolysis. Later that same year, Davy reported extraction of sodium from the similar substance caustic soda (NaOH, lye) by a similar technique, demonstrating the elements, and thus the salts, to be different.

Petalite ($LiAlSi_4O_{10}$) was discovered in 1800 by the Brazilian chemist José Bonifácio de Andrada in a mine on the island of Utö, Sweden. However, it was not until 1817 that Johan August Arfwedson, then working in the laboratory of the chemist Jöns Jacob Berzelius, detected the presence of a new element while analysing petalite ore. This new element was noted by him to form compounds similar to those of sodium and potassium, though its carbonate and hydroxide were less soluble in water and more alkaline than the other alkali metals.

He named the metal inside the material *lithium*. Lithium, sodium, and potassium were part of the discovery of periodicity, as they are among a series of triads of elements in the same group that were noted by Johann Wolfgang Döbereiner in 1850 as having similar properties.

Rubidium and caesium were the first elements to be discovered using the spectroscope, invented in 1859 by Robert Bunsen and Gustav Kirchhoff. The next year, they discovered caesium in the mineral water from Bad Dürkheim, Germany.

Their discovery of rubidium came the following year in Heidelberg, Germany, finding it in the mineral lepidolite. The names of rubidium and caesium come from the most prominent lines in their emission spectra: a bright red line for rubidium (from the Latin word *rubidus*, meaning dark red or bright red), and a sky-blue line for caesium (derived from the Latin word *caesius*, meaning sky-blue).

Around 1865 John Newlands produced a series of papers where he listed the elements in order of increasing atomic weight and similar physical and chemical properties that recurred at intervals of eight; he likened such periodicity to the octaves of music, where notes an octave apart have similar musical functions. His version put all the alkali metals then known (lithium to caesium), as well as copper, silver, and thallium (which show the +1 oxidation state characteristic of the alkali metals), together into a group. His table placed hydrogen with the halogens.

After 1869, Dmitri Mendeleev proposed his periodic table placing lithium at the top of a group with sodium, potassium, rubidium, caesium, and thallium. Two years later, Mendeleev revised his table, placing hydrogen in group 1 above lithium, and also moving thallium to the boron group. In this 1871 version, copper, silver, and gold were placed twice, once as part of group IB, and once as part of a "group VIII" encompassing today's groups 8 to 11. After the introduction of the 18-column table, the group IB elements were moved to their current position in the d-block, while alkali metals were left in *group IA*. Later the group's name was changed to *group 1* in 1988. The trivial name "alkali metals" comes from the fact that the hydroxides of the group 1 elements are all strong alkalis when dissolved in water.

There were at least four erroneous and incomplete discoveries before Marguerite Perey of the Curie Institute in Paris, France discovered francium in 1939 by purifying a sample of actinium-227, which had been reported to have a decay energy of 220 keV. However, Perey noticed decay particles with an energy level below 80 keV. Perey thought this decay activity might have been caused by a previously unidentified decay product, one that was separated during purification, but emerged again out of the pure actinium-227.

Various tests eliminated the possibility of the unknown element being thorium, radium, lead, bismuth, or thallium. The new product exhibited chemical properties of an alkali metal (such as coprecipitating with caesium salts), which led Perey to believe that it was element 87, caused by the alpha decay of actinium-227. Perey then attempted to determine the proportion of beta decay to alpha decay in actinium-227. Her first test put the alpha branching at 0.6%, a figure that she later revised to 1%.

$$ {}^{227}_{89}\mathrm{Ac} \xrightarrow[21.77\ \mathrm{y}]{\alpha\ (1.38\%)} {}^{223}_{87}\mathrm{Fr} \xrightarrow[22\ \mathrm{min}]{\beta^-} {}^{223}_{88}\mathrm{Ra} \xrightarrow[11.4\ \mathrm{d}]{\alpha} {}^{219}_{86}\mathrm{Rn} $$

The next element below francium (eka-francium) in the periodic table would be ununennium (Uue), element 119. The synthesis of ununennium was first attempted in 1985 by bombarding a target of einsteinium-254 with calcium-48 ions at the superHILAC accelerator at the Lawrence Berkeley National Laboratory in Berkeley, California. No atoms were identified, leading to a limiting yield of 300 nb.

$$^{254}_{99}Es + ^{48}_{20}Ca \rightarrow ^{302}_{119}Uue^{*} \rightarrow \textit{no atoms}$$

It is highly unlikely that this reaction will be able to create any atoms of ununennium in the near future, given the extremely difficult task of making sufficient amounts of einsteinium-254, which is favoured for production of ultraheavy elements because of its large mass, relatively long half-life of 270 days, and availability in significant amounts of several micrograms, to make a large enough target to increase the sensitivity of the experiment to the required level; einsteinium has not been found in nature and has only been produced in laboratories, and in quantities smaller than those needed for effective synthesis of superheavy elements. However, given that ununennium is only the first period 8 element on the extended periodic table, it may well be discovered in the near future through other reactions, and indeed an attempt to synthesise it is currently ongoing in Japan. Currently, none of the period 8 elements has been discovered yet, and it is also possible, due to drip instabilities, that only the lower period 8 elements, up to around element 128, are physically possible. No attempts at synthesis have been made for any heavier alkali metals: due to their extremely high atomic number, they would require new, more powerful methods and technology to make.

9.2 OCCURRENCE

9.2.1 In the Solar System

The Oddo–Harkins rule holds that elements with even atomic numbers are more common that those with odd atomic numbers, with the exception of hydrogen. This rule argues that elements with odd atomic numbers have one unpaired proton and are more likely to capture another, thus increasing their atomic number. In elements with even atomic numbers, protons are paired, with each member of the pair offsetting the spin of the other, enhancing stability. All the alkali metals have odd atomic numbers and they are not as common as the elements with even atomic numbers adjacent to them (the noble gases and the alkaline earth metals) in the Solar System. The heavier alkali metals are also less abundant than the lighter ones as the alkali metals from rubidium onward can only be synthesised in supernovae and not in stellar nucleosynthesis. Lithium is also much less abundant than sodium and potassium as it is poorly synthesised in both Big Bang nucleosynthesis and in stars: the Big Bang could only produce trace quantities of lithium, beryllium and boron due to the absence of a stable nucleus with 5 or 8 nucleons, and stellar nucleosynthesis could only pass this bottleneck by the triple-alpha process, fusing three helium nuclei to form carbon, and skipping over those three elements.

9.2.2 On Earth

The Earth formed from the same cloud of matter that formed the Sun, but the planets acquired different compositions during the formation and evolution of the solar system. In turn, the natural history of the Earth caused parts of this planet to have differing concentrations of the elements. The mass of the Earth is approximately 5.98×10^{24} kg. It is composed mostly of iron (32.1 per cent), oxygen (30.1 per cent), silicon (15.1 per cent), magnesium (13.9 per cent), sulfur (2.9 per cent), nickel (1.8 per cent), calcium (1.5 per cent), and aluminium (1.4 per cent); with the remaining 1.2 per cent consisting of trace amounts of other elements. Due to planetary differentiation, the core region is believed to be primarily composed of iron (88.8 per cent), with smaller amounts of nickel (5.8 per cent), sulfur (4.5 per cent), and less than %1 trace elements.

The alkali metals, due to their high reactivity, do not occur naturally in pure form in nature. They are lithophiles and therefore remain close to the Earth's surface because they combine readily with oxygen and so associate strongly with silica, forming relatively low-density minerals that do not sink down into the Earth's core. Potassium, rubidium and caesium are also incompatible elements due to their large ionic radii.

Sodium and potassium are very abundant on Earth, both being among the ten most common elements in Earth's crust; sodium makes up approximately 2.6% of the Earth's crust measured by weight, making it the sixth most abundant element overall and the most abundant alkali metal. Potassium makes up approximately 1.5% of the Earth's crust and is the seventh most abundant element. Sodium is found in many different minerals, of which the most common is ordinary salt (sodium chloride), which occurs in vast quantities dissolved in seawater. Other solid deposits include halite, amphibole, cryolite, nitratine, and zeolite. Many of these solid deposits occur as a result of ancient seas evaporating, which still occurs now in places such as Utah's Great Salt Lake and the Dead Sea.

Despite their near-equal abundance in Earth's crust, sodium is far more common than potassium in the ocean, both because potassium's larger size makes its salts less soluble, and because potassium is bound by silicates in soil and what potassium leaches is absorbed far more readily by plant life than sodium.

Despite its chemical similarity, lithium typically does not occur together with sodium or potassium due to its smaller size. Due to its relatively low reactivity, it can be found in seawater in large amounts; it is estimated that lithium concentration in seawater is approximately 0.14 to 0.25 parts per million (ppm) or 25 micromolar.

Its diagonal relationship with magnesium often allows it to replace magnesium in ferromagnesium minerals, where its crustal concentration is about 18 ppm, comparable to that of gallium and niobium. Commercially, the most important lithium mineral is spodumene, which occurs in large deposits worldwide.

Rubidium is approximately as abundant as zinc and more abundant than copper. It occurs naturally in the minerals leucite, pollucite, carnallite, zinnwaldite, and lepidolite, although none of these contain only rubidium and no other alkali metals. Caesium is more abundant than some commonly known elements, such as antimony, cadmium, tin, and tungsten, but is much less abundant than rubidium.

Francium-223, the only naturally occurring isotope of francium, is the product of the alpha decay of actinium-227 and can be found in trace amounts in uranium minerals. In a given sample of uranium, there is estimated to be only one francium atom for every 10^{18} uranium atoms. It has been calculated that there are at most 30 grams of francium in the earth's crust at any time, due to its extremely short half-life of 22 minutes.

9.3 PROPERTIES

9.3.1 Physical and Chemical

The physical and chemical properties of the alkali metals can be readily explained by their having an ns^1 valence electron configuration, which results in weak metallic bonding. Hence, all the alkali metals are soft and have low densities, melting and boiling points, as well as heats of sublimation, vaporisation, and dissociation. They all crystallise in the body-centered cubic crystal structure, and have distinctive flame colours because their outer s electron is very easily excited. Indeed, these flame test colours are the most common way of identifying them since all their salts with common ions are soluble.

The ns^1 configuration also results in the alkali metals having very large atomic and ionic radii, as well as very high thermal and electrical conductivity. Their chemistry is dominated by the loss of their lone valence electron in the outermost s-orbital to form the +1 oxidation state, due to the ease of ionising this electron and the very high second ionisation energy. Most of the chemistry has been observed only for the first five members of the group. The chemistry of francium is not well established due to its extreme radioactivity; thus, the presentation of its properties here is limited. What little is known about francium shows that it is very close in behaviour to caesium, as expected. The physical properties of francium are even sketchier because the bulk element has never been observed; hence any data that may be found in the literature are certainly speculative extrapolations.

The alkali metals are more similar to each other than the elements in any other group are to each other. Indeed, the similarity is so great that it is quite difficult to separate potassium, rubidium, and caesium, due to their similar ionic radii; lithium and sodium are more distinct. For instance, when moving down the table, all known alkali metals show increasing atomic radius, decreasing electronegativity, increasing reactivity, and decreasing melting and boiling points as well as heats of fusion and vaporisation. In general, their densities increase when moving down the table, with the exception that potassium is less dense than sodium. One of the very few properties of the alkali metals that does not display a very smooth trend is their reduction potentials: lithium's value is anomalous, being more negative than the others. This is because the Li^+ ion has a very high hydration energy in the gas phase: though the lithium ion disrupts the structure of water significantly, causing a higher change in entropy, this high hydration energy is enough to make the reduction potentials indicate it as being the most electropositive alkali metal, despite the difficulty of ionising it in the gas phase.

The stable alkali metals are all silver-coloured metals except for caesium, which has a pale golden tint: it is one of only three metals that are clearly coloured (the other two being copper and gold). Additionally, the heavy alkaline earth metals calcium, strontium, and barium, as well as the divalent lanthanides europium and ytterbium, are pale yellow, though the colour is much less prominent than it is for caesium. Their lustre tarnishes rapidly in air due to oxidation.

All the alkali metals are highly reactive and are never found in elemental forms in nature. Because of this, they are usually stored in mineral oil or kerosene (paraffin oil). They react aggressively with the halogens to form the alkali metal halides, which are white ionic crystalline compounds that are all soluble in water except lithium fluoride (LiF). The alkali metals also react with water to form strongly alkaline hydroxides and thus should be handled with great care. The heavier alkali metals react more vigorously than the lighter ones; for example, when dropped into water, caesium produces a larger explosion than potassium if the same number of moles of each metal is used. The alkali metals have the lowest first ionisation energies in their respective periods of the periodic table because of their low effective nuclear charge and the ability to attain a noble gas configuration by losing just one electron. Not only do the alkali metals react with water, but also with proton donors like alcohols and phenols, gaseous ammonia, and alkynes, the last demonstrating the phenomenal degree of their reactivity. Their great power as reducing agents makes them very useful in liberating other metals from their oxides or halides.

The second ionisation energy of all of the alkali metals is very high as it is in a full shell that is also closer to the nucleus; thus, they almost always lose a single electron, forming cations. The alkalides are an exception: they are unstable compounds which contain alkali metals in a −1 oxidation state, which is very unusual as before the discovery of the alkalides, the alkali metals were not expected to be able to form anions and were thought to be able to appear in salts only as cations. The alkalide anions have filled s-subshells, which gives them enough stability to exist. All the stable alkali metals except lithium are known to be able to form alkalides, and the alkalides have much theoretical interest due to their unusual stoichiometry and low ionisation potentials. Alkalides are chemically similar to the electrides, which are salts with trapped electrons acting as anions. A particularly striking example of an alkalide is "inverse sodium hydride", H^+Na^- (both ions being complexed), as opposed to the usual sodium hydride, Na^+H^-: it is unstable in isolation, due to its high energy resulting from the displacement of two electrons from hydrogen to sodium, although several derivatives are predicted to be metastable or stable.

In aqueous solution, the alkali metal ions form aqua ions of the formula $[M(H_2O)_n]^+$, where n is the solvation number. Their coordination numbers and shapes agree well with those expected from their ionic radii. In aqueous solution the water molecules directly attached to the metal ion are said to belong to the first coordination sphere, also known as the first, or primary, solvation shell. The bond between a water molecule and the metal ion is a dative covalent bond, with the oxygen atom donating both electrons to the bond. Each coordinated water molecule may be attached by hydrogen bonds to other water molecules. The latter are said to reside in the second coordination sphere. However, for the alkali metal cations, the second coordination sphere is not well-defined as the +1 charge on the cation is not high enough to polarise the water molecules in the primary solvation shell enough for them to form strong hydrogen bonds with those in the second coordination sphere, producing a more stable entity. The solvation number for Li^+ has been experimentally determined to be 4, forming the tetrahedral $[Li(H_2O)_4]^+$: while solvation numbers of 3 to 6 have been found for lithium aqua ions, solvation numbers less than 4 may be the result of the formation of contact ion pairs, and the higher solvation numbers may be interpreted in terms of water molecules that approach $[Li(H_2O)_4]^+$ through a face of the tetrahedron, though molecular dynamic simulations may indicate the existence of an octahedral hexaaqua ion. There are also probably six water molecules in the primary solvation sphere of the sodium ion, forming the octahedral $[Na(H_2O)_6]^+$ ion. While it was previously thought that the heavier alkali metals also formed octahedral hexaaqua ions, it has since been found that potassium and rubidium probably form the $[K(H_2O)_8]^+$ and $[Rb(H_2O)_8]^+$ ions, which have the square antiprismatic structure, and that caesium forms the 12-coordinate $[Cs(H_2O)_{12}]^+$ ion.

Lithium

The chemistry of lithium shows several differences from that of the rest of the group as the small Li^+ cation polarises anions and gives its compounds a more covalent character. Lithium and magnesium have a diagonal relationship due to their similar atomic radii, so that they show some similarities. For example, lithium forms a stable nitride, a property common among all the alkaline earth metals (magnesium's group) but unique among the alkali metals. In addition, among their respective groups, only lithium and magnesium form organometallic compounds with significant covalent character (e.g. LiMe and $MgMe_2$).

Lithium fluoride is the only alkali metal halide that is poorly soluble in water, and lithium hydroxide is the only alkali metal hydroxide that is not deliquescent. Conversely, lithium perchlorate and other lithium salts with large anions that cannot be polarised are much more stable than the analogous compounds of the other alkali metals, probably because Li^+ has a high solvation energy. This effect also means that most simple lithium salts are commonly encountered in hydrated form, because the anhydrous forms are extremely hygroscopic: this allows salts like lithium chloride and lithium bromide to be used in dehumidifiers and air-conditioners.

Francium

Francium is also predicted to show some differences due to its high atomic weight, causing its electrons to travel at considerable fractions of the speed of light and thus making relativistic effects more prominent. In contrast to the trend of decreasing electronegativities and ionisation energies of the alkali metals, francium's electronegativity and ionisation energy are predicted to be higher than caesium's due to the relativistic stabilisation of the 7s electrons; also, its atomic radius is expected to be abnormally low. Thus, contrary to expectation, caesium is the most reactive of the alkali metals, not francium. All known physical properties of francium also deviate from the clear trends going from lithium to caesium, such as the first ionisation energy, electron affinity, and anion polarisability, though due to the paucity of known data about francium many sources give extrapolated values, ignoring that relativistic effects make the trend from lithium to caesium become inapplicable at francium. Some of the few properties of francium that have been predicted taking relativity into account are the electron affinity (47.2 kJ/mol) and the enthalpy of dissociation of the Fr_2 molecule (42.1 kJ/mol). The CsFr molecule is polarised as Cs^+Fr^-, showing that the 7s subshell of francium is much more strongly affected by relativistic effects than the 6s subshell of caesium. Additionally, francium superoxide (FrO_2) is expected to have significant covalent character, unlike the other alkali metal superoxides, because of bonding contributions from the 6p electrons of francium.

9.3.2 Nuclear

All the alkali metals have odd atomic numbers; hence, their isotopes must be either odd–odd (both proton and neutron number are odd) or odd–even (proton number is odd, but neutron number is even). Odd–odd nuclei have even mass numbers, whereas odd–even nuclei have odd mass numbers. Odd–odd primordial nuclides are rare because most odd–odd nuclei are highly unstable with respect to beta decay, because the decay products are even–even, and are therefore more strongly bound, due to nuclear pairing effects.

Due to the great rarity of odd–odd nuclei, almost all the primordial isotopes of the alkali metals are odd–even (the exceptions being the light stable isotope lithium-6 and the long-lived radioisotope potassium-40). For a given odd mass number, there can be only a single beta-stable nuclide, since there is not a difference in binding energy between even–odd and odd–even comparable to that between even–even and odd–odd, leaving other nuclides of the same mass number (isobars) free to beta decay toward the lowest-mass nuclide. An effect of the instability of an odd number of either type of nucleons is that odd-numbered elements, such as the alkali metals, tend to have fewer stable isotopes than even-numbered elements. Of the 26 monoisotopic elements that have only a single stable isotope, all but one have an odd atomic number and all but one also have an even number of neutrons. Beryllium is the single exception to both rules, due to its low atomic number.

All of the alkali metals except lithium and caesium have at least one naturally occurring radioisotope: sodium-22 and sodium-24 are trace radioisotopes produced cosmogenically, potassium-40 and rubidium-87 have very long half-lives and thus occur naturally, and all isotopes of francium are radioactive. Caesium was also thought to be radioactive in the early 20th century, although it has no naturally occurring radioisotopes. (Francium had not been discovered yet at that time.) The natural long-lived radioisotope of potassium, potassium-40, makes up about 0.012% of natural potassium, and thus natural potassium is weakly radioactive. This natural radioactivity became a basis for a mistaken claim of the discovery for element 87 (the next alkali metal after caesium) in 1925. Natural rubidium is similarly slightly radioactive, with 27.83% being the long-lived radioisotope rubidium-87.

Caesium-137, with a half-life of 30.17 years, is one of the two principal medium-lived fission products, along with strontium-90, which are responsible for most of the radioactivity of spent nuclear fuel after several years of cooling, up to several hundred years after use. It constitutes most of the radioactivity still left from the Chernobyl accident. Caesium-137 undergoes high-energy beta decay and eventually becomes stable barium-137. It is a strong emitter of gamma radiation. Caesium-137 has a very low rate of neutron capture and cannot be feasibly disposed of in this way, but must be allowed to decay. Caesium-137 has been used as a tracer in hydrologic studies, analogous to the use of tritium. Small amounts of caesium-134 and caesium-137 were released into the environment during nearly all nuclear weapon tests and some nuclear accidents, most notably the Goiânia accident and the Chernobyl disaster. As of 2005, caesium-137 is the principal source of radiation in the zone of alienation around the Chernobyl nuclear power plant. Its chemical properties as one of the alkali metals make it one of the most problematic of the short-to-medium-lifetime fission products because it easily moves and spreads in nature due to the high water solubility of its salts, and is taken up by the body, which mistakes it for its essential congeners sodium and potassium.

9.4 PERIODIC TRENDS

The alkali metals are more similar to each other than the elements in any other group are to each other. For instance, when moving down the table, all known alkali metals show increasing atomic radius, decreasing electronegativity, increasing reactivity, and decreasing melting and boiling points as well as heats of fusion and vaporisation. In general, their densities increase when moving down the table, with the exception that potassium is less dense than sodium.

9.4.1 Atomic and Ionic Radii

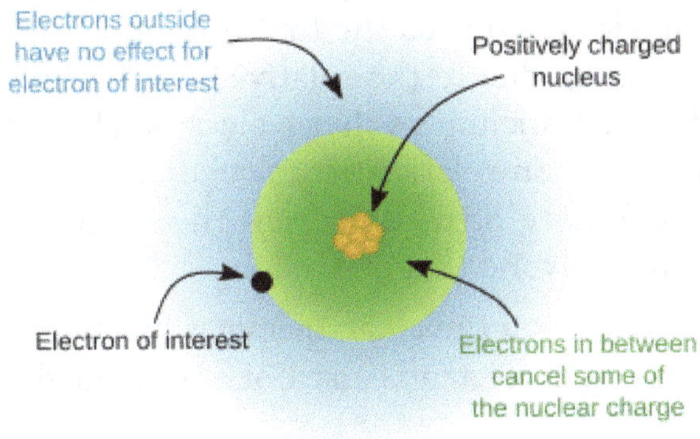

Fig 1: Effective Nuclear Charge on an Atomic Electron

The atomic radii of the alkali metals increase going down the group. Because of the shielding effect, when an atom has more than one electron shell, each electron feels electric repulsion from the other electrons as well as electric attraction from the nucleus. In the alkali metals, the outermost electron only feels a net charge of +1, as some of the nuclear charge (which is equal to the atomic number) is cancelled by the inner electrons; the number of inner electrons of an alkali metal is always one less than the nuclear charge. Therefore, the only factor which affects the atomic radius of the alkali metals is the number of electron shells. Since this number increases down the group, the atomic radius must also increase down the group.

The ionic radii of the alkali metals are much smaller than their atomic radii. This is because the outermost electron of the alkali metals is in a different electron shell than the inner electrons, and thus when it is removed the resulting atom has one fewer electron shell and is smaller. Additionally, the effective nuclear charge has increased, and thus the electrons are attracted more strongly towards the nucleus and the ionic radius decreases.

9.4.2 First Ionisation Energy

The first ionisation energy of an element or molecule is the energy required to move the most loosely held electron from one mole of gaseous atoms of the element or molecules to form one mole of gaseous ions with electric charge +1. The factors affecting the first ionisation energy are the nuclear charge, the amount of shielding by the inner electrons and the distance from the most loosely held electron from the nucleus, which is always an outer electron in main group elements. The first two factors change the effective nuclear charge the most loosely held electron feels. Since the outermost electron of alkali metals always feels the same effective nuclear charge (+1), the only factor which affects the first ionisation energy is the distance from the outermost electron to the nucleus.

Since this distance increases down the group, the outermost electron feels less attraction from the nucleus and thus the first ionisation energy decreases. This trend is broken in francium due to the relativistic stabilisation and contraction of the 7s orbital, bringing francium's valence electron closer to the nucleus than would be expected from non-relativistic calculations. This makes francium's outermost electron feel more attraction from the nucleus, increasing its first ionisation energy slightly beyond that of caesium.

The second ionisation energy of the alkali metals is much higher than the first as the second-most loosely held electron is part of a fully filled electron shell and is thus difficult to remove.

9.4.3 Reactivity

The reactivities of the alkali metals increase going down the group. This is the result of a combination of two factors: the first ionisation energies and atomisation energies of the alkali metals. Because the first ionisation energy of the alkali metals decreases down the group, it is easier for the outermost electron to be removed from the atom and participate in chemical reactions, thus increasing reactivity down the group. The atomisation energy measures the strength of the metallic bond of an element, which falls down the group as the atoms increase in radius and thus the metallic bond must increase in length, making the delocalised electrons further away from the attraction of the nuclei of the heavier alkali metals. Adding the atomisation and first ionisation energies gives a quantity closely related to (but not equal to) the activation energy of the reaction of an alkali metal with another substance.

This quantity decreases going down the group, and so does the activation energy; thus, chemical reactions can occur faster and the reactivity increases down the group.

9.4.4 Electronegativity

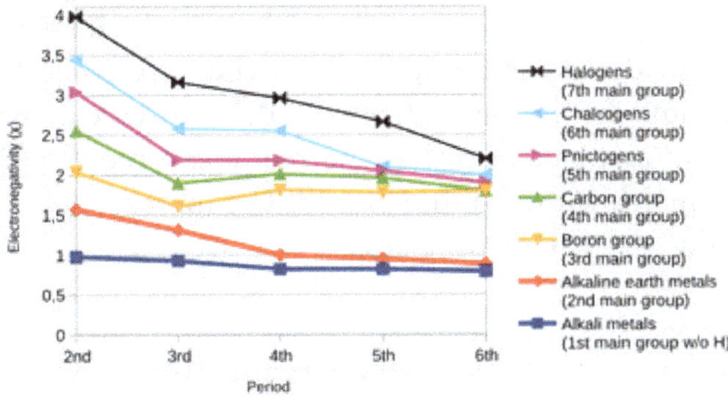

Fig 2: Periodic Variation of Pauling Electronegativities as One Descends the Main Groups of the Periodic Table from the Second to the Sixth Period

Electronegativity is a chemical property that describes the tendency of an atom or a functional group to attract electrons (or electron density) towards itself. If the bond between sodium and chlorine in sodium chloride were covalent, the pair of shared electrons would be attracted to the chlorine because the effective nuclear charge on the outer electrons is +7 in chlorine but is only +1 in sodium. The electron pair is attracted so close to the chlorine atom that they are practically transferred to the chlorine atom (an ionic bond). However, if the sodium atom was replaced by a lithium atom, the electrons will not be attracted as close to the chlorine atom as before because the lithium atom is smaller, making the electron pair more strongly attracted to the closer effective nuclear charge from lithium. Hence, the larger alkali metal atoms (further down the group) will be less electronegative as the bonding pair is less strongly attracted towards them. As mentioned previously, francium is expected to be an exception.

Because of the higher electronegativity of lithium, some of its compounds have a more covalent character. For example, lithium iodide (LiI) will dissolve in organic solvents, a property of most covalent compounds. Lithium fluoride (LiF) is the only alkali halide that is not soluble in water, and lithium hydroxide (LiOH) is the only alkali metal hydroxide that is not deliquescent.

9.4.5 Melting and Boiling Points

The melting point of a substance is the point where it changes state from solid to liquid while the boiling point of a substance (in liquid state) is the point where the vapour pressure of the liquid equals the environmental pressure surrounding the liquid and all the liquid changes state to gas. As a metal is heated to its melting point, the metallic bonds keeping the atoms in place weaken so that the atoms can move around, and the metallic bonds eventually break completely at the metal's boiling point. Therefore, the falling melting and boiling points of the alkali metals indicate that the strength of the metallic bonds of the alkali metals decreases down the group. This is because metal atoms are held together by the electromagnetic attraction from the positive ions to the delocalised electrons. As the atoms increase in size going down the group (because their atomic radius increases), the nuclei of the ions move further away from the delocalised electrons and hence the metallic bond becomes weaker so that the metal can more easily melt and boil, thus lowering the melting and boiling points. The increased nuclear charge is not a relevant factor due to the shielding effect.

9.4.6 Density

The alkali metals all have the same crystal structure (body-centred cubic) and thus the only relevant factors are the number of atoms that can fit into a certain volume and the mass of one of the atoms, since density is defined as mass per unit volume. The first factor depends on the volume of the atom and thus the atomic radius, which increases going down the group; thus, the volume of an alkali metal atom increases going down the group. The mass of an alkali metal atom also increases going down the group. Thus, the trend for the densities of the alkali metals depends on their atomic weights and atomic radii; if figures for these two factors are known, the ratios between the densities of the alkali metals can then be calculated. The resultant trend is that the densities of the alkali metals increase down the table, with an exception at potassium. Due to having the lowest atomic weight and the largest atomic radius of all the elements in their periods, the alkali metals are the least dense metals in the periodic table. Lithium, sodium, and potassium are the only three metals in the periodic table that are less dense than water: in fact, lithium is the least dense known solid at room temperature.

9.5 COMPOUNDS

The alkali metals form complete series of compounds with all usually encountered anions, which well illustrate group trends. These compounds can be described as involving the alkali metals losing electrons to acceptor species and forming monopositive ions. This description is most accurate for alkali halides and becomes less and less accurate as cationic and anionic charge increase, and as the anion becomes larger and more polarisable. For instance, ionic bonding gives way to metallic bonding along the series NaCl, Na_2O, Na_2S, Na_3P, Na_3As, Na_3Sb, Na_3Bi, Na.

9.5.1 Hydroxides

All the alkali metals react vigorously or explosively with cold water, producing an aqueous solution of a strongly basic alkali metal hydroxide and releasing hydrogen gas. This reaction becomes more vigorous going down the group: lithium reacts steadily with effervescence, but sodium and potassium can ignite, and rubidium and caesium sink in water and generate hydrogen gas so rapidly that shock waves form in the water that may shatter glass containers. When an alkali metal is dropped into water, it produces an explosion, of which there are two separate stages. The metal reacts with the water first, breaking the hydrogen bonds in the water and producing hydrogen gas; this takes place faster for the more reactive heavier alkali metals. Second, the heat generated by the first part of the reaction often ignites the hydrogen gas, causing it to burn explosively into the surrounding air. This secondary hydrogen gas explosion produces the visible flame above the bowl of water, lake or other body of water, not the initial reaction of the metal with water (which tends to happen mostly under water). The alkali metal hydroxides are the most basic known hydroxides.

Recent research has suggested that the explosive behavior of alkali metals in water is driven by a Coulomb explosion rather than solely by rapid generation of hydrogen itself. All alkali metals melt as a part of the reaction with water. Water molecules ionise the bare metallic surface of the liquid metal, leaving a positively charged metal surface and negatively charged water ions. The attraction between the charged metal and water ions will rapidly increase the surface area, causing an exponential increase of ionisation. When the repulsive forces within the liquid metal surface exceeds the forces of the surface tension, it vigorously explodes.

The hydroxides themselves are the most basic hydroxides known, reacting with acids to give salts and with alcohols to give oligomeric alkoxides. They easily react with carbon dioxide to form carbonates or bicarbonates, or with hydrogen sulfide to form sulfides or bisulfides, and may be used to separate thiols from petroleum. They react with amphoteric oxides: for example, the oxides of aluminium, zinc, tin, and lead react with the alkali metal hydroxides to give aluminates, zincates, stannates, and plumbates. Silicon dioxide is acidic, and thus the alkali metal hydroxides can also attack silicate glass.

9.5.2 Intermetallic Compounds

The alkali metals form many intermetallic compounds with each other and the elements from groups 2 to 13 in the periodic table of varying stoichiometries, such as the sodium amalgams with mercury, including Na_5Hg_8 and Na_3Hg. Some of these have ionic characteristics: taking the alloys with gold, the most electronegative of metals, as an example, NaAu and KAu are metallic, but RbAu and CsAu are semiconductors. NaK is an alloy of sodium and potassium that is very useful because it is liquid at room temperature, although precautions must be taken due to its extreme reactivity towards water and air. The eutectic mixture melts at -12.6 °C. An alloy of 41% caesium, 47% sodium, and 12% potassium has the lowest known melting point of any metal or alloy, -78 °C.

9.5.3 Compounds with the Group 13 Elements

The intermetallic compounds of the alkali metals with the heavier group 13 elements (aluminium, gallium, indium, and thallium), such as NaTl, are poor conductors or semiconductors, unlike the normal alloys with the preceding elements, implying that the alkali metal involved has lost an electron to the Zintl anions involved. Nevertheless, while the elements in group 14 and beyond tend to form discrete anionic clusters, group 13 elements tend to form polymeric ions with the alkali metal cations located between the giant ionic lattice. For example, NaTl consists of a polymeric anion $(—Tl^-—)_n$ with a covalent diamond cubic structure with Na^+ ions located between the anionic lattice. The larger alkali metals cannot fit similarly into an anionic lattice and tend to force the heavier group 13 elements to form anionic clusters.

Boron is a special case, being the only nonmetal in group 13. The alkali metal borides tend to be boron-rich, involving appreciable boron–boron bonding involving deltahedral structures, and are thermally unstable due to the alkali metals having a very high vapour pressure at elevated temperatures. This makes direct synthesis problematic because the alkali metals do not react with boron below 700 °C, and thus this must be accomplished in sealed containers with the alkali metal in excess. Furthermore, exceptionally in this group, reactivity with boron decreases down the group: lithium reacts completely at 700 °C, but sodium at 900 °C and potassium not until 1200 °C, and the reaction is instantaneous for lithium but takes hours for potassium. Rubidium and caesium borides have not even been characterised. Various phases are known, such as LiB_{10}, NaB_6, NaB_{15}, and KB_6. Under high pressure the boron–boron bonding in the lithium borides changes from following Wade's rules to forming Zintl anions like the rest of group 13.

9.5.4 Compounds with the Group 14 Elements

Lithium and sodium react with carbon to form acetylides, Li_2C_2 and Na_2C_2, which can also be obtained by reaction of the metal with acetylene. Potassium, rubidium, and caesium react with graphite; their atoms are intercalated between the hexagonal graphite layers, forming graphite intercalation compounds of formulae MC_{60} (dark grey, almost black), MC_{48} (dark grey, almost black), MC_{36} (blue), MC_{24} (steel blue), and MC_8 (bronze) (M = K, Rb, or Cs). These compounds are over 200 times more electrically conductive than pure graphite, suggesting that the valence electron of the alkali metal is transferred to the graphite layers (e.g. M^+C-8). Upon heating of KC_8, the elimination of potassium atoms results in the conversion in sequence to KC_{24}, KC_{36}, KC_{48} and finally KC_{60}. KC_8 is a very strong reducing agent and is pyrophoric and explodes on contact with water. While the larger alkali metals

(K, Rb, and Cs) initially form MC_8, the smaller ones initially form MC_6, and indeed they require reaction of the metals with graphite at high temperatures around 500 °C to form. Apart from this, the alkali metals are such strong reducing agents that they can even reduce buckminsterfullerene to produce solid fullerides M_nC_{60}; sodium, potassium, rubidium, and caesium can form fullerides where $n = 2, 3, 4$, or 6, and rubidium and caesium additionally can achieve $n = 1$.

When the alkali metals react with the heavier elements in the carbon group (silicon, germanium, tin, and lead), ionic substances with cage-like structures are formed, such as the silicides M_4Si_4 (M = K, Rb, or Cs), which contains M^+ and tetrahedral Si4-4 ions. The chemistry of alkali metal germanides, involving the germanide ion Ge^{4-} and other cluster (Zintl) ions such as Ge2-4, Ge4-9, Ge2-9, and $[(Ge_9)_2]^{6-}$, is largely analogous to that of the corresponding silicides. Alkali metal stannides are mostly ionic, sometimes with the stannide ion (Sn^{4-}), and sometimes with more complex Zintl ions such as Sn4-9, which appears in tetrapotassium nonastannide (K_4Sn_9). The monatomic plumbide ion (Pb^{4-}) is unknown, and indeed its formation is predicted to be energetically unfavourable; alkali metal plumbides have complex Zintl ions, such as Pb4-9. These alkali metal germanides, stannides, and plumbides may be produced by reducing germanium, tin, and lead with sodium metal in liquid ammonia.

9.5.5 Nitrides and Pnictides

Lithium, the lightest of the alkali metals, is the only alkali metal which reacts with nitrogen at standard conditions, and its nitride is the only stable alkali metal nitride. Nitrogen is an unreactive gas because breaking the strong triple bond in the dinitrogen molecule (N_2) requires a lot of energy. The formation of an alkali metal nitride would consume the ionisation energy of the alkali metal (forming M^+ ions), the energy required to break the triple bond in N_2 and the formation of N^{3-} ions, and all the energy released from the formation of an alkali metal nitride is from the lattice energy of the alkali metal nitride.

The lattice energy is maximised with small, highly charged ions; the alkali metals do not form highly charged ions, only forming ions with a charge of +1, so only lithium, the smallest alkali metal, can release enough lattice energy to make the reaction with nitrogen exothermic, forming lithium nitride. The reactions of the other alkali metals with nitrogen would not release enough lattice energy and would thus be endothermic, so they do not form nitrides at standard conditions. Sodium nitride (Na_3N) and potassium nitride (K_3N), while existing, are extremely unstable, being prone to decomposing back into their constituent elements, and cannot be produced by reacting the elements with each other at standard conditions. Steric hindrance forbids the existence of rubidium or caesium nitride. However, sodium and potassium form colourless azide salts involving the linear N-3 anion; due to the large size of the alkali metal cations, they are thermally stable enough to be able to melt before decomposing.

All the alkali metals react readily with phosphorus and arsenic to form phosphides and arsenides with the formula M_3Pn (where M represents an alkali metal and Pn represents a pnictogen – phosphorus, arsenic, antimony, or bismuth). This is due to the greater size of the P^{3-} and As^{3-} ions, so that less lattice energy needs to be released for the salts to form. These are not the only phosphides and arsenides of the alkali metals: for example, potassium has nine different known phosphides, with formulae K_3P, K_4P_3, K_5P_4, KP, K_4P_6, K_3P_7, K_3P_{11}, $KP_{10.3}$, and KP_{15}. While most metals form arsenides, only the alkali and alkaline earth metals form mostly ionic arsenides.

The structure of Na_3As is complex with unusually short Na–Na distances of 328–330 pm which are shorter than in sodium metal, and this indicates that even with these electropositive metals the bonding cannot be straightforwardly ionic. Other alkali metal arsenides not conforming to the formula M_3As are known, such as LiAs, which has a metallic lustre and electrical conductivity indicating the presence of some metallic bonding. The antimonides are unstable and reactive as the Sb^{3-} ion is a strong reducing agent; reaction of them with acids form the toxic and unstable gas stibine (SbH_3). Indeed, they have some metallic properties, and the alkali metal antimonides of stoichiometry MSb involve antimony atoms bonded in a spiral Zintl structure. Bismuthides are not even wholly ionic; they are intermetallic compounds containing partially metallic and partially ionic bonds.

9.5.6 Oxides and Chalcogenides

All the alkali metals react vigorously with oxygen at standard conditions. They form various types of oxides, such as simple oxides (containing the O^{2-} ion), peroxides (containing the O2-2 ion, where there is a single bond between the two oxygen atoms), superoxides (containing the O-2 ion), and many others. Lithium burns in air to form lithium oxide, but sodium reacts with oxygen to form a mixture of sodium oxide and sodium peroxide.

Potassium forms a mixture of potassium peroxide and potassium superoxide, while rubidium and caesium form the superoxide exclusively. Their reactivity increases going down the group: while lithium, sodium and potassium merely burn in air, rubidium and caesium are pyrophoric (spontaneously catch fire in air).

The smaller alkali metals tend to polarise the larger anions (the peroxide and superoxide) due to their small size. This attracts the electrons in the more complex anions towards one of its constituent oxygen atoms, forming an oxide ion and an oxygen atom. This causes lithium to form the oxide exclusively on reaction with oxygen at room temperature. This effect becomes drastically weaker for the larger sodium and potassium, allowing them to form the less stable peroxides. Rubidium and caesium, at the bottom of the group, are so large that even the least stable superoxides can form. Because the superoxide releases the most energy when formed, the superoxide is preferentially formed for the larger alkali metals where the more complex anions are not polarised.

The oxides and peroxides for these alkali metals do exist, but do not form upon direct reaction of the metal with oxygen at standard conditions. In addition, the small size of the Li^+ and O^{2-} ions contributes to their forming a stable ionic lattice structure. Under controlled conditions, however, all the alkali metals, with the exception of francium, are known to form their oxides, peroxides, and superoxides.

The alkali metal peroxides and superoxides are powerful oxidising agents. Sodium peroxide and potassium superoxide react with carbon dioxide to form the alkali metal carbonate and oxygen gas, which allows them to be used in submarine air purifiers; the presence of water vapour, naturally present in breath, makes the removal of carbon dioxide by potassium superoxide even more efficient. All the stable alkali metals except lithium can form red ozonides (MO_3) through low-temperature reaction of the powdered anhydrous hydroxide with ozone: the ozonides may be then extracted using liquid ammonia. They slowly decompose at standard conditions to the superoxides and oxygen, and hydrolyse immediately to the hydroxides when in contact with water. Potassium, rubidium, and caesium also form sesquioxides M_2O_3, which may be better considered peroxide disuperoxides, $[(M^+)_4(O2-2)(O-2)_2]$.

Rubidium and caesium can form a great variety of suboxides with the metals in formal oxidation states below +1. Rubidium can form Rb_6O and Rb_9O_2 (copper-coloured) upon oxidation in air, while caesium forms an immense variety of oxides, such as the ozonide CsO_3 and several brightly coloured suboxides, such as Cs_7O (bronze), Cs_4O (red-violet), $Cs_{11}O_3$ (violet), Cs_3O (dark green), CsO, Cs_3O_2, as well as Cs_7O_2. The last of these may be heated under vacuum to generate Cs_2O.

The alkali metals can also react analogously with the heavier chalcogens (sulfur, selenium, tellurium, and polonium), and all the alkali metal chalcogenides are known (with the exception of francium's). Reaction with an excess of the chalcogen can similarly result in lower chalcogenides, with chalcogen ions containing chains of the chalcogen atoms in question.

For example, sodium can react with sulfur to form the sulfide (Na_2S) and various polysulfides with the formula Na_2S_x (*x* from 2 to 6), containing the $S^{2-}{}_x$ ions. Due to the basicity of the Se^{2-} and Te^{2-} ions, the alkali metal selenides and tellurides are alkaline in solution; when reacted directly with selenium and tellurium, alkali metal polyselenides and polytellurides are formed along with the selenides and tellurides with the $Se^{2-}{}_x$ and $Te^{2-}{}_x$ ions. They may be obtained directly from the elements in liquid ammonia or when air is not present, and are colourless, water-soluble compounds that air oxidises quickly back to selenium or tellurium. The alkali metal polonides are all ionic compounds containing the Po^{2-} ion; they are very chemically stable and can be produced by direct reaction of the elements at around 300–400 °C.

9.5.7 Halides, Hydrides, and Pseudohalides

The alkali metals are among the most electropositive elements on the periodic table and thus tend to bond ionically to the most electronegative elements on the periodic table, the halogens (fluorine, chlorine, bromine, iodine, and astatine), forming salts known as the alkali metal halides. The reaction is very vigorous and can sometimes result in explosions.

All twenty stable alkali metal halides are known; the unstable ones are not known, with the exception of sodium astatide, because of the great instability and rarity of astatine and francium. The most well-known of the twenty is certainly sodium chloride, otherwise known as common salt. All of the stable alkali metal halides have the formula MX where M is an alkali metal and X is a halogen. They are all white ionic crystalline solids that have high melting points. All the alkali metal halides are soluble in water except for lithium fluoride (LiF), which is insoluble in water due to its very high lattice enthalpy. The high lattice enthalpy of lithium fluoride is due to the small sizes of the Li^+ and F^- ions, causing the electrostatic interactions between them to be strong: a similar effect occurs for magnesium fluoride, consistent with the diagonal relationship between lithium and magnesium.

The alkali metals also react similarly with hydrogen to form ionic alkali metal hydrides, where the hydride anion acts as a pseudohalide: these are often used as reducing agents, producing hydrides, complex metal hydrides, or hydrogen gas. Other pseudohalides are also known, notably the cyanides. These are isostructural to the respective halides except for lithium cyanide, indicating that the cyanide ions may rotate freely. Ternary alkali metal halide oxides, such as Na_3ClO, K_3BrO (yellow), Na_4Br_2O, Na_4I_2O, and K_4Br_2O, are also known. The polyhalides are rather unstable, although those of rubidium and caesium are greatly stabilised by the feeble polarising power of these extremely large cations.

9.5.8 Coordination Complexes

Alkali metal cations do not usually form coordination complexes with simple Lewis bases due to their low charge of just +1 and their relatively large size; thus the Li^+ ion forms most complexes and the heavier alkali metal ions form less and less (though exceptions occur for weak complexes). Lithium in particular has a very rich coordination chemistry in which it exhibits coordination numbers from 1 to 12, although octahedral hexacoordination is its preferred mode. In aqueous solution, the alkali metal ions exist as octahedral hexahydrate complexes $[M(H_2O)_6]^+$, with the exception of the lithium ion, which due to its small size forms tetrahedral tetrahydrate complexes $[Li(H_2O)_4]^+$; the alkali metals form these complexes because their ions are attracted by electrostatic forces of attraction to the polar water molecules. Because of this, anhydrous salts containing alkali metal cations are often used as desiccants. Alkali metals also readily form complexes with crown ethers (e.g. 12-crown-4 for Li^+, 15-crown-5 for Na^+, 18-crown-6 for K^+, and 21-crown-7 for Rb^+) and cryptands due to electrostatic attraction.

9.5.9 Ammonia Solutions

The alkali metals dissolve slowly in liquid ammonia, forming ammoniacal solutions of solvated metal cation M^+ and solvated electron e^-, which react to form hydrogen gas and the alkali metal amide (MNH_2, where M represents an alkali metal): this was first noted by Humphry Davy in 1809 and rediscovered by W. Weyl in 1864. The process may be speeded up by a catalyst. Similar solutions are formed by the heavy divalent alkaline earth metals calcium, strontium, barium, as well as the divalent lanthanides, europium and ytterbium. The amide salt is quite insoluble and readily precipitates out of solution, leaving intensely coloured ammonia solutions of the alkali metals.

In 1907, Charles A. Kraus identified the colour as being due to the presence of solvated electrons, which contribute to the high electrical conductivity of these solutions. At low concentrations (below 3 M), the solution is dark blue and has ten times the conductivity of aqueous sodium chloride; at higher concentrations (above 3 M), the solution is copper-coloured and has approximately the conductivity of liquid metals like mercury. In addition to the alkali metal amide salt and solvated electrons, such ammonia solutions also contain the alkali metal cation (M^+), the neutral alkali metal atom (M), diatomic alkali metal molecules (M_2) and alkali metal anions (M^-). These are unstable and eventually become the more thermodynamically stable alkali metal amide and hydrogen gas. Solvated electrons are powerful reducing agents and are often used in chemical synthesis.

9.5.10 Organometallic

Organolithium

Being the smallest alkali metal, lithium forms the widest variety of and most stable organometallic compounds, which are bonded covalently. Organolithium compounds are electrically non-conducting volatile solids or liquids that melt at low temperatures, and tend to form oligomers with the structure $(RLi)_x$ where R is the organic group. As the electropositive nature of lithium puts most of the charge density of the bond on the carbon atom, effectively creating a carbanion, organolithium compounds are extremely powerful bases and nucleophiles. For use as bases, butyllithiums are often used and are commercially available. An example of an organolithium compound is methyllithium ($(CH_3Li)_x$), which exists in tetrameric ($x = 4$, tetrahedral) and hexameric ($x = 6$, octahedral) forms. Organolithium compounds, especially *n*-butyllithium, are useful reagents in organic synthesis, as might be expected given lithium's diagonal relationship with magnesium, which plays an important role in the Grignard reaction. For example, alkyllithiums and aryllithiums may be used to synthesise aldehydes and ketones by reaction with metal carbonyls.

The reaction with nickel tetracarbonyl, for example, proceeds through an unstable acyl nickel carbonyl complex which then undergoes electrophilic substitution to give the desired aldehyde (using H^+ as the electrophile) or ketone (using an alkyl halide) product.

$$LiR + Ni(CO)_4 \longrightarrow Li^+[RCONi(CO)_3]^-$$

$$Li^+[RCONi(CO)_3]^- \xrightarrow[\text{solvent}]{H^+} Li^+ + RCHO + [(\text{solvent})Ni(CO)_3]$$

$$Li^+[RCONi(CO)_3]^- \xrightarrow[\text{solvent}]{R'Br} Li^+ + RR'CO + [(\text{solvent})Ni(CO)_3]$$

Alkyllithiums and aryllithiums may also react with *N,N*-disubstituted amides to give aldehydes and ketones, and symmetrical ketones by reacting with carbon monoxide.

They thermally decompose to eliminate a β-hydrogen, producing alkenes and lithium hydride: another route is the reaction of ethers with alkyl- and aryllithiums that act as strong bases. In non-polar solvents, aryllithiums react as the carbanions they effectively are, turning carbon dioxide to aromatic carboxylic acids ($ArCO_2H$) and aryl ketones to tertiary carbinols ($Ar'_2C(Ar)OH$). Finally, they may be used to synthesise other organometallic compounds through metal-halogen exchange.

Heavier Alkali Metals

Unlike the organolithium compounds, the organometallic compounds of the heavier alkali metals are predominantly ionic. The application of organosodium compounds in chemistry is limited in part due to competition from organolithium compounds, which are commercially available and exhibit more convenient reactivity.

The principal organosodium compound of commercial importance is sodium cyclopentadienide. Sodium tetraphenylborate can also be classified as an organosodium compound since in the solid state sodium is bound to the aryl groups. Organometallic compounds of the higher alkali metals are even more reactive than organosodium compounds and of limited utility.

A notable reagent is Schlosser's base, a mixture of *n*-butyllithium and potassium *tert*-

butoxide. This reagent reacts with propene to form the compound allylpotassium (KCH_2CHCH_2). *cis*-2-Butene and *trans*-2-butene equilibrate when in contact with alkali metals. Whereas isomerisation is fast with lithium and sodium, it is slow with the heavier alkali metals.

The heavier alkali metals also favour the sterically congested conformation. Several crystal structures of organopotassium compounds have been reported, establishing that they, like the sodium compounds, are polymeric. Organosodium, organopotassium, organorubidium and organocaesium compounds are all mostly ionic and are insoluble (or nearly so) in nonpolar solvents.

Alkyl and aryl derivatives of sodium and potassium tend to react with air. They cause the cleavage of ethers, generating alkoxides. Unlike alkyllithium compounds, alkylsodiums and alkylpotassiums cannot be made by reacting the metals with alkyl halides because Wurtz coupling occurs:

$$RM + R'X \longrightarrow R\text{–}R' + MX$$

As such, they have to be made by reacting alkylmercury compounds with sodium or potassium metal in inert hydrocarbon solvents. While methylsodium forms tetramers like methyllithium, methylpotassium is more ionic and has the nickel arsenide structure with discrete methyl anions and potassium cations.

The alkali metals and their hydrides react with acidic hydrocarbons, for example cyclopentadienes and terminal alkynes, to give salts.

Liquid ammonia, ether, or hydrocarbon solvents are used, the most common of which being tetrahydrofuran. The most important of these compounds is sodium cyclopentadienide, NaC_5H_5, an important precursor to many transition metal cyclopentadienyl derivatives. Similarly, the alkali metals react with cyclooctatetraene in tetrahydrofuran to give alkali metal cyclooctatetraenides; for example, dipotassium cyclooctatetraenide ($K_2C_8H_8$) is an important precursor to many metal cyclooctatetraenyl derivatives, such as uranocene. The large and very weakly polarising alkali metal cations can stabilise large, aromatic, polarisable radical anions, such as the dark-green sodium naphthalenide, $Na^+[C_{10}H_8\bullet]^-$, a strong reducing agent.

9.6 REPRESENTATIVE REACTIONS OF ALKALI METALS

9.6.1 Reaction with Oxygen

Upon reacting with oxygen, alkali metals form oxides, peroxides, superoxides and suboxides. However, the first three are more common. The table below shows the types of compounds formed in reaction with oxygen. The compound in brackets represents the minor product of combustion.

Alkali metal	Oxide	Peroxide	Superoxide
Li	Li_2O	(Li_2O_2)	
Na	(Na_2O)	Na_2O_2	
K			KO_2
Rb			RbO_2
Cs			CsO_2

The alkali metal peroxides are ionic compounds that are unstable in water. The peroxide anion is weakly bound to the cation, and it is hydrolysed, forming stronger covalent bonds.

$$Na_2O_2 + 2H_2O \rightarrow 2NaOH + H_2O_2$$

The other oxygen compounds are also unstable in water.

$$2KO_2 + 2H_2O \rightarrow 2KOH + H_2O_2 + O_2$$

$$Li_2O + H_2O \rightarrow 2LiOH$$

9.6.2 Reaction with Sulfur

With sulfur, they form sulfides and polysulfides.

$$2Na + 1/8S_8 \rightarrow Na_2S + 1/8S_8 \rightarrow Na_2S_2...Na_2S_7$$

Because alkali metal sulfides are essentially salts of a weak acid and a strong base, they form basic solutions.

$$S^{2-} + H_2O \rightarrow HS^- + HO^-$$

$$HS^- + H_2O \rightarrow H_2S + HO^-$$

9.6.3 Reaction with Nitrogen

Lithium is the only metal that combines directly with nitrogen at room temperature.

$$3Li + 1/2N_2 \rightarrow Li_3N$$

Li_3N can react with water to liberate ammonia.

$$Li_3N + 3H_2O \rightarrow 3LiOH + NH_3$$

9.6.4 Reaction with Hydrogen

With hydrogen, alkali metals form saline hydrides that hydrolyse in water.

$$2\,Na + H_2 \xrightarrow{\Delta} 2\,NaH$$

$$2\,NaH + 2\,H_2O \longrightarrow 2\,NaOH + H_2\uparrow$$

9.6.5 Reaction with Carbon

Lithium is the only metal that reacts directly with carbon to give dilithium acetylide. Na and K can react with acetylene to give acetylides.

$$2Li + 2C \longrightarrow Li_2C_2$$

$$2Na + 2C_2H_2 \xrightarrow{150\ ^\circ C} 2NaC_2H + H_2$$

$$2Na + 2NaC_2H \xrightarrow{220\ ^\circ C} 2Na_2C_2 + H_2$$

9.6.6 Reaction with Water

On reaction with water, they generate hydroxide ions and hydrogen gas. This reaction is vigorous and highly exothermic and the hydrogen resulted may ignite in air or even explode in the case of Rb and Cs.

$$Na + H_2O \rightarrow NaOH + 1/2H_2$$

9.6.7 Reaction with other Salts

The alkali metals are very good reducing agents. They can reduce metal cations that are less electropositive. Titanium is produced industrially by the reduction of titanium tetrachloride with Na at 400 °C (van Arkel–de Boer process).

$$TiCl_4 + 4Na \rightarrow 4NaCl + Ti$$

9.6.8 Reaction with Organohalide Compounds

Alkali metals react with halogen derivatives to generate hydrocarbon via the Wurtz reaction.

$$2CH_3\text{-}Cl + 2Na \rightarrow H_3C\text{-}CH_3 + 2NaCl$$

9.6.9 Alkali Metals in Liquid Ammonia

Alkali metals dissolve in liquid ammonia or other donor solvents like aliphatic amines or hexamethylphosphoramide to give blue solutions. These solutions are believed to contain free electrons.

$$Na + xNH_3 \rightarrow Na^+ + e(NH_3)_x^-$$

Due to the presence of solvated electrons, these solutions are very powerful reducing agents used in organic synthesis.

1)

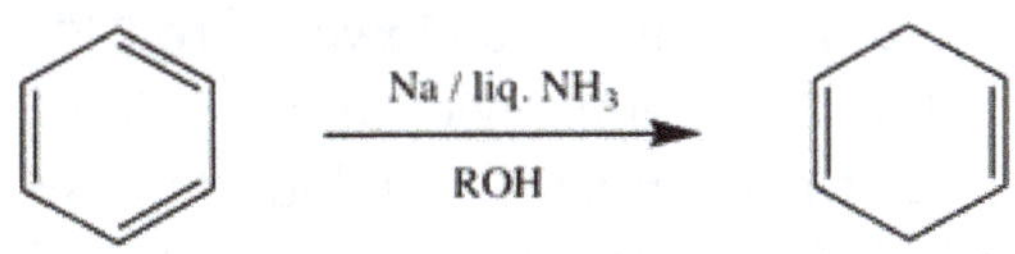

2)

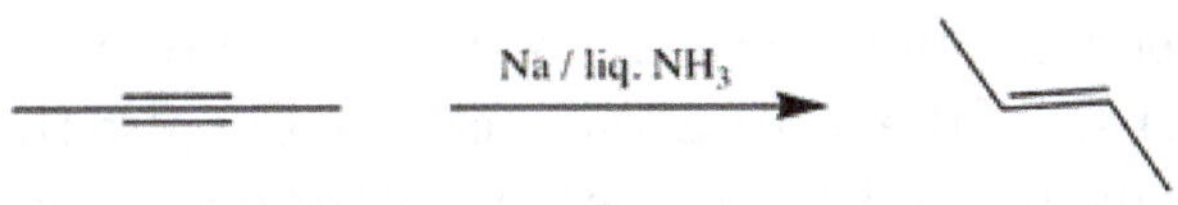

Fig 3: Reduction Reactions Using Sodium in Liquid Ammonia

Reaction 1) is known as Birch reduction. Other reductions that can be carried by these solutions are:

$$S_8 + 2e^- \longrightarrow S_8^{2-}$$
$$Fe(CO)_5 + 2e^- \longrightarrow Fe(CO)_4^{2-} + CO$$

9.7 EXTENSIONS

Although francium is the heaviest alkali metal that has been discovered, there has been some theoretical work predicting the physical and chemical characteristics of hypothetical heavier alkali metals. Being the first period 8 element, the undiscovered element ununennium (element 119) is predicted to be the next alkali metal after francium and behave much like their lighter congeners; however, it is also predicted to differ from the lighter alkali metals in some properties. Its chemistry is predicted to be closer to that of potassium or rubidium instead of caesium or francium.

This is unusual as periodic trends, ignoring relativistic effects would predict ununennium to be even more reactive than caesium and francium. This lowered reactivity is due to the relativistic stabilisation of ununennium's valence electron, increasing ununennium's first ionisation energy and decreasing the metallic and ionic radii; this effect is already seen for francium. This assumes that ununennium will behave chemically as an alkali metal, which, although likely, may not be true due to relativistic effects.

The relativistic stabilisation of the 8s orbital also increases ununennium's electron affinity far beyond that of caesium and francium; indeed, ununennium is expected to have an electron affinity higher than all the alkali metals lighter than it. Relativistic effects also cause a very large drop in the polarisability of ununennium. On the other hand, ununennium is predicted to continue the trend of melting points decreasing going down the group, being expected to have a melting point between 0 °C and 30°C.

The stabilisation of ununennium's valence electron and thus the contraction of the 8s orbital cause its atomic radius to be lowered to 240 pm, very close to that of rubidium (247 pm), so that the chemistry of ununennium in the +1 oxidation state should be more similar to the chemistry of rubidium than to that of francium. On the other hand, the ionic radius of the Uue^+ ion is predicted to be larger than that of Rb^+, because the 7p orbitals are destabilised and are thus larger than the p-orbitals of the lower shells. Ununennium may also show the +3 and +5 oxidation states, which are not seen in any other alkali metal, in addition to the +1 oxidation state that is characteristic of the other alkali metals and is also the main oxidation state of all the known alkali metals: this is because of the destabilisation and expansion of the $7p_{3/2}$ spinor, causing its outermost electrons to have a lower ionisation energy than what would otherwise be expected. Indeed, many ununennium compounds are expected to have a large covalent character, due to the involvement of the $7p_{3/2}$ electrons in the bonding.

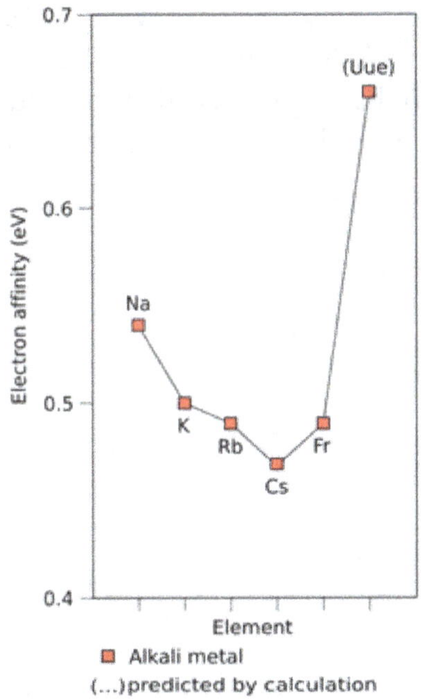

Fig 4: Empirical (Na–Fr) and Predicted (Uue) Electron Affinity of the Alkali Metals from the Third to the Eighth Period, Measured in Electron Volts

Not as much work has been done predicting the properties of the alkali metals beyond ununennium. Although a simple extrapolation of the periodic table (by the Aufbau principle) would put element 169, unhexennium, under ununennium, Dirac-Fock calculations predict that the next element after ununennium with alkali-metal-like properties may be element 165, unhexpentium, which is predicted to have the electron configuration [Og] $5g^{18}$ $6f^{14}$ $7d^{10}$ $8s^2$ $8p_{1/2}^{\ 2}$ $9s^1$. This element would be intermediate in properties between an alkali metal and a group 11 element, and while its physical and atomic properties would be closer to the former, its chemistry may be closer to that of the latter.

Further calculations show that unhexpentium would follow the trend of increasing ionisation energy beyond caesium, having an ionisation energy comparable to that of sodium, and that it should also continue the trend of decreasing atomic radii beyond caesium, having an atomic radius comparable to that of potassium. However, the 7d electrons of unhexpentium may also be able to participate in chemical reactions along with the 9s electron, possibly allowing oxidation states beyond +1, whence the likely transition metal behaviour of unhexpentium. Due to the alkali and alkaline earth metals both being s-block elements, these predictions for the trends and properties of ununennium and unhexpentium also mostly hold quite similarly for the corresponding alkaline earth metals unbinilium (Ubn) and unhexhexium (Uhh).

Unsepttrium, element 173, may be an even better heavier homologue of ununennium; with a predicted electron configuration of [Usb] $6g^1$, it returns to the alkali-metal-like situation of having one easily removed electron far above a closed p-shell in energy, and is expected to be even more reactive than caesium.

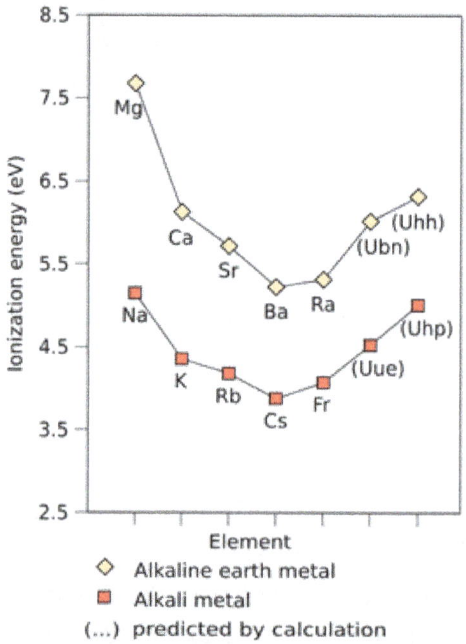

Fig 5: Empirical (Na–Fr, Mg–Ra) and Predicted (Uue–Uhp, Ubn–Uhh) Ionisation Energy of the Alkali and Alkaline Earth Metals from the Third to the Ninth Period, Measured in Electron Volts

The probable properties of further alkali metals beyond unsepttrium have not been explored yet as of 2019, and they may or may not be able to exist. In periods 8 and above of the periodic table, relativistic and shell-structure effects become so strong that extrapolations from lighter congeners become completely inaccurate.

In addition, the relativistic and shell-structure effects (which stabilise the s-orbitals and destabilise and expand the d-, f-, and g-orbitals of higher shells) have opposite effects, causing even larger difference between relativistic and non-relativistic calculations of the properties of elements with such high atomic numbers.

Interest in the chemical properties of ununennium, unhexpentium, and unsepttrium stems from the fact that they are located close to the expected locations of islands of stability, centered at elements 122 (^{306}Ubb) and 164 (^{482}Uhq).

9.8 PSEUDO-ALKALI METALS

Many other substances are similar to the alkali metals in their tendency to form monopositive cations. Analogously to the pseudohalogens, they have sometimes been called "pseudo-alkali metals". These substances include some elements and many more polyatomic ions; the polyatomic ions are especially similar to the alkali metals in their large size and weak polarising power.

9.8.1 Hydrogen

The element hydrogen, with one electron per neutral atom, is usually placed at the top of Group 1 of the periodic table because of its electron configuration. But hydrogen is not normally considered to be an alkali metal. Metallic hydrogen, which only exists at very high pressures, is known for its electrical and magnetic properties, not its chemical properties. Under typical conditions, pure hydrogen exists as a diatomic gas consisting of two atoms per molecule (H_2); however, the alkali metals form diatomic molecules (such as dilithium, Li_2) only at high temperatures, when they are in the gaseous state.

Hydrogen, like the alkali metals, has one valence electron and reacts easily with the halogens, but the similarities mostly end there because of the small size of a bare proton H^+ compared to the alkali metal cations. Its placement above lithium is primarily due to its electron configuration. It is sometimes placed above fluorine due to their similar chemical properties, though the resemblance is likewise not absolute.

The first ionisation energy of hydrogen (1312.0 kJ/mol) is much higher than that of the alkali metals. As only one additional electron is required to fill in the outermost shell of the hydrogen atom, hydrogen often behaves like a halogen, forming the negative hydride ion, and is very occasionally considered to be a halogen on that basis. (The alkali metals can also form negative ions, known as alkalides, but these are little more than laboratory curiosities, being unstable.) An argument against this placement is that formation of hydride from hydrogen is endothermic, unlike the exothermic formation of halides from halogens. The radius of the H^- anion also does not fit the trend of increasing size going down the halogens: indeed, H^- is very diffuse because its single proton cannot easily control both electrons. It was expected for some time that liquid hydrogen would show metallic properties; while this has been shown to not be the case, under extremely high pressures, such as those found at the cores of Jupiter and Saturn, hydrogen does become metallic and behaves like an alkali metal; in this phase, it is known as metallic hydrogen. The electrical resistivity of liquid metallic hydrogen at 3000 K is approximately equal to that of liquid rubidium and caesium at 2000 K at the respective pressures when they undergo a nonmetal-to-metal transition.

The $1s^1$ electron configuration of hydrogen, while analogous to that of the alkali metals (ns^1), is unique because there is no 1p subshell. Hence it can lose an electron to form the hydron H^+, or gain one to form the hydride ion H^-. In the former case it resembles superficially the alkali metals; in the latter case, the halogens, but the differences due to the lack of a 1p subshell are important enough that neither group fits the properties of hydrogen well. Group 14 is also a good fit in terms of thermodynamic properties such as ionisation energy and electron affinity, but hydrogen cannot be tetravalent.

Thus none of the three placements are entirely satisfactory, although group 1 is the most common placement (if one is chosen) because of the electron configuration and the fact that the hydron is by far the most important of all monatomic hydrogen species, being the foundation of acid-base chemistry. As an example of hydrogen's unorthodox properties stemming from its unusual electron configuration and small size, the hydrogen ion is very small (radius around 150 fm compared to the 50–220 pm size of most other atoms and ions) and so is nonexistent in condensed systems other than in association with other atoms or molecules.

Indeed, transferring of protons between chemicals is the basis of acid-base chemistry. Also unique is hydrogen's ability to form hydrogen bonds, which are an effect of charge-transfer, electrostatic, and electron correlative contributing phenomena. While analogous lithium bonds are also known, they are mostly electrostatic. Nevertheless, hydrogen can take on the same structural role as the alkali metals in some molecular crystals, and has a close relationship with the lightest alkali metals (especially lithium).

9.8.2 Ammonium and Derivatives

The ammonium ion (NH+4) has very similar properties to the heavier alkali metals, acting as an alkali metal intermediate between potassium and rubidium, and is often considered a close relative. For example, most alkali metal salts are soluble in water, a property which ammonium salts share.

Ammonium is expected to behave stably as a metal (NH+4 ions in a sea of delocalised electrons) at very high pressures (though less than the typical pressure where transitions from insulating to metallic behaviour occur around, 100 GPa), and could possibly occur inside the ice giants Uranus and Neptune, which may have significant impacts on their interior magnetic fields. It has been estimated that the transition from a mixture of ammonia and dihydrogen molecules to metallic ammonium may occur at pressures just below 25 GPa. Under standard conditions, ammonium can form a metallic amalgam with mercury.

Other "pseudo-alkali metals" include the alkylammonium cations, in which some of the hydrogen atoms in the ammonium cation are replaced by alkyl or aryl groups.

In particular, the quaternary ammonium cations (NR+4) are very useful since they are permanently charged, and they are often used as an alternative to the expensive Cs^+ to stabilise very large and very easily polarisable anions such as HI−2. Tetraalkylammonium hydroxides, like alkali metal hydroxides, are very strong bases that react with atmospheric carbon dioxide to form carbonates. Furthermore, the nitrogen atom may be replaced by a phosphorus, arsenic, or antimony atom (the heavier nonmetallic pnictogens), creating a phosphonium (PH+4) or arsonium (AsH+4) cation that can itself be substituted similarly; while stibonium (SbH+4) itself is not known, some of its organic derivatives are characterised.

9.8.3 Cobaltocene and Derivatives

Cobaltocene, $Co(C_5H_5)_2$, is a metallocene, the cobalt analogue of ferrocene. It is a dark purple solid. Cobaltocene has 19 valence electrons, one more than usually found in organotransition metal complexes, such as its very stable relative, ferrocene, in accordance with the 18-electron rule. This additional electron occupies an orbital that is antibonding with respect to the Co–C bonds. Consequently, many chemical reactions of $Co(C_5H_5)_2$ are characterized by its tendency to lose this "extra" electron, yielding a very stable 18-electron cation known as cobaltocenium. Many cobaltocenium salts coprecipitate with caesium salts, and cobaltocenium hydroxide is a strong base that absorbs atmospheric carbon dioxide to form cobaltocenium carbonate. Like the alkali metals, cobaltocene is a strong reducing agent, and decamethylcobaltocene is stronger still due to the combined inductive effect of the ten methyl groups. Cobalt may be substituted by its heavier congener rhodium to give rhodocene, an even stronger reducing agent. Iridocene (involving iridium) would presumably be still more potent, but is not very well-studied due to its instability.

9.8.4 Thallium

Thallium is the heaviest stable element in group 13 of the periodic table. At the bottom of the periodic table, the inert-pair effect is quite strong, because of the relativistic stabilisation of the 6s orbital and the decreasing bond energy as the atoms increase in size so that the amount of energy released in forming two more bonds is not worth the high ionisation energies of the 6s electrons.

It displays the +1 oxidation state that all the known alkali metals display, and thallium compounds with thallium in its +1 oxidation state closely resemble the corresponding potassium or silver compounds stoichiometrically due to the similar ionic radii of the Tl^+ (164 pm), K^+ (152 pm) and Ag^+ (129 pm) ions. It was sometimes considered an alkali metal in continental Europe (but not in England) in the years immediately following its discovery, and was placed just after caesium as the sixth alkali metal in Dmitri Mendeleev's 1869 periodic table and Julius Lothar Meyer's

1868 periodic table. Mendeleev's 1871 periodic table and Meyer's 1870 periodic table put thallium in its current position in the boron group and left the space below caesium blank. However, thallium also displays the oxidation state +3, which no known alkali metal displays (although ununennium, the undiscovered seventh alkali metal, is predicted to possibly display the +3 oxidation state). The sixth alkali metal is now considered to be francium. While Tl^+ is stabilised by the inert-pair effect, this inert pair of 6s electrons is still able to participate chemically, so that these electrons are stereochemically active in aqueous solution. Additionally, the thallium halides (except TlF) are quite insoluble in water, and TlI has an unusual structure because of the presence of the stereochemically active inert pair in thallium.

9.8.5 Copper, Silver, and Gold

The group 11 metals (or coinage metals), copper, silver, and gold, are typically categorised as transition metals given they can form ions with incomplete d-shells. Physically, they have the relatively low melting points and high electronegativity values associated with post-transition metals. "The filled *d* subshell and free *s* electron of Cu, Ag, and Au contribute to their high electrical and thermal conductivity. Transition metals to the left of group 11 experience interactions between *s* electrons and the partially filled *d* subshell that lower electron mobility." Chemically, the group 11 metals behave like main-group metals in their +1 valence states, and are hence somewhat related to the alkali metals: this is one reason for their previously being labelled as "group IB", paralleling the alkali metals' "group IA". They are occasionally classified as post-transition metals. Their spectra are analogous to those of the alkali metals. Their monopositive ions are paramagnetic and contribute no colour to their salts, like those of the alkali metals.

In Mendeleev's 1871 periodic table, copper, silver, and gold are listed twice, once under group VIII (with the iron triad and platinum group metals), and once under group IB. Group IB was nonetheless parenthesised to note that it was tentative. Mendeleev's main criterion for group assignment was the maximum oxidation state of an element: on that basis, the group 11 elements could not be classified in group IB, due to the existence of copper(II) and gold(III) compounds being known at that time. However, eliminating group IB would make group I the only main group (group VIII was labelled a transition group) to lack an A–B bifurcation. Soon afterward, a majority of chemists chose to classify these elements in group IB and remove them from group VIII for the resulting symmetry: this was the predominant classification until the rise of the modern medium-long 18-column periodic table, which separated the alkali metals and group 11 metals.

The coinage metals were traditionally regarded as a subdivision of the alkali metal group, due to them sharing the characteristic s^1 electron configuration of the alkali metals (group 1: p^6s^1; group 11: $d^{10}s^1$). However, the similarities are largely confined to the stoichiometries of the +1 compounds of both groups, and not their chemical properties. This stems from the filled d subshell providing a much weaker shielding effect on the outermost s electron than the filled p subshell, so that the coinage metals have much higher first ionisation energies and smaller ionic radii than do the corresponding alkali metals.

Furthermore, they have higher melting points, hardnesses, and densities, and lower reactivities and solubilities in liquid ammonia, as well as having more covalent character in their compounds. Finally, the alkali metals are at the top of the electrochemical series, whereas the coinage metals are almost at the very bottom. The coinage metals' filled d shell is much more easily disrupted than the alkali metals' filled p shell, so that the second and third ionisation energies are lower, enabling higher oxidation states than +1 and a richer coordination chemistry, thus giving the group 11 metals clear transition metal character. Particularly noteworthy is gold forming ionic compounds with rubidium and caesium, in which it forms the auride ion (Au^-) which also occurs in solvated form in liquid ammonia solution: here gold behaves as a pseudohalogen because its $5d^{10}6s^1$ configuration has one electron less than the quasi-closed shell $5d^{10}6s^2$ configuration of mercury.

9.9 PRODUCTION AND ISOLATION

The production of pure alkali metals is somewhat complicated due to their extreme reactivity with commonly used substances, such as water. From their silicate ores, all the stable alkali metals may be obtained the same way: sulfuric acid is first used to dissolve the desired alkali metal ion and aluminium(III) ions from the ore (leaching), whereupon basic precipitation removes aluminium ions from the mixture by precipitating it as the hydroxide. The remaining insoluble alkali metal carbonate is then precipitated selectively; the salt is then dissolved in hydrochloric acid to produce the chloride.

The result is then left to evaporate and the alkali metal can then be isolated. Lithium and sodium are typically isolated through electrolysis from their liquid chlorides, with calcium chloride typically added to lower the melting point of the mixture. The heavier alkali metals, however, are more typically isolated in a different way, where a reducing agent (typically sodium for potassium and magnesium or calcium for the heaviest alkali metals) is used to reduce the alkali metal chloride. The liquid or gaseous product (the alkali metal) then undergoes fractional distillation for purification.

Most routes to the pure alkali metals require the use of electrolysis due to their high reactivity; one of the few which does not is the pyrolysis of the corresponding alkali metal azide, which yields the metal for sodium, potassium, rubidium, and caesium and the nitride for lithium.

Lithium salts have to be extracted from the water of mineral springs, brine pools, and brine deposits. The metal is produced electrolytically from a mixture of fused lithium chloride and potassium chloride.

Sodium occurs mostly in seawater and dried seabed, but is now produced through electrolysis of sodium chloride by lowering the melting point of the substance to below 700 °C through the use of a Downs cell. Extremely pure sodium can be produced through the thermal decomposition of sodium azide. Potassium occurs in many minerals, such as sylvite (potassium chloride). Previously, potassium was generally made from the electrolysis of potassium chloride or potassium hydroxide, found extensively in places such as Canada, Russia, Belarus, Germany, Israel, United States, and Jordan, in a method similar to how sodium was produced in the late 1800s and early 1900s.

It can also be produced from seawater. However, these methods are problematic because the potassium metal tends to dissolve in its molten chloride and vaporises significantly at the operating temperatures, potentially forming the explosive superoxide. As a result, pure potassium metal is now produced by reducing molten potassium chloride with sodium metal at 850 °C.

$$Na\ (g) + KCl\ (l) \rightleftharpoons NaCl\ (l) + K\ (g)$$

Although sodium is less reactive than potassium, this process works because at such high temperatures potassium is more volatile than sodium and can easily be distilled off, so that the equilibrium shifts towards the right to produce more potassium gas and proceeds almost to completion.

Metals like sodium are obtained by electrolysis of molten salts. Rb & Cs obtained mainly as by products of Li processing. To make pure caesium, ores of caesium and rubidium are crushed and heated to 650 °C with sodium metal, generating an alloy that can then be separated via a fractional distillation technique. Because metallic caesium is too reactive to handle, it is normally offered as caesium azide (CsN3). Caesium hydroxide is formed when caesium interacts aggressively with water and ice (CsOH).

Rubidium is the 16th most abundant element in the earth's crust; however, it is quite rare. Some minerals found in North America, South Africa, Russia, and Canada contain rubidium. Some potassium minerals (lepidolites, biotites, feldspar, carnallite) contain it, together with caesium. Pollucite, carnallite, leucite, and lepidolite are all minerals that contain rubidium. As a by-product of lithium extraction, it is commercially obtained from lepidolite. Rubidium is also found in potassium rocks and brines, which is a commercial supply. The majority of rubidium is now obtained as a byproduct of refining lithium. Rubidium is used in vacuum tubes as a getter, a material that combines with and removes trace gases from vacuum tubes.

For several years in the 1950s and 1960s, a by-product of the potassium production called Alkarb was a main source for rubidium. Alkarb contained 21% rubidium while the rest was potassium and a small fraction of caesium. Today the largest producers of caesium, for example the Tanco Mine in Manitoba, Canada, produce rubidium as by-product from pollucite.

Today, a common method for separating rubidium from potassium and caesium is the fractional crystallisation of a rubidium and caesium alum $(Cs, Rb)Al(SO_4)_2{\cdot}12H_2O$, which yields pure rubidium alum after approximately 30 recrystallisations. The limited applications and the lack of a mineral rich in rubidium limit the production of rubidium compounds to 2 to 4 tonnes per year. Caesium, however, is not produced from the above reaction. Instead, the mining of pollucite ore is the main method of obtaining pure caesium, extracted from the ore mainly by three methods: acid digestion, alkaline decomposition, and direct reduction. Both metals are produced as by-products of lithium production: after 1958, when interest in lithium's thermonuclear properties increased sharply, the production of rubidium and caesium also increased correspondingly. Pure rubidium and caesium metals are produced by reducing their chlorides with calcium metal at 750 °C and low pressure.

As a result of its extreme rarity in nature, most francium is synthesised in the nuclear reaction $^{197}Au + {}^{18}O \longrightarrow {}^{210}Fr + 5\ n$, yielding francium-209, francium-210, and francium-211. The greatest quantity of francium ever assembled to date is about 300,000 neutral atoms, which were synthesised using the nuclear reaction given above. When the only natural isotope francium-223 is specifically required, it is produced as the alpha daughter of actinium-227, itself produced synthetically from the neutron irradiation of natural radium-226, one of the daughters of natural uranium-238.

9.10 APPLICATIONS

Lithium, sodium, and potassium have many useful applications, while rubidium and caesium are very notable in academic contexts but do not have many applications yet. Lithium is the key ingredient for a range of lithium-based batteries, and lithium oxide can help process silica. Lithium stearate is a thickener and can be used to make lubricating greases; it is produced from lithium hydroxide, which is also used to absorb carbon dioxide in space capsules and submarines. Lithium chloride is used as a brazing alloy for aluminium parts. In medicine, some lithium salts are used as mood-stabilising pharmaceuticals. Metallic lithium is used in alloys with magnesium and aluminium to give very tough and light alloys.

Sodium compounds have many applications, the most well-known being sodium chloride as table salt. Sodium salts of fatty acids are used as soap. Pure sodium metal also has many applications, including use in sodium-vapour lamps, which produce very efficient light compared to other types of lighting, and can help smooth the surface of other metals. Being a strong reducing agent, it is often used to reduce many other metals, such as titanium and zirconium, from their chlorides. Furthermore, it is very useful as a heat-exchange liquid in fast breeder nuclear reactors due to its low melting point, viscosity, and cross-section towards neutron absorption. Sodium-ion batteries may provide cheaper alternatives to their equivalent lithium-based cells. Both sodium and potassium are commonly used as GRAS counterions to create more water-soluble and hence more bioavailable salt forms of acidic pharmaceuticals.

Potassium compounds are often used as fertilisers as potassium is an important element for plant nutrition. Potassium hydroxide is a very strong base, and is used to control the pH of various substances. Potassium nitrate and potassium permanganate are often used as powerful oxidising agents.

Potassium superoxide is used in breathing masks, as it reacts with carbon dioxide to give potassium carbonate and oxygen gas. Pure potassium metal is not often used, but its alloys with sodium may substitute for pure sodium in fast breeder nuclear reactors.

Rubidium and caesium are often used in atomic clocks. Caesium atomic clocks are extraordinarily accurate; if a clock had been made at the time of the dinosaurs, it would be off by less than four seconds (after 80 million years). For that reason, caesium atoms are used as the definition of the second. Rubidium ions are often used in purple fireworks, and caesium is often used in drilling fluids in the petroleum industry.

Francium has no commercial applications, but because of francium's relatively simple atomic structure, among other things, it has been used in spectroscopy experiments, leading to more information regarding energy levels and the coupling constants between subatomic particles. Studies on the light emitted by laser-trapped francium-210 ions have provided accurate data on transitions between atomic energy levels, similar to those predicted by quantum theory.

9.11 BIOLOGICAL ROLE AND PRECAUTIONS

9.11.1 Metals

Pure alkali metals are dangerously reactive with air and water and must be kept away from heat, fire, oxidising agents, acids, most organic compounds, halocarbons, plastics, and moisture. They also react with carbon dioxide and carbon tetrachloride, so that normal fire extinguishers are counterproductive when used on alkali metal fires. Some Class D dry powder extinguishers designed for metal fires are effective, depriving the fire of oxygen and cooling the alkali metal.

Experiments are usually conducted using only small quantities of a few grams in a fume hood. Small quantities of lithium may be disposed of by reaction with cool water, but the heavier alkali metals should be dissolved in the less reactive isopropanol. The alkali metals must be stored under mineral oil or an inert atmosphere. The inert atmosphere used may be argon or nitrogen gas, except for lithium, which reacts with nitrogen. Rubidium and caesium must be kept away from air, even under oil, because even a small amount of air diffused into the oil may trigger formation of the dangerously explosive peroxide; for the same reason, potassium should not be stored under oil in an oxygen-containing atmosphere for longer than 6 months.

9.11.2 Ions

The bioinorganic chemistry of the alkali metal ions has been extensively reviewed. Solid state crystal structures have been determined for many complexes of alkali metal ions in small peptides, nucleic acid constituents, carbohydrates and ionophore complexes.

Lithium naturally only occurs in traces in biological systems and has no known biological role, but does have effects on the body when ingested. Lithium carbonate is used as a mood stabiliser in psychiatry to treat bipolar disorder (manic-depression) in daily doses of about 0.5 to 2 grams, although there are side-effects. Excessive ingestion of lithium causes drowsiness, slurred speech and vomiting, among other symptoms, and poisons the central nervous system, which is dangerous as the required dosage of lithium to treat bipolar disorder is only slightly lower than the toxic dosage. Its biochemistry, the way it is handled by the human body and studies using rats and goats suggest that it is an essential trace element, although the natural biological function of lithium in humans has yet to be identified.

Sodium and potassium occur in all known biological systems, generally functioning as electrolytes inside and outside cells. Sodium is an essential nutrient that regulates blood volume, blood pressure, osmotic equilibrium and pH; the minimum physiological requirement for sodium is 500 milligrams per day. Sodium chloride (also known as common salt) is the principal source of sodium in the diet, and is used as seasoning and preservative, such as for pickling and jerky; most of it comes from processed foods. The Dietary Reference Intake for sodium is 1.5 grams per day, but most people in the United States consume more than 2.3 grams per day, the minimum amount that promotes hypertension; this in turn causes 7.6 million premature deaths worldwide.

Potassium is the major cation (positive ion) inside animal cells, while sodium is the major cation outside animal cells. The concentration differences of these charged particles causes a difference in electric potential between the inside and outside of cells, known as the membrane potential. The balance between potassium and sodium is maintained by ion transporter proteins in the cell membrane. The cell membrane potential created by potassium and sodium ions allows the cell to generate an action potential—a "spike" of electrical discharge. The ability of cells to produce electrical discharge is critical for body functions such as neurotransmission, muscle contraction, and heart function.

Disruption of this balance may thus be fatal: for example, ingestion of large amounts of potassium compounds can lead to hyperkalemia strongly influencing the cardiovascular system. Potassium chloride is used in the United States for lethal injection executions.

Due to their similar atomic radii, rubidium and caesium in the body mimic potassium and are taken up similarly. Rubidium has no known biological role, but may help stimulate metabolism, and, similarly to caesium, replace potassium in the body causing potassium deficiency.

Partial substitution is quite possible and rather non-toxic: a 70 kg person contains on average 0.36 g of rubidium, and an increase in this value by 50 to 100 times did not show negative effects in test persons. Rats can survive up to 50% substitution of potassium by rubidium.

Rubidium (and to a much lesser extent caesium) can function as temporary cures for hypokalemia; while rubidium can adequately physiologically substitute potassium in some systems, caesium is never able to do so. There is only very limited evidence in the form of deficiency symptoms for rubidium being possibly essential in goats; even if this is true, the trace amounts usually present in food are more than enough.

Caesium compounds are rarely encountered by most people, but most caesium compounds are mildly toxic. Like rubidium, caesium tends to substitute potassium in the body, but is significantly larger and is therefore a poorer substitute. Excess caesium can lead to hypokalemia, arrhythmia, and acute cardiac arrest, but such amounts would not ordinarily be encountered in natural sources. As such, caesium is not a major chemical environmental pollutant.

The median lethal dose (LD_{50}) value for caesium chloride in mice is 2.3 g per kilogram, which is comparable to the LD_{50} values of potassium chloride and sodium chloride. Caesium chloride has been promoted as an alternative cancer therapy, but has been linked to the deaths of over 50 patients, on whom it was used as part of a scientifically unvalidated cancer treatment.

Radioisotopes of caesium require special precautions: the improper handling of caesium-137 gamma ray sources can lead to release of this radioisotope and radiation injuries.

Perhaps the best-known case is the Goiânia accident of 1987, in which an improperly-disposed-of radiation therapy system from an abandoned clinic in the city of Goiânia, Brazil, was scavenged from a junkyard, and the glowing caesium salt sold to curious, uneducated buyers. This led to four deaths and serious injuries from radiation exposure.

Together with caesium-134, iodine-131, and strontium-90, caesium-137 was among the isotopes distributed by the Chernobyl disaster which constitute the greatest risk to health. Radioisotopes of francium would presumably be dangerous as well due to their high decay energy and short half-life, but none have been produced in large enough amounts to pose any serious risk.

CHAPTER 10
THE CHLOR-ALKALI INDUSTRY

The chlor-alkali industry is one of the most essential sectors in the chemical industry, producing chemicals like chlorine, sodium hydroxide (caustic soda), and hydrogen. These chemicals are fundamental to a wide range of applications, from water treatment and PVC production to pharmaceuticals and food processing.

10.1 INTRODUCTION TO THE CHLOR-ALKALI INDUSTRY

The chlor-alkali industry is an important branch of the chemical industry that primarily focuses on the electrolysis of sodium chloride (salt) to produce chlorine (Cl_2), sodium hydroxide (NaOH), and hydrogen gas (H_2). These three products are generated simultaneously in the process, and each has significant industrial applications. Chlorine is a powerful disinfectant and a raw material in the production of plastics, solvents, and chemicals. Sodium hydroxide is widely used in the production of paper, textiles, and detergents, while hydrogen has applications in chemical synthesis, refining, and fuel cells.

10.2 CHLOR-ALKALI PRODUCTION PROCESSES

The chlor-alkali industry primarily relies on three types of electrolysis processes to produce chlorine, sodium hydroxide, and hydrogen. These processes are the diaphragm cell process, the membrane cell process, and the mercury cell process.

10.2.1 Diaphragm Cell Process

In the diaphragm cell process, a porous diaphragm made of asbestos or polymer separates the anode and cathode. Saltwater (brine) flows through the anode compartment, where chlorine gas is released. The sodium ions then migrate through the diaphragm to the cathode side, where they combine with hydroxide ions to form sodium hydroxide. The diaphragm prevents the mixing of chlorine and sodium hydroxide, but it often contains some salt impurities in the final product.

10.2.2 Membrane Cell Process

The membrane cell process uses an ion-selective membrane to separate the anode and cathode compartments. Chlorine gas is produced at the anode, while sodium ions move through the membrane to combine with hydroxide ions at the cathode, forming sodium hydroxide. This process is considered more environmentally friendly than the diaphragm cell process because it avoids asbestos and produces a higher purity caustic soda. Additionally, it requires less energy, which makes it the most widely used method today.

10.2.3 Mercury Cell Process

The mercury cell process, also known as the Castner-Kellner process, uses a mercury cathode to produce a sodium-mercury amalgam at the cathode. This amalgam then reacts with water in a separate chamber to form sodium hydroxide and hydrogen gas.

Although this process can yield very pure caustic soda, it has raised significant environmental concerns due to the use of mercury, which is highly toxic. Many countries have phased out or restricted the use of mercury cells due to environmental and health risks.

10.3 KEY PRODUCTS OF THE CHLOR-ALKALI INDUSTRY

The chlor-alkali process results in the production of three main chemicals: chlorine, sodium hydroxide, and hydrogen. Each of these has numerous industrial applications that contribute significantly to various sectors.

10.3.1 Chlorine

Chlorine is one of the most widely used chemicals in the world. It is essential for the production of polyvinyl chloride (PVC), which is used in building materials, pipes, and medical devices. Chlorine is also a key component in the manufacture of solvents, agricultural chemicals, and refrigerants. Furthermore, chlorine is widely used for water treatment and disinfection, helping to ensure safe drinking water and to sanitize swimming pools.

10.3.2 Sodium Hydroxide (Caustic Soda)

Sodium hydroxide is a strong alkaline compound used in a variety of industrial processes. It is vital in the paper and pulp industry, where it helps to break down wood into paper pulp. Caustic soda is also used in the production of soaps, detergents, textiles, and aluminum. Its ability to neutralize acids makes it valuable for pH control in various chemical processes. In addition, sodium hydroxide is widely used in the treatment of wastewater to remove heavy metals and impurities.

10.3.3 Hydrogen

Hydrogen, the third byproduct of the chlor-alkali process, is an important chemical in its own right. It is used in the production of ammonia for fertilizers, the hydrogenation of oils and fats, and as a fuel in the refining industry. Additionally, hydrogen is being explored as a clean energy source in fuel cells and is gaining attention as a potential solution for reducing carbon emissions in transportation and industry.

10.4 ENVIRONMENTAL IMPACT OF THE CHLOR-ALKALI INDUSTRY

The chlor-alkali industry has traditionally been associated with several environmental challenges. Among the most significant concerns are the energy-intensive nature of the process, the use of hazardous substances, and the potential for harmful emissions.

10.4.1 Energy Consumption

The electrolysis process used in chlor-alkali production requires substantial amounts of electricity, making it one of the most energy-intensive processes in the chemical industry. This energy demand has both economic and environmental impacts, as it often relies on fossil fuels, contributing to greenhouse gas emissions.

10.4.2 Mercury Pollution

The mercury cell process, once a common method of chlor-alkali production, has led to concerns about mercury contamination. Mercury is highly toxic and can accumulate in living organisms, causing serious health problems. Many countries have implemented regulations to phase out mercury-based processes in favor of more environmentally friendly alternatives, such as the membrane cell process.

10.4.3 Chlorine and Byproduct Disposal

Chlorine production and handling can pose environmental risks due to its toxicity and potential to form harmful compounds. Chlorine gas is corrosive and can cause respiratory problems if inhaled. Furthermore, improper disposal of byproducts and wastewater from chlor-alkali plants can lead to water and soil contamination.

10.5 TECHNOLOGICAL ADVANCEMENTS AND SUSTAINABLE PRACTICES

In recent years, the chlor-alkali industry has made strides toward reducing its environmental impact by adopting new technologies and sustainable practices. These advancements include energy-efficient membrane cells, renewable energy integration, and improved waste management techniques.

10.5.1 Energy-Efficient Membrane Cells

The membrane cell process is now the preferred method for chlor-alkali production due to its lower energy consumption and reduced environmental footprint. Modern membrane cells are designed to minimize energy requirements, reducing the overall cost and environmental impact of the process.

10.5.2 Renewable Energy Integration

To reduce the carbon footprint of the chlor-alkali industry, some facilities are beginning to integrate renewable energy sources, such as solar and wind power, into their operations. This approach helps to offset the significant energy demands of the process and reduces dependence on fossil fuels.

10.5.3 Waste Management and Recycling

Improved waste management practices, including the recycling of brine and the capture of byproducts, have also been implemented in the chlor-alkali industry. For example, capturing and reusing hydrogen as a fuel source can help reduce waste and provide a sustainable energy alternative.

10.6 APPLICATIONS AND ECONOMIC IMPORTANCE

The products of the chlor-alkali industry are integral to many sectors of the global economy. Chlorine and caustic soda are essential for manufacturing everything from pharmaceuticals and textiles to electronics and plastics. The industry also provides key raw materials for wastewater treatment, food processing, and energy production, supporting numerous downstream industries and contributing to economic growth.

10.7 CONCLUSION

The chlor-alkali industry plays a crucial role in the global economy, producing essential chemicals with a wide range of applications. However, the industry also faces significant environmental challenges, particularly related to energy consumption and pollution. Through technological advancements, such as the adoption of membrane cell processes and the integration of renewable energy, the chlor-alkali industry is working to minimize its environmental impact and enhance sustainability. As the industry continues to evolve, balancing economic importance with environmental responsibility will be key to its future success.

CHAPTER 11

GROUP 2—THE ALKALINE EARTH ELEMENTS

The alkaline earth elements are a group of metallic elements found in Group 2 of the periodic table. This group consists of six elements: beryllium (Be), magnesium (Mg), calcium (Ca), strontium (Sr), barium (Ba), and radium (Ra). Known for their unique properties and chemical behavior, the alkaline earth metals play a vital role in both natural processes and various industrial applications.

11.1 GENERAL CHARACTERISTICS OF THE ALKALINE EARTH ELEMENTS

The elements in Group 2 share several common physical and chemical characteristics. These similarities stem from their electron configurations, which feature two valence electrons in the outermost shell. The general electron configuration for these elements is [noble gas] ns^2. Due to this configuration, alkaline earth metals tend to lose two electrons to form doubly charged cations with a +2 oxidation state, such as Mg^{2+} and Ca^{2+}. This loss of electrons results in stable ionic compounds with strong ionic bonds.

Physically, the alkaline earth metals are all relatively soft, silver-colored metals with relatively low densities. Their metallic character and reactivity increase as one moves down the group, with beryllium being the least reactive and radium the most reactive. Unlike Group 1 alkali metals, alkaline earth metals have higher melting and boiling points, reflecting the stronger metallic bonding due to their smaller atomic radii and higher charge density.

11.2 INDIVIDUAL PROPERTIES OF THE ALKALINE EARTH ELEMENTS

Each of the six alkaline earth elements has distinct characteristics and uses, though they all share the same +2 oxidation state and similar chemical reactivity.

11.2.1 Beryllium (Be)

Beryllium is a relatively rare, lightweight metal known for its high strength and rigidity. It has a high melting point and is highly resistant to corrosion. Beryllium is also nonmagnetic and has a low density, making it valuable in aerospace applications. Due to its toxicity, however, beryllium handling requires strict safety measures.

11.2.2 Magnesium (Mg)

Magnesium is a highly abundant and essential metal that is lightweight, strong, and corrosion-resistant. It burns with a bright white flame, making it useful in flares and fireworks.

Magnesium alloys are widely used in the aerospace, automotive, and electronics industries due to their strength-to-weight ratio. Additionally, magnesium is biologically important and plays a crucial role in cellular processes, such as photosynthesis in plants.

11.2.3 Calcium (Ca)

Calcium is another highly abundant element in the Earth's crust and is essential for living organisms.

It is a major component of bones and teeth in humans and animals and is crucial in muscle function, nerve transmission, and blood clotting. Calcium compounds, such as calcium carbonate ($CaCO_3$), are commonly used in construction, agriculture, and the production of cement and lime.

11.2.4 Strontium (Sr)

Strontium is a soft, silvery metal with properties similar to those of calcium. Strontium compounds are used in fireworks to produce a bright red flame, as well as in glass manufacturing and ceramic production. Strontium-90, a radioactive isotope, has applications in medical treatments and serves as a heat source in some radioisotope thermoelectric generators.

11.2.5 Barium (Ba)

Barium is a reactive metal used in various applications, particularly in the form of barium sulfate ($BaSO_4$). Due to its high density and opacity to X-rays, barium sulfate is used in medical imaging as a contrast agent for X-ray scans of the digestive system.

Barium compounds are also used in the production of paints, glass, and ceramics, as well as in oil and gas drilling fluids.

11.2.6 Radium (Ra)

Radium is a highly radioactive metal discovered in uranium ores. Due to its radioactivity, radium was once used in luminous paints for watch dials and instrument panels, although it has since been replaced by safer alternatives. Radium is primarily of historical interest and is used in limited medical applications for radiation therapy.

11.3 CHEMICAL PROPERTIES AND REACTIVITY

The alkaline earth metals are less reactive than the Group 1 alkali metals but still exhibit strong reactivity, particularly with nonmetals. Their reactivity increases as one moves down the group, with beryllium being the least reactive and radium the most.

- **Reaction with Water:** Magnesium, calcium, strontium, and barium react with water to produce hydrogen gas and a corresponding hydroxide, although the reactivity with water increases down the group. Beryllium does not react with water, even at elevated temperatures, due to a protective oxide layer that forms on its surface.
- **Reaction with Halogens:** All alkaline earth metals react readily with halogens to form ionic halides, such as calcium chloride ($CaCl_2$) and magnesium chloride ($MgCl_2$).
- **Reaction with Oxygen:** When heated in the presence of oxygen, the alkaline earth metals form oxides, such as magnesium oxide (MgO) and calcium oxide (CaO). These oxides are basic and react with water to produce hydroxides.

11.4 INDUSTRIAL AND TECHNOLOGICAL APPLICATIONS

The alkaline earth elements have a wide range of industrial applications due to their unique properties. These applications vary across sectors, including construction, aerospace, electronics, and medicine.

11.4.1 Construction and Building Materials

Calcium compounds, particularly calcium carbonate and calcium oxide, are fundamental to the construction industry. They are used to produce cement, concrete, and lime, which are essential for building structures, roads, and infrastructure.

11.4.2 Aerospace and Automotive Industries

Magnesium alloys are prized for their low density and high strength, making them ideal for applications where weight reduction is critical, such as in aircraft and automotive manufacturing. Beryllium is also used in aerospace applications, though more selectively due to its high cost and toxicity.

11.4.3 Electronics and Technology

Barium compounds are utilized in the electronics industry, particularly in the production of cathode ray tubes, which were once common in television and computer monitors. Barium titanate, for example, is a dielectric material used in capacitors and other electronic components.

11.4.4 Medical and Biological Applications

Magnesium and calcium are essential to human health and play significant roles in biological processes.

Magnesium supplements are commonly used to prevent and treat magnesium deficiencies, while calcium supplements support bone health. Barium sulfate is used as a contrast agent for X-rays of the gastrointestinal tract, allowing doctors to observe and diagnose conditions in the digestive system.

11.5 ENVIRONMENTAL AND BIOLOGICAL SIGNIFICANCE

The alkaline earth metals are not only important industrially but also play crucial roles in the environment and biological systems.

- **Biological Roles:** Magnesium and calcium are essential nutrients for plants, animals, and humans. Calcium is a primary component of bones and teeth, while magnesium is involved in enzyme activity and DNA synthesis. Both are necessary for muscle function and cardiovascular health.
- **Environmental Roles:** Alkaline earth metals influence soil quality and water chemistry. Calcium carbonate, for instance, is found in limestone and chalk, which are essential components of soil that regulate pH levels. Additionally, calcium and magnesium ions are significant contributors to water hardness, which impacts water treatment and industrial processes.

11.6 HEALTH AND SAFETY CONSIDERATIONS

While magnesium and calcium are beneficial to human health, some of the alkaline earth metals, such as beryllium and radium, pose health risks. Beryllium is highly toxic if inhaled and can cause lung disease.

Similarly, radium's radioactivity poses serious health risks and has led to strict regulation in its handling and use.

11.6.1 Beryllium Toxicity

Beryllium exposure can lead to a chronic lung disease known as berylliosis. The metal's toxicity has necessitated stringent safety protocols in industries where beryllium is used, such as aerospace and nuclear sectors.

11.6.2 Radioactivity of Radium

Radium, being highly radioactive, can cause radiation sickness and increase cancer risk. Its historical use in luminous paints led to severe health issues among workers, sparking awareness of the dangers of radiation exposure.

11.7 CONCLUSION

The alkaline earth elements are an integral part of the periodic table, each with distinct properties and applications that benefit numerous industries and biological systems.

From the essential roles of magnesium and calcium in human health to the industrial applications of beryllium, magnesium, and barium, these metals continue to influence technology, medicine, and environmental science. While some alkaline earth elements pose health risks, responsible use and handling allow their properties to be harnessed safely.

As research and technology advance, the applications and understanding of these elements will continue to expand, further cementing their importance in both scientific and everyday contexts.

PART THREE
THE P-BLOCK ELEMENTS

CHAPTER 12
THE GROUP 13 ELEMENTS

Group 13 elements of the periodic table, also known as the boron group, consist of five elements: boron (B), aluminum (Al), gallium (Ga), indium (In), and thallium (Tl). These elements share similar properties due to their common valence electron configuration, which is characterized by three electrons in their outermost shell.

The Group 13 elements are situated in the p-block of the periodic table, making them intriguing in both their chemistry and industrial applications.

12.1 GENERAL CHARACTERISTICS OF GROUP 13 ELEMENTS

The Group 13 elements have three valence electrons with an electron configuration of ns^2np^1, giving them a +3 oxidation state when they lose all three electrons. However, the stability of this oxidation state varies across the group. These elements demonstrate a transition from non-metallic to metallic behavior as we move down the group, with boron being a metalloid, while aluminum, gallium, indium, and thallium are metals. Group 13 elements have moderate to low melting and boiling points.

Boron has the highest melting and boiling points, while the remaining elements show much lower values. Atomic and ionic radii increase down the group due to additional electron shells, and ionization energies tend to decrease down the group, with boron having the highest ionization energy and thallium the lowest.

12.2 INDIVIDUAL PROPERTIES OF GROUP 13 ELEMENTS

Each Group 13 element possesses unique properties, despite sharing general characteristics.

12.2.1 Boron (B)

Boron is a metalloid with unique properties, occupying an intermediate position between metals and non-metals. It forms covalent compounds and is known for its hardness and high melting point.

Boron does not occur in nature in its free form; rather, it is found in compounds like borax (sodium borate) and kernite. Boron is relatively inert at room temperature but becomes more reactive when heated. It has significant applications in ceramics, glass, and as a doping agent in the semiconductor industry.

12.2.2 Aluminum (Al)

Aluminum is the most abundant metal in the Earth's crust and is widely used due to its low density, high strength-to-weight ratio, and resistance to corrosion. It is a highly reactive metal that forms a thin oxide layer on its surface, protecting it from further oxidation. Aluminum is primarily found in bauxite ore and is refined through the Hall-Héroult process. It is used extensively in the automotive, aerospace, and packaging industries due to its excellent mechanical properties.

12.2.3 Gallium (Ga)

Gallium is a soft, silvery metal that melts at temperatures slightly above room temperature. It is not found as a free element in nature but is often obtained as a byproduct of aluminum and zinc production. Gallium is essential in the electronics industry, particularly in the form of gallium arsenide (GaAs) and gallium nitride (GaN), which are used in high-speed electronic devices and LED technology. Gallium's unique melting point makes it useful in temperature measurement and as a component in low-melting alloys.

12.2.4 Indium (In)

Indium is a soft, malleable metal with a relatively low melting point. It is often used in the electronics industry for its excellent electrical conductivity. Indium is also found in low-melting alloys and as a coating for bearings. Its compounds, like indium tin oxide (ITO), are used in touch screens, LCD displays, and photovoltaic cells. Indium is relatively rare, found in small amounts in zinc ores and some iron, lead, and copper ores.

12.2.5 Thallium (Tl)

Thallium is a toxic, soft metal with limited industrial uses due to its toxicity. It is highly reactive and forms a protective oxide layer when exposed to air. Thallium compounds are used in specialized applications such as optical lenses, infrared detectors, and some medical treatments. Due to its toxicity, thallium is handled with care, and its applications are limited compared to other Group 13 elements.

12.3 CHEMICAL BEHAVIOR OF GROUP 13 ELEMENTS

The Group 13 elements exhibit similar chemical behaviors, yet each element's reactivity varies due to differences in atomic size, ionization energy, and electron configurations.

12.3.1 Oxidation States

The most common oxidation state for Group 13 elements is +3, corresponding to the loss of all three valence electrons. However, as the elements become heavier, the +1 oxidation state becomes more stable due to the inert pair effect, particularly for thallium. For instance, while boron and aluminum predominantly exhibit a +3 oxidation state, thallium often forms compounds in the +1 state, such as TlCl.

12.3.2 Reaction with Oxygen

The Group 13 elements react with oxygen to form oxides. Boron forms boron trioxide (B_2O_3), a covalent oxide that acts as a glass-forming agent in borosilicate glass.

Aluminum forms aluminum oxide (Al_2O_3), a stable and protective layer that prevents further oxidation. Gallium, indium, and thallium oxides exhibit a range of properties from amphoteric to basic.

12.3.3 Reaction with Halogens

Group 13 elements react with halogens to form trihalides, such as aluminum chloride ($AlCl_3$) and boron trifluoride (BF_3). These trihalides have covalent bonds and tend to act as Lewis acids due to their electron deficiency. For example, boron trifluoride is a strong Lewis acid used in organic synthesis.

12.3.4 Hydrolysis and Acid-Base Behavior

The oxides and hydroxides of Group 13 elements exhibit varying degrees of amphoteric behavior. Boron oxide is acidic, while aluminum oxide is amphoteric, reacting with both acids and bases. As we move down the group, the oxides and hydroxides become more basic. Thallium oxide, for instance, is more basic and reacts readily with acids to form thallium salts.

12.4 INDUSTRIAL AND TECHNOLOGICAL APPLICATIONS

The Group 13 elements have numerous applications across various industries, owing to their distinct properties.

12.4.1 Boron Applications

Boron compounds are used in the production of heat-resistant glass, detergents, and as additives in fiberglass. Boron's role in the semiconductor industry as a doping agent for silicon makes it essential in the electronics field. Boron compounds are also employed in neutron-absorbing materials for nuclear reactors due to boron's high neutron absorption cross-section.

12.4.2 Aluminum Applications

Aluminum is one of the most widely used metals globally, with applications in transportation, packaging, construction, and consumer goods. Aluminum's light weight and corrosion resistance make it ideal for airplane and automobile manufacturing. Its reflective properties are also exploited in solar panels and insulation materials.

12.4.3 Gallium Applications

Gallium is crucial in semiconductor technology, particularly in high-frequency transistors, solar cells, and LED lights. Gallium arsenide (GaAs) is a widely used compound in optoelectronics, and gallium nitride (GaN) is significant in power electronics due to its wide bandgap properties.

12.4.4 Indium Applications

Indium is a key component in indium tin oxide (ITO), which is used in the production of touchscreens, LCD displays, and photovoltaic cells. Indium's low melting point also makes it valuable in low-melting alloys and as a coating for other metals.

12.4.5 Thallium Applications

Thallium's uses are limited due to its toxicity, but it is applied in certain specialized fields. It is used in optical lenses, infrared detectors, and in some medical imaging procedures.

12.5 ENVIRONMENTAL AND BIOLOGICAL CONSIDERATIONS

- **Boron:** Essential for plant growth in small quantities, boron is relatively nontoxic but can cause adverse effects if present in excess.
- **Aluminum:** Although aluminum is not considered toxic, high exposure can have detrimental effects on soil and aquatic ecosystems, leading to soil acidification and toxicity to aquatic life.
- **Gallium and Indium:** These elements are relatively rare in nature, and their extraction can have environmental consequences. They are not known to be toxic to humans but must be carefully managed to prevent environmental harm.
- **Thallium:** Highly toxic to both humans and wildlife, thallium's presence in the environment is carefully regulated. It can cause significant health problems if ingested or inhaled.

12.6 CONCLUSION

The Group 13 elements play vital roles in various industries and have diverse applications due to their unique properties. From boron's role in electronics and glass manufacturing to aluminum's extensive use in construction and transportation, these elements are integral to modern life. As technology advances, the demand for gallium, indium, and even the rarer thallium continues to grow in fields like electronics, semiconductors, and medicine. Understanding the properties, reactivity, and applications of Group 13 elements not only highlights their industrial importance but also emphasizes the need for responsible handling and environmental stewardship.

CHAPTER 13
THE GROUP 14 ELEMENTS

The Group 14 elements, also known as the carbon group or the crystallogens, are a set of elements located in the p-block of the periodic table. This group includes carbon (C), silicon (Si), germanium (Ge), tin (Sn), and lead (Pb). Each of these elements has four valence electrons, allowing them to form four covalent bonds with other elements. The Group 14 elements display a wide range of physical and chemical properties, with carbon being a non-metal, silicon and germanium metalloids, and tin and lead metals.

13.1 GENERAL CHARACTERISTICS OF GROUP 14 ELEMENTS

The Group 14 elements have the electron configuration ns^2np^2, giving them four electrons in their outermost shell. This configuration allows them to form a wide variety of compounds by achieving both +4 and +2 oxidation states, although the stability of these states varies across the group. As one moves down the group, the elements exhibit an increase in metallic character, transitioning from non-metal (carbon) to metal (tin and lead).

These elements display a range of physical properties. For example, carbon has the highest melting and boiling points, while lead has the lowest. This variance is due to the types of bonding and structures these elements can form. The atomic and ionic radii of Group 14 elements increase as one moves down the group due to the addition of electron shells, while ionization energies and electronegativity decrease.

13.2 INDIVIDUAL PROPERTIES OF GROUP 14 ELEMENTS

Each Group 14 element possesses unique properties that determine its behavior and applications.

13.2.1 Carbon (C)

Carbon is the most versatile element in the periodic table due to its ability to form strong covalent bonds with itself and other elements, creating a wide array of compounds. Carbon occurs in several allotropes, such as graphite, diamond, and fullerenes, each with distinct properties. Diamond, for example, is one of the hardest known materials, while graphite is a good conductor of electricity and is used as a lubricant. Carbon is the fundamental building block of all organic compounds and is crucial to life on Earth. It also has applications in energy, nanotechnology, and materials science.

13.2.2 Silicon (Si)

Silicon is a metalloid with properties intermediate between metals and non-metals. It is the second most abundant element in the Earth's crust, predominantly found in the form of silicon dioxide (SiO_2) and silicates. Silicon's most notable application is in the electronics industry, where it serves as a semiconductor material for devices such as transistors, solar cells, and integrated circuits. Silicon is also used in the manufacturing of glass, ceramics, and silicone compounds, which are essential in construction and personal care products.

13.2.3 Germanium (Ge)

Germanium is a brittle, gray-white metalloid that shares properties with silicon. It was initially used in early semiconductor devices before silicon became the dominant material. Today, germanium is still used in certain high-frequency electronics, fiber-optic systems, and infrared optics. Germanium is less abundant than silicon but can be found in zinc ores and certain coals. Despite its limited availability, germanium remains valuable in the electronics and telecommunications industries.

13.2.4 Tin (Sn)

Tin is a soft, malleable metal with a low melting point, widely used in alloys such as bronze and solder. It is resistant to corrosion, making it ideal for coating other metals to prevent rust, as seen in tin-plated steel. Tin also has important applications in the production of glass, where molten tin is used in the float glass process. Additionally, tin is used in a variety of chemical compounds, such as tin(II) fluoride, which is an ingredient in toothpaste, and organotin compounds, which are used as stabilizers in plastics.

13.2.5 Lead (Pb)

Lead is a heavy, soft metal that has been used since ancient times. It is known for its high density, low melting point, and resistance to corrosion. Lead has been traditionally used in pipes, paints, and gasoline additives, although its use has declined due to health concerns. Today, lead is primarily used in batteries, particularly lead-acid batteries for vehicles. It is also employed in radiation shielding, as it effectively absorbs X-rays and gamma rays. However, due to its toxicity, lead usage is heavily regulated.

13.3 CHEMICAL BEHAVIOR OF GROUP 14 ELEMENTS

The Group 14 elements exhibit similar chemical behaviors, although their reactivity varies.

13.3.1 Oxidation States

The most common oxidation states for Group 14 elements are +4 and +2. Carbon, silicon, and germanium primarily exhibit the +4 oxidation state, forming compounds like carbon dioxide (CO_2) and silicon dioxide (SiO_2). Tin and lead can display both +4 and +2 oxidation states, with the +2 state becoming more stable as one moves down the group due to the inert pair effect. For example, lead is more commonly found in the +2 state, as in lead(II) oxide (PbO), while tin frequently forms compounds in both states, such as tin(II) chloride ($SnCl_2$) and tin(IV) chloride ($SnCl_4$).

13.3.2 Reactivity with Oxygen

The Group 14 elements react with oxygen to form oxides. Carbon, for instance, forms carbon dioxide (CO_2) and carbon monoxide (CO) depending on the availability of oxygen. Silicon forms silicon dioxide (SiO_2), which is the primary component of sand. Germanium and tin form oxides with varying levels of stability, while lead forms lead(II) oxide (PbO) as its most common oxide. These oxides vary from acidic to amphoteric, with carbon dioxide being an acidic oxide, while tin and lead oxides are amphoteric.

13.3.3 Reactivity with Halogens

Group 14 elements react with halogens to form tetrahalides (EX_4) and, for tin and lead, dihalides (EX_2). Carbon tetrahalides like carbon tetrachloride (CCl_4) are commonly used as solvents, while silicon tetrahalides serve as intermediates in the synthesis of silicones. Tin and lead halides exhibit interesting reactivity, with lead(II) chloride ($PbCl_2$) being a common compound. These halides are useful in organic synthesis, as catalysts, and in various industrial applications.

13.4 INDUSTRIAL AND TECHNOLOGICAL APPLICATIONS

The Group 14 elements have extensive applications across multiple industries.

13.4.1 Carbon Applications

Carbon's allotropes have various uses, from diamond in cutting tools to graphite in batteries and lubricants. Carbon-based materials such as carbon fiber and carbon nanotubes are used in aerospace, sports equipment, and nanotechnology. Carbon compounds like methane (CH_4) serve as fuels, while carbon dioxide is vital for photosynthesis and industrial processes.

13.4.2 Silicon Applications

Silicon's role as a semiconductor material is fundamental to modern electronics, including computers, smartphones, and solar panels. Silicon compounds like silica are essential in construction and glassmaking. Additionally, silicones, derived from silicon, are used in sealants, adhesives, and personal care products.

13.4.3 Germanium Applications

Germanium is valuable in fiber optics, infrared optics, and as an alloying agent. Its use in semiconductors, particularly for high-speed electronics, is also significant. Germanium is employed in radiation detectors and certain chemotherapy drugs.

13.4.4 Tin Applications

Tin's corrosion resistance makes it suitable for coating other metals, such as in tin cans. It is used in solders, alloys, and as a stabilizer in plastics. Tin compounds are also employed in glass production and as flame retardants.

13.4.5 Lead Applications

Lead's primary use is in lead-acid batteries for vehicles and backup power systems. It is also used in radiation shielding, ammunition, and certain glass products. Lead compounds are used in pigments and glassmaking, although they are less common due to toxicity concerns.

13.5 ENVIRONMENTAL AND HEALTH CONSIDERATIONS

The environmental and health impacts of Group 14 elements vary. Carbon dioxide is a greenhouse gas, contributing to climate change. Lead is toxic and has been phased out in many products due to health risks, particularly in children. Silicon and tin are relatively non-toxic, while germanium poses minimal environmental risk. Proper handling and disposal are essential for managing these elements safely.

13.6 CONCLUSION

The Group 14 elements are integral to various fields, from electronics to energy and materials science. Their unique properties, from carbon's versatile bonding to silicon's semiconducting capabilities, make them essential to modern technology. Despite environmental challenges associated with certain elements like lead, these elements' applications continue to expand. As technology advances, the role of Group 14 elements will remain crucial in developing sustainable, innovative solutions.

CHAPTER 14

THE GROUP 15 ELEMENTS

The Group 15 elements, also known as the nitrogen group or pnictogens, occupy a significant place in the p-block of the periodic table. This group includes nitrogen (N), phosphorus (P), arsenic (As), antimony (Sb), and bismuth (Bi). Each of these elements has five valence electrons, resulting in a general electronic configuration of ns^2np^3. This unique configuration allows them to form a wide variety of compounds by achieving oxidation states from -3 to +5. These elements display a gradual transition from non-metallic to metallic properties as one moves down the group, with nitrogen and phosphorus being non-metals, arsenic and antimony metalloids, and bismuth a metal.

14.1 GENERAL CHARACTERISTICS OF GROUP 15 ELEMENTS

The Group 15 elements share the ns^2np^3 electron configuration, giving them five valence electrons. This setup results in a strong tendency to form covalent bonds, especially in the case of nitrogen and phosphorus. The reactivity and physical characteristics of these elements vary, with a distinct shift from non-metals at the top of the group to metals at the bottom.

The atomic and ionic radii of the Group 15 elements increase as one moves down the group due to the addition of electron shells, which leads to a decrease in ionization energies and electronegativity. The melting and boiling points of these elements also increase down the group, with nitrogen being a gas at room temperature and bismuth a solid. Nitrogen and phosphorus are essential for life, playing crucial roles in biological processes, while arsenic, antimony, and bismuth are primarily used in industrial applications.

14.2 INDIVIDUAL PROPERTIES OF GROUP 15 ELEMENTS

Each Group 15 element exhibits unique properties that influence its chemical behavior and uses in various industries.

14.2.1 Nitrogen (N)

Nitrogen is a diatomic, colorless, odorless gas that makes up about 78% of the Earth's atmosphere. Its strong triple bond ($N \equiv N$) makes it relatively inert under standard conditions, but it can react under specific circumstances to form compounds like ammonia (NH_3), nitric acid (HNO_3), and various oxides of nitrogen.

Nitrogen is essential for life, as it is a key component of amino acids, proteins, and DNA. In industrial applications, nitrogen is used in fertilizers, explosives, and the production of synthetic materials.

14.2.2 Phosphorus (P)

Phosphorus exists in several allotropes, with white, red, and black phosphorus being the most common. White phosphorus is highly reactive and toxic, while red phosphorus is more stable and commonly used in matches, fireworks, and smoke bombs. Phosphorus is a crucial element for life, forming part of DNA, RNA, and ATP, the energy currency of cells. In industry, phosphorus is essential for fertilizers and is also used in the production of steel, glass, and various chemicals.

14.2.3 Arsenic (As)

Arsenic is a metalloid with several allotropes, although the gray metallic form is the most stable. Arsenic is toxic and has been historically used as a poison. It also has applications in the electronics industry, particularly in semiconductors, due to its electrical properties. Arsenic is found in compounds such as arsenic trioxide (As_2O_3), which is used in pesticides and wood preservatives. Due to its toxicity, arsenic is highly regulated, especially in applications involving water and food.

14.2.4 Antimony (Sb)

Antimony is a brittle, silver-white metalloid that is primarily used as an alloying agent. It enhances the hardness and durability of metals, making it useful in applications like lead-acid batteries and solder. Antimony compounds are also used as flame retardants, pigments, and in the glass and ceramics industries. Antimony exists in two stable isotopes and is often found in the mineral stibnite (Sb_2S_3).

14.2.5 Bismuth (Bi)

Bismuth is a dense, brittle metal with a low melting point and is considered the heaviest stable element in Group 15. Unlike arsenic and antimony, bismuth is relatively non-toxic and is used in a variety of applications, including cosmetics, pharmaceuticals, and as a replacement for lead in environmentally friendly products. Bismuth has a high density and low thermal conductivity, making it useful for alloys and fire-detection devices.

14.3 CHEMICAL BEHAVIOR OF GROUP 15 ELEMENTS

The Group 15 elements exhibit diverse chemical behaviors due to their varied physical properties and bonding capabilities.

14.3.1 Oxidation States

Group 15 elements can exhibit a wide range of oxidation states, from -3 to +5. Nitrogen can achieve all these oxidation states, while the heavier elements tend to show fewer stable oxidation states due to the inert pair effect. Nitrogen commonly forms -3 oxidation state compounds, like ammonia (NH_3), and +5 oxidation state compounds, such as nitric acid (HNO_3). The +3 and +5 states are most stable for phosphorus and arsenic, while antimony and bismuth favor +3 states, with bismuth primarily forming +3 compounds due to its metallic nature.

14.3.2 Reactivity with Hydrogen

Group 15 elements form hydrides of the general formula EH_3. Ammonia (NH_3) is the most significant hydride, characterized by its basicity and widespread use in agriculture and industry. Phosphine (PH_3) is less stable and toxic, while arsine (AsH_3) and stibine (SbH_3) are highly toxic and unstable, decomposing quickly. Bismuth hydride (BiH_3) is extremely unstable and decomposes rapidly.

14.3.3 Reactivity with Oxygen

Group 15 elements form oxides with oxygen, exhibiting a variety of oxidation states. Nitrogen forms nitrogen dioxide (NO_2) and dinitrogen pentoxide (N_2O_5), which are both important in atmospheric chemistry and industrial processes. Phosphorus forms tetraphosphorus decaoxide (P_4O_{10}), a crucial component in fertilizers. Arsenic, antimony, and bismuth form oxides with lower oxidation states, such as arsenic trioxide (As_2O_3) and bismuth oxide (Bi_2O_3), which are used in glass and ceramics.

14.3.4 Reactivity with Halogens

The Group 15 elements react with halogens to form compounds with the general formula EX_3 and EX_5. Nitrogen forms nitrogen trichloride (NCl_3), a highly unstable and explosive compound. Phosphorus reacts with chlorine to form phosphorus trichloride (PCl_3) and phosphorus pentachloride (PCl_5), which are important industrial chemicals. The heavier elements, like arsenic and antimony, form trihalides and pentahalides used in various applications, including flame retardants and electronics.

14.4 INDUSTRIAL AND TECHNOLOGICAL APPLICATIONS

The Group 15 elements have diverse applications across various industries, from agriculture to electronics.

14.4.1 Nitrogen Applications

Nitrogen is essential for plant growth, and its compounds, such as ammonia and nitrates, are crucial in fertilizers. Nitrogen is also used in the production of explosives, such as TNT, and in the chemical industry for producing nitric acid and other compounds. Additionally, nitrogen gas is used to create inert atmospheres in industries ranging from food preservation to pharmaceuticals.

14.4.2 Phosphorus Applications

Phosphorus is vital for agriculture, as it is a primary component of fertilizers. Phosphorus compounds are also used in matches, fireworks, and pesticides. Red phosphorus is essential in safety matches and certain chemical reactions, while white phosphorus is used in military applications and as a component in some incendiary devices.

14.4.3 Arsenic Applications

Despite its toxicity, arsenic has specific industrial uses. It is used in the production of semiconductors and certain alloys, as well as in the treatment of wood and as a component in some pesticides. However, due to its health risks, the use of arsenic is highly regulated and limited to specialized applications.

14.4.4 Antimony Applications

Antimony is primarily used as an alloying element, particularly in lead-based alloys for batteries and solder. Antimony compounds are also used in flame retardants, glass, and ceramics. The use of antimony in electronics and flame retardants is widespread, particularly in the automotive and construction industries.

14.4.5 Bismuth Applications

Bismuth is used in pharmaceuticals, such as bismuth subsalicylate for treating digestive disorders. It is also used as a replacement for lead in alloys, cosmetics, and medical devices. Bismuth's non-toxic nature makes it a preferred alternative in applications where environmental safety is a concern.

14.5 ENVIRONMENTAL AND HEALTH CONSIDERATIONS

The environmental and health impacts of Group 15 elements vary widely. Nitrogen compounds contribute to air and water pollution, particularly from fertilizers. Phosphorus compounds can lead to eutrophication in aquatic systems. Arsenic is a well-known carcinogen, and its presence in water sources is a global health concern. Antimony has moderate toxicity and environmental impact, while bismuth is considered relatively safe.

14.6 CONCLUSION

The Group 15 elements are essential to many aspects of human life and industry. Their unique chemical behaviors and properties, from nitrogen's role in biological systems to antimony's use in flame retardants, make them invaluable. However, the environmental and health implications of these elements must be carefully managed. With advances in technology and growing awareness of sustainability, the applications of Group 15 elements continue to evolve, reflecting their significance in modern science and industry.

CHAPTER 15
GROUP 16—THE CHALCOGENS

The Group 16 elements, also known as the chalcogens, are crucial members of the p-block in the periodic table. This group includes oxygen (O), sulfur (S), selenium (Se), tellurium (Te), and polonium (Po). These elements are characterized by their six valence electrons, with a general electron configuration of ns^2np^4, giving them a tendency to gain or share electrons in chemical reactions to achieve a stable octet.

The term "chalcogen" is derived from Greek, meaning "ore former," as many metal ores contain oxygen or sulfur.

15.1 GENERAL CHARACTERISTICS OF GROUP 16 ELEMENTS

The Group 16 elements are distinguished by their ns^2np^4 electron configuration, which makes them two electrons short of a complete octet. This results in a tendency for these elements to form -2 oxidation states by gaining two electrons. However, they can also exhibit positive oxidation states due to the availability of empty d-orbitals, especially in the heavier elements like sulfur, selenium, and tellurium. As one moves down the group, the atomic and ionic radii increase due to the addition of electron shells. Consequently, ionization energies and electronegativities decrease down the group, which also leads to a gradual transition from non-metallic to more metallic behavior. Oxygen and sulfur are non-metals, selenium and tellurium are metalloids, and polonium is a metal. Each of these elements has unique properties that affect their chemical behavior and uses.

15.2 INDIVIDUAL PROPERTIES OF GROUP 16 ELEMENTS

Each chalcogen has distinct characteristics, making them suitable for a range of applications across various industries.

15.2.1 Oxygen (O)

Oxygen is a diatomic, colorless, and odorless gas that constitutes about 21% of the Earth's atmosphere. It is highly reactive and forms compounds with almost all other elements. Oxygen is essential for life, as it plays a crucial role in cellular respiration for animals and photosynthesis for plants. It also forms oxides and is involved in combustion reactions. Industrially, oxygen is used in steelmaking, medical applications, and as an oxidizer in rocket fuel.

15.2.2 Sulfur (S)

Sulfur is a yellow, non-metallic solid that is abundant in nature. It is commonly found in volcanic areas and hot springs. Sulfur can exist in several allotropes, with the most stable form being rhombic sulfur. It has a unique smell, often associated with rotten eggs, due to hydrogen sulfide (H_2S). Sulfur is widely used in the production of sulfuric acid, which is a critical industrial chemical. It is also used in vulcanization of rubber, fertilizers, and in the production of pesticides.

15.2.3 Selenium (Se)

Selenium is a metalloid that exhibits both metallic and non-metallic properties. It exists in several allotropic forms, including red, gray, and black selenium. Selenium is an essential trace element for animals and plays a role in antioxidant enzymes. In industry, selenium is used in the production of glass, where it helps remove greenish tints, and in electronics, particularly in photocopiers and solar cells. It is also used as a dietary supplement and in the manufacture of pigments.

15.2.4 Tellurium (Te)

Tellurium is a rare metalloid with metallic characteristics. It is brittle and has a silvery-white appearance. Tellurium is commonly found in association with gold and other metals. It is used as an alloying agent to improve the machinability of metals like steel and copper. Tellurium compounds are used in thermoelectric devices, which convert temperature differences into electric voltage. It also finds applications in semiconductors and solar panels.

15.2.5 Polonium (Po)

Polonium is a radioactive metal and the heaviest element in Group 16. Discovered by Marie Curie, it has no stable isotopes and is highly radioactive, with applications primarily limited to specialized scientific research. Polonium-210 is used as a heat source in space satellites due to its ability to release significant heat energy. Due to its intense radioactivity, polonium is also a source of neutron radiation when combined with beryllium.

15.3 CHEMICAL BEHAVIOR OF GROUP 16 ELEMENTS

The chemical behavior of the chalcogens is influenced by their tendency to form various oxidation states, especially -2, +4, and +6. The reactivity generally decreases down the group.

15.3.1 Oxidation States

Oxygen commonly exhibits a -2 oxidation state due to its high electronegativity. Sulfur, selenium, and tellurium can form compounds in oxidation states of +4 and +6, with sulfur being capable of forming compounds in every oxidation state from

-2 to +6. This versatility is seen in compounds such as sulfur dioxide (SO_2), sulfur trioxide (SO_3), and sulfides. Tellurium also forms similar compounds but is less reactive than sulfur and selenium.

15.3.2 Reactivity with Hydrogen

The Group 16 elements react with hydrogen to form hydrides, with the general formula H_2E, where E represents the chalcogen. Water (H_2O) is the most well-known hydride of oxygen, essential for all life forms. Sulfur, selenium, and tellurium form hydrogen sulfide (H_2S), hydrogen selenide (H_2Se), and hydrogen telluride (H_2Te), respectively. These hydrides are toxic and have unpleasant odors. The stability of these hydrides decreases as one moves down the group, with H_2O being the most stable and H_2Te the least stable.

15.3.3 Reactivity with Oxygen

The Group 16 elements readily form oxides when they react with oxygen. Oxygen can form simple oxides like sulfur dioxide (SO_2) and more complex compounds like sulfur trioxide (SO_3). Selenium and tellurium oxides, such as selenium dioxide (SeO_2) and tellurium dioxide (TeO_2), are important in glass production and other industries. Polonium also forms polonium dioxide (PoO_2), although its radioactive nature limits its applications.

15.3.4 Reactivity with Halogens

The chalcogens form compounds with halogens, resulting in halides such as sulfur hexafluoride (SF_6) and selenium tetrafluoride (SeF_4). These halides vary in stability and application. SF_6 is a potent greenhouse gas, used in the electrical industry as an insulating material. Sulfur also forms halides like sulfur dichloride (SCl_2), used in organic synthesis.

15.4 INDUSTRIAL AND TECHNOLOGICAL APPLICATIONS

The Group 16 elements are widely used across various industries due to their unique chemical properties and reactivity.

15.4.1 Oxygen Applications

Oxygen is essential in the steel industry, where it helps remove impurities from molten iron. It is also used in welding and cutting metals, as well as in medical applications for patients who require oxygen therapy. Oxygen finds extensive use in the aerospace industry, particularly in rocket propulsion systems.

15.4.2 Sulfur Applications

Sulfur is primarily used to produce sulfuric acid, which is vital in manufacturing fertilizers, detergents, and explosives. Sulfur compounds are used in the vulcanization of rubber and as fungicides in agriculture. The cosmetic and pharmaceutical industries use sulfur for skin treatments and in medications.

15.4.3 Selenium Applications

Selenium is used in the glass industry to remove impurities and improve clarity. It is also used in photocopiers and as a component in solar cells. Selenium supplements are available due to its role as an essential micronutrient, helping in antioxidant activities in the human body.

15.4.4 Tellurium Applications

Tellurium is used in alloying applications to improve the machinability of metals like copper and stainless steel. It is also used in the production of thermoelectric materials, which are used in cooling devices and power generation. The semiconductor industry utilizes tellurium in the manufacture of certain solar panels and electronic components.

15.4.5 Polonium Applications

Polonium's applications are limited due to its radioactivity. It is used as a heat source in space applications and in scientific research. Polonium is also used in certain devices that eliminate static electricity in industries that handle sensitive materials.

15.5 ENVIRONMENTAL AND HEALTH CONSIDERATIONS

The chalcogens play significant roles in environmental chemistry. Oxygen is a crucial component of the atmosphere and is essential for life. However, sulfur oxides contribute to air pollution and acid rain. Selenium and tellurium compounds can be toxic in high concentrations, although selenium is a necessary micronutrient in small amounts. Polonium is highly radioactive and poses significant health risks upon exposure, requiring stringent safety measures in handling.

15.6 CONCLUSION

Group 16 elements, or the chalcogens, are essential components of the p-block with diverse chemical properties and applications. Oxygen and sulfur are crucial for biological and industrial processes, while selenium, tellurium, and polonium find uses in specialized applications. Although some of these elements can pose environmental and health risks, they are integral to modern technology and industry. As we continue to innovate and seek sustainable solutions, the study and application of the chalcogens will remain vital in fields ranging from energy production to environmental protection.

CHAPTER 16

GROUP 17—THE HALOGENS

Group 17 of the periodic table consists of five elements: fluorine (F), chlorine (Cl), bromine (Br), iodine (I), and astatine (At). These elements are collectively known as the halogens, derived from the Greek word "halos," meaning salt, and "gen," meaning to produce. This reflects their ability to form salts, particularly when reacting with alkali metals. The halogens are among the most reactive elements, and their unique properties have wide-ranging applications in industry, health, and everyday life.

16.1 GENERAL CHARACTERISTICS OF HALOGENS

The halogens are highly reactive non-metals found in the p-block of the periodic table. Each halogen has seven valence electrons (ns^2np^5 configuration), meaning they are one electron short of achieving a stable octet. This electron configuration gives halogens a strong tendency to gain an electron during chemical reactions, resulting in a -1 oxidation state. However, halogens, especially the heavier ones, can also exhibit positive oxidation states in compounds. The elements in Group 17 are diatomic molecules (F_2, Cl_2, Br_2, I_2, At_2) in their natural form, and their physical states vary from gases (fluorine, chlorine) to liquids (bromine) to solids (iodine, astatine) at room temperature. Halogens are highly electronegative, with fluorine being the most electronegative element in the periodic table. This property is responsible for their strong reactivity and ability to form compounds with a wide variety of elements.

16.2 PHYSICAL PROPERTIES AND TRENDS IN GROUP 17

The halogens exhibit a number of physical properties that follow predictable trends as one moves down the group from fluorine to astatine. These trends are based on the increasing atomic and ionic radii, which result from the addition of electron shells.

16.2.1 Atomic and Ionic Radii: Understanding Size and Charge in Atoms and Ions

As one moves down the group, the atomic and ionic radii increase due to the addition of extra electron shells. This leads to a decrease in the effective nuclear charge experienced by the outermost electrons, making them more distant from the nucleus and thus increasing the size of the atom. Fluorine has the smallest atomic radius, while astatine has the largest.

16.2.2 Ionization Energy and Electronegativity

Ionization energy and electronegativity both decrease as one moves down the group. Fluorine, being the smallest and most electronegative, has the highest ionization energy. As the size of the atom increases, it becomes easier to remove an electron, leading to lower ionization energies for bromine, iodine, and astatine. The electronegativity trend follows a similar pattern, with fluorine being the most electronegative element in the periodic table, while astatine is the least electronegative halogen.

16.2.3 Melting and Boiling Points

The melting and boiling points of the halogens increase down the group. This is because the intermolecular van der Waals forces, which hold the molecules together, become stronger as the size and mass of the molecules increase. Fluorine and chlorine are gases at room temperature, bromine is a liquid, and iodine and astatine are solids.

16.3 CHEMICAL PROPERTIES AND REACTIVITY OF HALOGENS

The halogens are highly reactive elements, with reactivity decreasing as one moves down the group. Their reactivity is due to their high electronegativity and the desire to gain one electron to achieve a stable octet configuration.

16.3.1 Oxidation States

The halogens typically exhibit a -1 oxidation state in their compounds, especially in reactions with metals or hydrogen. However, the heavier halogens (chlorine, bromine, iodine) can exhibit positive oxidation states in compounds, ranging from +1 to +7. This occurs when these elements form compounds with more electronegative elements such as oxygen or fluorine.

16.3.2 Reactivity with Metals and Non-Metals

Halogens readily react with metals to form ionic halides. For example, sodium (Na) reacts with chlorine (Cl_2) to form sodium chloride (NaCl), a common salt. The reactivity of the halogens decreases down the group, with fluorine being the most reactive and astatine the least. Fluorine can react even with noble gases like xenon, a characteristic not shared by the other halogens. Halogens also react with non-metals to form covalent compounds, such as hydrogen halides (HF, HCl, HBr, HI).

16.3.3 Displacement Reactions

A key characteristic of halogens is their ability to displace less reactive halogens from halide solutions. For example, chlorine can displace bromine from a bromide salt solution, and bromine can displace iodine from an iodide solution. This reactivity trend is used in industrial and laboratory processes to extract and purify halogens.

16.4 INDUSTRIAL AND TECHNOLOGICAL APPLICATIONS OF HALOGENS

Halogens are essential to a variety of industrial and technological applications due to their reactivity and ability to form stable compounds. Each halogen has unique uses, from water purification to the production of pharmaceuticals and electronic components.

16.4.1 Fluorine

Fluorine is widely used in the production of fluorinated compounds, including polytetrafluoroethylene (PTFE), commonly known as Teflon, which is used as a non-stick coating for cookware. Fluorine is also used in the production of uranium hexafluoride (UF_6), which is necessary for the nuclear fuel cycle. Fluoride compounds, such as sodium fluoride, are added to toothpaste and drinking water to prevent tooth decay.

16.4.2 Chlorine

Chlorine is perhaps the most commercially significant halogen. It is used extensively in water treatment and disinfection, helping to ensure safe drinking water worldwide. Chlorine is also used in the production of polyvinyl chloride (PVC), a widely used plastic, and in the manufacture of solvents, insecticides, and pharmaceuticals. In addition, chlorine is involved in the production of bleach and other household cleaning products.

16.4.3 Bromine

Bromine is primarily used in the production of flame retardants, which are added to materials to reduce their flammability. It is also used in water treatment, particularly in swimming pools and hot tubs, where it serves as a disinfectant. Bromine compounds are also used in the production of certain dyes, photographic chemicals, and pharmaceuticals.

16.4.4 Iodine

Iodine is an essential element for human health, as it is required for the production of thyroid hormones. Iodine deficiency can lead to goiter and other health issues, so iodine is added to table salt in many countries to prevent deficiency. Iodine is also used as a disinfectant, in medical imaging (iodine-based contrast agents), and in the production of pharmaceuticals and dyes.

16.4.5 Astatine

Astatine is a rare, radioactive halogen with very limited commercial applications. Its isotopes are studied for potential use in cancer treatment, specifically in targeted alpha-particle therapy, where it may help destroy cancerous cells with minimal damage to surrounding healthy tissue.

16.5 ENVIRONMENTAL AND HEALTH IMPACTS OF HALOGENS

While halogens are essential for various industrial applications, they also have significant environmental and health impacts.

16.5.1 Environmental Concerns

Halogen-containing compounds, particularly chlorofluorocarbons (CFCs), have been linked to the depletion of the Earth's ozone layer. This led to the banning of CFCs under the Montreal Protocol, an international treaty designed to protect the ozone layer. Additionally, brominated flame retardants have been found to persist in the environment and accumulate in living organisms, raising concerns about their long-term effects on ecosystems.

16.5.2 Health Considerations

Exposure to high levels of halogens or halogen-containing compounds can pose health risks. Chlorine gas, for example, is highly toxic and was used as a chemical weapon during World War I. Fluorine gas is also highly corrosive and dangerous to handle. In contrast, iodine is essential for human health, but excessive exposure can lead to thyroid problems. Halogenated organic compounds, such as dioxins and PCBs, are persistent pollutants that can accumulate in the food chain and cause adverse health effects.

16.6 CONCLUSION

Group 17 elements, the halogens, are among the most reactive elements in the periodic table, and their chemical and physical properties follow distinct trends down the group. From the highly reactive fluorine to the rare and radioactive astatine, the halogens have a wide range of applications, from water purification to the production of plastics, flame retardants, and pharmaceuticals. However, their reactivity also poses environmental and health challenges, necessitating careful management and regulation of halogen-containing compounds. The halogens' unique combination of reactivity, versatility, and importance to modern industry underscores their significance in both chemistry and daily life.

CHAPTER 17

GROUP 18—THE NOBLE GASES

Group 18 of the periodic table, known as the noble gases, comprises the elements helium (He), neon (Ne), argon (Ar), krypton (Kr), xenon (Xe), and radon (Rn). These elements are found in the p-block and share a unique set of properties, making them distinct from other elements in the periodic table. Noble gases are characterized by their complete valence electron shells, rendering them mostly inert and stable.

This chemical stability, along with other unique properties, has led to a wide range of applications, from lighting to medical technology.

17.1 GENERAL CHARACTERISTICS OF NOBLE GASES

Noble gases are colorless, odorless, tasteless, and nonflammable under standard conditions. They exist as monoatomic gases with complete valence electron shells, which generally results in a lack of reactivity. The electronic configuration of noble gases is ns^2np^6 (except for helium, which is $1s^2$), giving them eight valence electrons. This stable octet configuration means noble gases do not readily gain, lose, or share electrons, making them relatively inert.

This inertness is a hallmark of the noble gases and contributes to their various practical applications. While noble gases are traditionally viewed as chemically inert, several of them, particularly xenon, have been found to form compounds under certain conditions. These compounds, however, are generally unstable and rare.

17.2 PHYSICAL PROPERTIES AND TRENDS OF NOBLE GASES

The physical properties of noble gases, including atomic size, boiling and melting points, and density, follow distinct trends as one moves down the group from helium to radon. These trends are influenced by atomic number, mass, and other periodic factors.

17.2.1 Atomic and Ionic Radii

The atomic radius of noble gases increases as one moves down the group. This is because each successive element has an additional electron shell, increasing the distance between the nucleus and the outermost electrons. Helium, with only one electron shell, has the smallest atomic radius, while radon, with six shells, has the largest.

17.2.2 Melting and Boiling Points

Noble gases have very low melting and boiling points due to weak intermolecular forces, specifically London dispersion forces. These forces increase with atomic size and mass, so the melting and boiling points of noble gases rise slightly as one moves down the group. For instance, helium remains a liquid only under extreme cooling, while xenon and radon have slightly higher melting and boiling points but are still gases at room temperature.

17.2.3 Density

The density of noble gases also increases as one moves down the group. This is due to the increase in atomic mass and the relatively low volume occupied by each atom. Helium, the lightest noble gas, is significantly less dense than air, which is why it is used in balloons. Radon, on the other hand, is one of the densest gases known.

17.3 CHEMICAL PROPERTIES AND REACTIVITY OF NOBLE GASES

The noble gases are known for their low reactivity. Their filled valence electron shells provide stability, which makes them unlikely to participate in chemical reactions under standard conditions. However, under certain extreme conditions, some noble gases can form compounds, especially with highly electronegative elements.

17.3.1 Inertness and Stability

Noble gases are generally inert due to their stable electron configurations. This stability is most pronounced in lighter noble gases like helium and neon, which rarely, if ever, form compounds.

For much of the 19th and 20th centuries, noble gases were thought to be entirely unreactive, hence their designation as "noble."

17.3.2 Formation of Compounds

While noble gases are typically inert, heavier noble gases like xenon and krypton can form compounds, particularly with fluorine and oxygen. Xenon, for example, can form xenon hexafluoride (XeF_6) and xenon trioxide (XeO_3). Krypton can form krypton difluoride (KrF_2) under specific conditions. These compounds are typically formed under high pressure or with the aid of a catalyst and are generally unstable.

17.4 DISCOVERY AND HISTORY OF NOBLE GASES

The noble gases were discovered over several decades in the late 19th century. The discovery of these elements expanded the periodic table and contributed to the understanding of atomic structure.

17.4.1 Helium

Helium was the first noble gas to be discovered, but it was initially detected in the sun rather than on Earth. In 1868, astronomers Pierre Janssen and Joseph Norman Lockyer observed a yellow spectral line in sunlight, which did not correspond to any known element. This observation led to the identification of helium, named after the Greek word for the sun, "helios." Helium was later isolated on Earth in 1895.

17.4.2 Argon

Argon was the first noble gas to be discovered on Earth. In 1894, scientists Lord Rayleigh and Sir William Ramsay identified argon as an inert component of air that could not be explained by any known elements at the time. This discovery led Ramsay and others to further investigate the existence of other noble gases in the atmosphere.

17.4.3 Neon, Krypton, and Xenon

Following the discovery of argon, Ramsay and Morris Travers continued to isolate additional noble gases from air. In 1898, they discovered neon, krypton, and xenon. These elements, like argon, were found to be chemically inert, reinforcing the concept of a new group of elements in the periodic table.

17.4.4 Radon

Radon, the heaviest noble gas, was discovered in 1900 by Friedrich Ernst Dorn while studying the decay of radium. Radon is radioactive and occurs naturally as a decay product of uranium and thorium in rocks and soil.

17.5 INDUSTRIAL AND TECHNOLOGICAL APPLICATIONS OF NOBLE GASES

The unique properties of noble gases, including their inertness, low reactivity, and specific physical characteristics, make them suitable for a wide variety of applications in industry, technology, and medicine.

17.5.1 Helium

Helium has a wide range of applications due to its low density, non-flammability, and low boiling point. It is used in balloons, as a coolant for superconducting magnets in MRI machines, and in cryogenics. Helium is also used as a shielding gas in welding and as a component of breathing gases for deep-sea divers.

17.5.2 Neon

Neon is primarily known for its use in neon lighting, where it emits a bright red-orange glow. Neon signs are widely used for advertising and decorative purposes. Neon is also used in high-voltage indicators and vacuum tubes.

17.5.3 Argon

Argon is used extensively as a shielding gas in welding, especially for metals that are sensitive to atmospheric gases.

It is also used in incandescent and fluorescent light bulbs to prevent the oxidation of filament materials. Additionally, argon is used in certain types of lasers and in the preservation of historical documents and materials.

17.5.4 Krypton

Krypton is used in certain types of lighting, particularly in high-performance light bulbs and flash photography. Krypton is also used in some types of lasers, including krypton-ion lasers, which emit light at specific wavelengths useful in scientific research.

17.5.5 Xenon

Xenon is used in high-intensity discharge lamps, such as those found in car headlights, and in specialized lighting for photography and film production. Xenon is also used as a general anesthetic and in ion propulsion systems for spacecraft due to its high atomic mass.

17.5.6 Radon

Radon has limited industrial applications due to its radioactivity. However, it has been studied for potential use in cancer treatment through targeted radiation therapy. Radon is also used in geological research to measure and predict earthquakes.

17.6 ENVIRONMENTAL AND HEALTH CONCERNS OF NOBLE GASES

While most noble gases are non-toxic and relatively safe, radon poses significant health risks due to its radioactivity. Radon gas can accumulate in buildings, particularly in basements, and prolonged exposure has been linked to lung cancer.

Helium, although safe for most applications, can pose suffocation risks if it displaces oxygen in confined spaces. Additionally, the extraction and use of noble gases, especially helium, can impact the environment if not managed sustainably.

PART FOUR

THE D-BLOCK ELEMENTS

CHAPTER 18

AN INTRODUCTION TO THE TRANSITION ELEMENTS

In chemistry, a transition metal (or transition element) is a chemical element in the d-block of the periodic table (groups 3 to 12), though the elements of group 12 (and less often group 3) are sometimes excluded. The lanthanide and actinide elements (the f-block) are called inner transition metals and are sometimes considered to be transition metals as well.

Since they are metals, they are lustrous and have good electrical and thermal conductivity. Most (with the exception of group 11 and group 12) are hard and strong, and have high melting and boiling temperatures. They form compounds in any of two or more different oxidation states and bind to a variety of ligands to form coordination complexes that are often coloured. They form many useful alloys and are often employed as catalysts in elemental form or in compounds such as coordination complexes and oxides. Most are strongly paramagnetic because of their unpaired d electrons, as are many of their compounds. All of the elements that are ferromagnetic near room temperature are transition metals (iron, cobalt and nickel) or inner transition metals (gadolinium).

English chemist Charles Rugeley Bury (1890–1968) first used the word *transition* in this context in 1921, when he referred to a *transition series of elements* during the change of an inner layer of electrons (for example $n = 3$ in the 4th row of the periodic table) from a stable group of 8 to one of 18, or from 18 to 32. These elements are now known as the d-block.

18.1 DEFINITION AND CLASSIFICATION

The 2011 IUPAC *Principles of Chemical Nomenclature* describe a "transition metal" as any element in groups 3 to 12 on the periodic table. This corresponds exactly to the d-block elements, and many scientists use this definition. In actual practice, the f-block lanthanide and actinide series are called "inner transition metals". The 2005 *Red Book* allows for the group 12 elements to be excluded, but not the 2011 *Principles.*

The IUPAC *Gold Book* defines a transition metal as "an element whose atom has a partially filled d sub-shell, or which can give rise to cations with an incomplete d sub-shell", but this definition is taken from an old edition of the *Red Book* and is no longer present in the current edition. In the d-block, the atoms of the elements have between zero and ten d electrons.

Table 1: Transition Metals in the D-Block

Group	3	4	5	6	7	8	9	10	11	12
Period 4	$_{21}$Sc	$_{22}$Ti	$_{23}$V	$_{24}$Cr	$_{25}$Mn	$_{26}$Fe	$_{27}$Co	$_{28}$Ni	$_{29}$Cu	$_{30}$Zn
5	$_{39}$Y	$_{40}$Zr	$_{41}$Nb	$_{42}$Mo	$_{43}$Tc	$_{44}$Ru	$_{45}$Rh	$_{46}$Pd	$_{47}$Ag	$_{48}$Cd
6	$_{71}$Lu	$_{72}$Hf	$_{73}$Ta	$_{74}$W	$_{75}$Re	$_{76}$Os	$_{77}$Ir	$_{78}$Pt	$_{79}$Au	$_{80}$Hg
7	$_{103}$Lr	$_{104}$Rf	$_{105}$Db	$_{106}$Sg	$_{107}$Bh	$_{108}$Hs	$_{109}$Mt	$_{110}$Ds	$_{111}$Rg	$_{112}$CN

Published texts and periodic tables show variation regarding the heavier members of group 3. The common placement of lanthanum and actinium in these positions is not supported by physical, chemical, and electronic evidence, which overwhelmingly favour putting lutetium and lawrencium in those places. Some authors prefer to leave the spaces below yttrium blank as a third option, but there is confusion on whether this format implies that group 3 contains only scandium and yttrium, or if it also contains all the lanthanides and actinides; additionally, it creates a 15-element-wide f-block, when quantum mechanics dictates that the f-block should only be 14 elements wide. The form with lutetium and lawrencium in group 3 is supported by a 1988 IUPAC report on physical, chemical, and electronic grounds, and again by a 2021 IUPAC preliminary report as it is the only form that allows simultaneous:

- Preservation of the sequence of increasing atomic numbers,
- A 14-element-wide f-block, and
- Avoidance of the split in the d-block.

Argumentation can still be found in the contemporary literature purporting to defend the form with lanthanum and actinium in group 3, but many authors consider it to be logically inconsistent (a particular point of contention being the differing treatment of actinium and thorium, which both can use 5f as a valence orbital but have no 5f occupancy as single atoms); the majority of investigators considering the problem agree with the updated form with lutetium and lawrencium.

The group 12 elements zinc, cadmium, and mercury are sometimes excluded from the transition metals. This is because they have the electronic configuration [ ] $d^{10}s^2$, where the d shell is complete, and they still have a complete d shell in all their known oxidation states. The group 12 elements Zn, Cd and Hg may therefore, under certain criteria, be classed as post-transition metals in this case. However, it is often convenient to include these elements in a discussion of the transition elements. For example, when discussing the crystal field stabilization energy of first-row transition elements, it is convenient to also include the elements calcium and zinc, as both Ca^{2+} and Zn^{2+} have a value of zero, against which the value for other transition metal ions may be compared. Another example occurs in the Irving-Williams series of stability constants of complexes. Moreover, Zn, Cd, and Hg can use their d orbitals for bonding even though they are not known in oxidation states that would formally require breaking open the d-subshell, which sets them apart from the p-block elements.

The 2007 (though disputed and so far not reproduced independently) synthesis of mercury(IV) fluoride (HgF_4) has been taken by some to reinforce the view that the group 12 elements should be considered transition metals, but some authors still consider this compound to be exceptional.

Copernicium is expected to be able to use its d electrons for chemistry as its 6d subshell is destabilised by strong relativistic effects due to its very high atomic number, and as such is expected to have transition-metal-like behaviour and show higher oxidation states than +2 (which are not definitely known for the lighter group 12 elements). Even in bare dications, Cn^{2+} is predicted to be $6d^8 7s^2$, unlike Hg^{2+} which is $5d^{10} 6s^0$.

Although meitnerium, darmstadtium, and roentgenium are within the d-block and are expected to behave as transition metals analogous to their lighter congeners iridium, platinum, and gold, this has not yet been experimentally confirmed. Whether copernicium behaves more like mercury or has properties more similar to those of the noble gas radon is not clear. Relative inertness of Cn would come from the relativistically expanded $7s–7p_{1/2}$ energy gap, which is already adumbrated in the $6s–6p_{1/2}$ gap for Hg, weakening metallic bonding and causing its well-known low melting and boiling points.

Transition metals with lower or higher group numbers are described as 'earlier' or 'later', respectively. When described in a two-way classification scheme, early transition metals are on the left side of the d-block from group 3 to group 7. Late transition metals are on the right side of the d-block, from group 8 to 11 (or 12, if they are counted as transition metals). In an alternative three-way scheme, groups 3, 4, and 5 are classified as early transition metals, 6, 7, and 8 are classified as middle transition metals, and 9, 10, and 11 (and sometimes group 12) are classified as late transition metals.

The heavy group 2 elements calcium, strontium, and barium do not have filled d-orbitals as single atoms, but are known to have d-orbital bonding participation in some compounds, and for that reason have been called "honorary" transition metals. Probably the same is true of radium.

The f-block elements La–Yb and Ac–No have chemical activity of the (n-1)d shell, but importantly also have chemical activity of the (n-2)f shell that is absent in d-block elements. Hence they are often treated separately as inner transition elements.

18.2 ELECTRONIC CONFIGURATION

The general electronic configuration of the d-block atoms is [noble gas]$(n - 1)d^{0-10}ns^{0-2}np^{0-1}$. Here "[noble gas]" is the electronic configuration of the last noble gas preceding the atom in question, and n is the highest principal quantum number of an occupied orbital in that atom. For example, Ti ($Z = 22$) is in period 4 so that $n = 4$, the first 18 electrons have the same configuration of Ar at the end of period 3, and the overall configuration is $[Ar]3d^24s^2$. The period 6 and 7 transition metals also add core $(n - 2)f^{14}$ electrons, which are omitted from the tables below. The p orbitals are almost never filled in free atoms (the one exception being lawrencium due to relativistic effects that become important at such high Z), but they can contribute to the chemical bonding in transition metal compounds. The Madelung rule predicts that the inner d orbital is filled after the valence-shell s orbital. The typical electronic structure of transition metal atoms is then written as [noble gas] $ns^2(n - 1)d^m$. This rule is approximate, but holds for most of the transition metals. Even when it fails for the neutral ground state, it accurately describes a low-lying excited state.

The d subshell is the next-to-last subshell and is denoted as $(n - 1)$d subshell. The number of s electrons in the outermost s subshell is generally one or two except palladium (Pd), with no electron in that s sub shell in its ground state. The s subshell in the valence shell is represented as the ns subshell, e.g. 4s. In the periodic table, the transition metals are present in ten groups (3 to 12). The elements in group 3 have an $ns^2(n - 1)d^1$ configuration, except for lawrencium (Lr): its $7s^27p^1$ configuration exceptionally does not fill the 6d orbitals at all. The first transition series is present in the 4th period, and starts after Ca ($Z = 20$) of group 2 with the configuration $[Ar]4s^2$, or scandium (Sc), the first element of group 3 with atomic number $Z = 21$ and configuration $[Ar]4s^23d^1$, depending on the definition used. As we move from left to right, electrons are added to the same d subshell till it is complete. Since the electrons added fill the $(n - 1)$d orbitals, the properties of the d-block elements are quite different from those of s and p block elements in which the filling occurs either in s or in p orbitals of the valence shell. The electronic configuration of the individual elements present in all the d-block series are given below:

Table 2: First (3d) D-Block Series (Sc–Zn)

Group	3	4	5	6	7	8	9	10	11	12
Atomic number	21	22	23	24	25	26	27	28	29	30
Element	Sc	Ti	V	Cr	Mn	Fe	Co	Ni	Cu	Zn
Electron configu-ration	$3d^14s^2$	$3d^24s^2$	$3d^34s^2$	$3d^54s^1$	$3d^54s^2$	$3d^64s^2$	$3d^74s^2$	$3d^84s^2$	$3d^{10}4s^1$	$3d^{10}4s^2$

Table 3: Second (4d) D-Block Series (Y–Cd)

Atomic Number	39	40	41	42	43	44	45	46	47	48
Element	Y	Zr	Nb	Mo	Tc	Ru	Rh	Pd	Ag	Cd
Electron configuration	$4d^15s^2$	$4d^25s^2$	$4d^45s^1$	$4d^55s^1$	$4d^55s^2$	$4d^75s^1$	$4d^85s^1$	$4d^{10}5s^0$	$4d^{10}5s^1$	$4d^{10}5s^2$

Table 4: Third (5d) D-Block Series (Lu–Hg)

Atomic Number	71	72	73	74	75	76	77	78	79	80
Element	Lu	Hf	Ta	W	Re	Os	Ir	Pt	Au	Hg
Electron configuration	$5d^16s^2$	$5d^26s^2$	$5d^36s^2$	$5d^46s^2$	$5d^56s^2$	$5d^66s^2$	$5d^76s^2$	$5d^96s^1$	$5d^{10}6s^1$	$5d^{10}6s^2$

Table 5: Fourth (6d) D-Block Series (Lr–Cn): (Configurations Predicted for Mt–Cn)

Atomic Number	103	104	105	106	107	108	109	110	111	112
Element	Lr	Rf	Db	Sg	Bh	Hs	Mt	Ds	Rg	Cn
Electron configuration	$7s^27p^1$	$6d^27s^2$	$6d^37s^2$	$6d^47s^2$	$6d^57s^2$	$6d^67s^2$	$6d^77s^2$	$6d^87s^2$	$6d^97s^2$	$6d^{10}7s^2$

A careful look at the electronic configuration of the elements reveals that there are certain exceptions to the Madelung rule. For Cr as an example the rule predicts the configuration $3d^44s^2$, but the observed atomic spectra show that the real ground state is $3d^54s^1$.

To explain such exceptions, it is necessary to consider the effects of increasing nuclear charge on the orbital energies, as well as the electron–electron interactions including both Coulomb repulsion and exchange energy. The exceptions are in any case not very relevant for chemistry because the energy difference between them and the expected configuration is always quite low.

The (n - 1)d orbitals that are involved in the transition metals are very significant because they influence such properties as magnetic character, variable oxidation states, formation of coloured compounds etc. The valence s and p orbitals (ns and np) have very little contribution in this regard since they hardly change in the moving from left to the right in a transition series. In transition metals, there are greater horizontal similarities in the properties of the elements in a period in comparison to the periods in which the d orbitals are not involved. This is because in a transition series, the valence shell electronic configuration of the elements do not change. However, there are some group similarities as well.

18.3 CHARACTERISTIC PROPERTIES

There are a number of properties shared by the transition elements that are not found in other elements, which results from the partially filled d shell. These include:

- The formation of compounds whose colour is due to d–d electronic transitions.
- The formation of compounds in many oxidation states, due to the relatively low energy gap between different possible oxidation states.
- The formation of many paramagnetic compounds due to the presence of unpaired d electrons. A few compounds of main-group elements are also paramagnetic (e.g., nitric oxide, oxygen).

Most transition metals can be bound to a variety of ligands, allowing for a wide variety of transition metal complexes.

18.3.1 Coloured Compounds

Colour in transition-series metal compounds is generally due to electronic transitions of two principal types.

- **Charge Transfer Transitions:** An electron may jump from a predominantly ligand orbital to a predominantly metal orbital, giving rise to a ligand-to-metal charge-transfer (LMCT) transition. These can most easily occur when the metal is in a high oxidation state. For example, the colour of chromate, dichromate and permanganate ions is due to LMCT transitions. Another example is that mercuric iodide, HgI_2, is red because of a LMCT transition.

A metal-to-ligand charge transfer (MLCT) transition will be most likely when the metal is in a low oxidation state and the ligand is easily reduced.

In general charge transfer transitions result in more intense colours than d–d transitions.

- **D–D Transitions:** An electron jumps from one d orbital to another. In complexes of the transition metals the d orbitals do not all have the same energy. The pattern of splitting of the d orbitals can be calculated using crystal field theory. The extent of the splitting depends on the particular metal, its oxidation state and the nature of the ligands. The actual energy levels are shown on Tanabe–Sugano diagrams.

In centrosymmetric complexes, such as octahedral complexes, d–d transitions are forbidden by the Laporte rule and only occur because of vibronic coupling in which a molecular vibration occurs together with a d–d transition. Tetrahedral complexes have somewhat more intense colour because mixing d and p orbitals is possible when there is no centre of symmetry, so transitions are not pure d–d transitions. The molar absorptivity (ε) of bands caused by d–d transitions are relatively low, roughly in the range 5-500 $M^{-1}cm^{-1}$ (where M = mol dm^{-3}). Some d–d transitions are spin forbidden. An example occurs in octahedral, high-spin complexes of manganese(II), which has a d^5 configuration in which all five electrons have parallel spins; the colour of such complexes is much weaker than in complexes with spin-allowed transitions. Many compounds of manganese(II) appear almost colourless. The spectrum of $[Mn(H_2O)_6]^{2+}$ shows a maximum molar absorptivity of about 0.04 $M^{-1}cm^{-1}$ in the visible spectrum.

18.3.2 Oxidation States

Main-group elements in groups 13 to 18 also exhibit multiple oxidation states. The "common" oxidation states of these elements typically differ by two instead of one. For example, compounds of gallium in oxidation states +1 and +3 exist in which there is a single gallium atom. Compounds of Ga(II) would have an unpaired electron and would behave as a free radical and generally be destroyed rapidly, but some stable radicals of Ga(II) are known. Gallium also has a formal oxidation state of +2 in dimeric compounds, such as $[Ga_2Cl_6]^{2-}$, which contain a Ga-Ga bond formed from the unpaired electron on each Ga atom. Thus the main difference in oxidation states, between transition elements and other elements is that oxidation states are known in which there is a single atom of the element and one or more unpaired electrons.

The maximum oxidation state in the first row transition metals is equal to the number of valence electrons from titanium (+4) up to manganese (+7), but decreases in the later elements. In the second row, the maximum occurs with ruthenium (+8), and in the third row, the maximum occurs with iridium (+9). In compounds such as $[MnO_4]^-$ and OsO_4, the elements achieve a stable configuration by covalent bonding.

The lowest oxidation states are exhibited in metal carbonyl complexes such as $Cr(CO)_6$ (oxidation state zero) and $[Fe(CO)_4]^{2-}$ (oxidation state -2) in which the 18-electron rule is obeyed. These complexes are also covalent.

Ionic compounds are mostly formed with oxidation states +2 and +3. In aqueous solution, the ions are hydrated by (usually) six water molecules arranged octahedrally.

18.3.3 Magnetism

Transition metal compounds are paramagnetic when they have one or more unpaired d electrons. In octahedral complexes with between four and seven d electrons both high spin and low spin states are possible. Tetrahedral transition metal complexes such as $[FeCl_4]^{2-}$ are high spin because the crystal field splitting is small so that the energy to be gained by virtue of the electrons being in lower energy orbitals is always less than the energy needed to pair up the spins. Some compounds are diamagnetic. These include octahedral, low-spin, d^6 and square-planar d^8 complexes. In these cases, crystal field splitting is such that all the electrons are paired up.

Ferromagnetism occurs when individual atoms are paramagnetic and the spin vectors are aligned parallel to each other in a crystalline material. Metallic iron and the alloy alnico are examples of ferromagnetic materials involving transition metals. Antiferromagnetism is another example of a magnetic property arising from a particular alignment of individual spins in the solid state.

18.3.4 Catalytic Properties

The transition metals and their compounds are known for their homogeneous and heterogeneous catalytic activity. This activity is ascribed to their ability to adopt multiple oxidation states and to form complexes. Vanadium(V) oxide (in the contact process), finely divided iron (in the Haber process), and nickel (in catalytic hydrogenation) are some of the examples. Catalysts at a solid surface (nanomaterial-based catalysts) involve the formation of bonds between reactant molecules and atoms of the surface of the catalyst (first row transition metals utilize 3d and 4s electrons for bonding). This has the effect of increasing the concentration of the reactants at the catalyst surface and also weakening of the bonds in the reacting molecules (the activation energy is lowered). Also because the transition metal ions can change their oxidation states, they become more effective as catalysts.

An interesting type of catalysis occurs when the products of a reaction catalyse the reaction producing more catalyst (autocatalysis). One example is the reaction of oxalic acid with acidified potassium permanganate (or manganate (VII)). Once a little Mn^{2+} has been produced, it can react with MnO_4^- forming Mn^{3+}. This then reacts with $C_2O_4^-$ ions forming Mn^{2+} again.

18.3.5 Physical Properties

As implied by the name, all transition metals are metals and thus conductors of electricity. In general, transition metals possess a high density and high melting points and boiling points. These properties are due to metallic bonding by delocalized d electrons, leading to cohesion which increases with the number of shared electrons. However the group 12 metals have much lower melting and boiling points since their full d subshells prevent d–d bonding, which again tends to differentiate them from the accepted transition metals. Mercury has a melting point of -38.83°C (-37.89°F) and is a liquid at room temperature.

CHAPTER 19
GROUP 3—THE SCANDIUM GROUP

The elements in Group 3 of the d-block of the periodic table, commonly referred to as the Scandium Group, include scandium (Sc), yttrium (Y), and typically the lanthanides lanthanum (La) and actinides such as actinium (Ac). These elements, which begin the d-block, are relatively rare and have properties that distinguish them from other d-block metals. The Scandium Group is crucial in various industries due to its unique properties, particularly its role in alloy production, electronics, and medical applications.

19.1 OVERVIEW OF GROUP 3 ELEMENTS

Group 3 elements are transition metals with a partially filled d-orbital. The electronic configuration of these elements usually involves the filling of d and f orbitals, which contribute to their unique physical and chemical properties. Group 3 elements are notable for their lightness and relatively low atomic numbers compared to other transition metals.

19.1.1 Scandium (Sc)

Scandium, with atomic number 21, is a soft, silvery-white metal. It is lightweight and has a high melting point, making it useful in various industrial applications. The electron configuration of scandium is [Ar] $3d^1\ 4s^2$, indicating that it has one electron in the d-orbital, which is characteristic of transition metals.

19.1.2 Yttrium (Y)

Yttrium, with atomic number 39, is another silvery transition metal that is relatively rare. Its electron configuration is [Kr] $4d^1\ 5s^2$. Yttrium is heavier than scandium but shares similar chemical properties. It is used in various applications due to its stability and the way it interacts with other elements, particularly in electronics and ceramic materials.

19.1.3 Lanthanum (La) and Actinium (Ac)

Lanthanum and actinium, sometimes included in Group 3, are the first elements in the lanthanide and actinide series, respectively. They share some properties with scandium and yttrium, such as similar oxidation states and electron configurations that allow them to act as bridging elements between the main d-block and f-block of the periodic table. Lanthanum, with atomic number 57, has the electron configuration [Xe] $5d^1\ 6s^2$, while actinium, with atomic number 89, is represented as [Rn] $6d^1\ 7s^2$.

19.2 PHYSICAL PROPERTIES OF GROUP 3 ELEMENTS

The physical properties of Group 3 elements include similarities in their metallic luster, malleability, and high melting and boiling points. Despite these similarities, each element has unique characteristics that make it valuable in specific industrial and scientific applications.

19.2.1 Density and Atomic Radius

Scandium and yttrium are relatively light metals, with densities lower than many other transition metals. As one moves down the group, the atomic radius increases, which is consistent with the trends observed in the periodic table. Scandium has an atomic radius of 162 pm, yttrium has an atomic radius of 180 pm, and lanthanum's atomic radius is approximately 188 pm.

19.2.2 Melting and Boiling Points

Group 3 elements generally have high melting and boiling points, reflecting the strong metallic bonds and electron configurations. Scandium has a melting point of 1,541 °C and a boiling point of 2,836 °C, while yttrium has a melting point of 1,522 °C and a boiling point of 3,345 °C. Lanthanum has a slightly lower melting point at 920 °C, with a boiling point of 3,469 °C.

19.2.3 Conductivity and Magnetism

Group 3 elements are good conductors of electricity due to their metallic structure and free-flowing electrons. They are typically paramagnetic, meaning they are attracted to magnetic fields, which has implications in various technological applications.

19.3 CHEMICAL PROPERTIES AND REACTIVITY OF GROUP 3 ELEMENTS

The chemical properties of Group 3 elements are influenced by their electron configurations. They commonly exhibit a +3 oxidation state, reflecting the loss of three outermost electrons. This oxidation state is stable for scandium, yttrium, lanthanum, and actinium, which makes them chemically similar.

19.3.1 Oxidation States

Group 3 elements typically exhibit the +3 oxidation state. Scandium, yttrium, lanthanum, and actinium are known to form compounds in this oxidation state due to the loss of their two 4s electrons and one 3d or 5d electron, leading to stable configurations. This makes them highly reactive with nonmetals such as oxygen and halogens, forming stable oxides and halides.

19.3.2 Reactivity with Oxygen

Group 3 elements readily react with oxygen to form oxides. For example, scandium forms Sc_2O_3, yttrium forms Y_2O_3, and lanthanum forms La_2O_3.

These oxides are stable, white solids that are insoluble in water but can dissolve in acids. They are often used in ceramics, glass, and electronics due to their high melting points and chemical stability.

19.3.3 Reactivity with Halogens

When Group 3 elements react with halogens, they form halides. For instance, scandium reacts with chlorine to form scandium chloride ($ScCl_3$). These halides are usually ionic in nature and dissolve in water, making them useful in various chemical processes.

19.4 OCCURRENCE AND EXTRACTION OF GROUP 3 ELEMENTS

The elements in Group 3 are relatively rare in the Earth's crust but are often found in minerals and ores. Scandium is typically found in minerals like thortveitite and in trace amounts in rare earth ores.

Yttrium is often found in monazite and bastnasite ores, which are also sources of rare earth elements. Lanthanum and actinium are also found in trace amounts in various minerals. The extraction process for these elements involves mining the ores, followed by chemical processes such as solvent extraction and reduction. These processes are complex and energy-intensive, given the scarcity and reactivity of Group 3 elements.

19.5 INDUSTRIAL AND TECHNOLOGICAL APPLICATIONS OF GROUP 3 ELEMENTS

Due to their unique properties, Group 3 elements have numerous applications in industries ranging from aerospace to electronics.

19.5.1 Aerospace Alloys

Scandium is highly valued in the aerospace industry due to its light weight and high strength. When alloyed with aluminum, scandium increases the strength and corrosion resistance of the metal, making it ideal for aircraft components and high-performance sporting equipment.

19.5.2 Electronics and Phosphors

Yttrium has several applications in electronics, particularly in the production of phosphors used in color televisions and LEDs. Yttrium is also used in the production of yttrium iron garnet, a material that has applications in microwave and laser technologies.

19.5.3 Medical Applications

Yttrium-90, a radioactive isotope of yttrium, is used in medical treatments, particularly in cancer therapy for targeted radiation. Lanthanum compounds are used in kidney disease treatments to reduce phosphate levels in patients.

19.6 ENVIRONMENTAL IMPACT AND HEALTH CONSIDERATIONS

While Group 3 elements have many useful applications, there are some environmental and health considerations. Mining and extraction of these metals can result in significant environmental degradation if not managed properly. In addition, certain compounds, particularly those of yttrium, can be toxic if inhaled or ingested, posing health risks to workers in industries where these elements are processed. Scandium and yttrium are not generally hazardous, but proper handling and disposal practices are essential in industrial settings.

19.7 CONCLUSION

The Group 3 elements—scandium, yttrium, and sometimes lanthanum and actinium—occupy a unique position in the periodic table. Their properties, such as high melting points, light weight, and low reactivity, make them invaluable in various industrial applications, particularly in aerospace, electronics, and medical fields. Despite their rarity, the distinctive characteristics of Group 3 elements continue to make them essential components in modern technology. As extraction techniques improve and new applications are discovered, the significance of the Scandium Group in industry is expected to grow further.

CHAPTER 20
GROUP 4—THE TITANIUM GROUP

Group 4 of the periodic table, also known as the Titanium Group, consists of titanium (Ti), zirconium (Zr), hafnium (Hf), and rutherfordium (Rf). These elements are classified as transition metals and are known for their strength, high melting points, and corrosion resistance. They play crucial roles in various industries, including aerospace, nuclear power, and medical technologies.

20.1 OVERVIEW OF GROUP 4 ELEMENTS

The Group 4 elements exhibit characteristics typical of transition metals, such as forming multiple oxidation states and high density. These elements are known for their stability, which is attributed to their partially filled d orbitals. As one moves down the group from titanium to rutherfordium, the atomic size increases, and the elements display greater stability and density.

20.1.1 Titanium (Ti)

Titanium, with atomic number 22, is a lightweight, silver-gray metal. Known for its strength, corrosion resistance, and relatively low density, it is widely used in aerospace and medical applications. The electron configuration of titanium is [Ar] $3d^2\ 4s^2$, indicating two electrons in the d orbital.

20.1.2 Zirconium (Zr)

Zirconium, with atomic number 40, is a lustrous, gray-white metal with an electron configuration of [Kr] $4d^2\ 5s^2$. It shares similar chemical properties with titanium and is frequently used in nuclear reactors due to its low neutron-absorption cross-section, which makes it ideal for fuel cladding in nuclear power plants.

20.1.3 Hafnium (Hf)

Hafnium, with atomic number 72, is a dense, silvery metal that is often found with zirconium in nature due to their similar properties. Its electron configuration is [Xe] $4f^{14}\ 5d^2\ 6s^2$. Hafnium is used in the electronics industry and as a neutron absorber in nuclear reactors due to its high thermal neutron-capture cross-section.

20.1.4 Rutherfordium (Rf)

Rutherfordium, with atomic number 104, is a synthetic element that is not found naturally. Due to its short half-life, it has limited practical applications, and research on this element remains largely theoretical. The electron configuration of rutherfordium is predicted to be [Rn] $5f^{14}$ $6d^2$ $7s^2$, consistent with the general electronic structure of Group 4 elements.

20.2 PHYSICAL PROPERTIES OF GROUP 4 ELEMENTS

The physical properties of Group 4 elements vary, but they are generally characterized by high melting points, high density, and significant tensile strength. These qualities make them valuable in high-performance and extreme environments.

20.2.1 Melting and Boiling Points

Group 4 elements possess high melting and boiling points, indicative of strong metallic bonding.

Titanium has a melting point of 1,668°C and a boiling point of 3,287°C, while zirconium has a melting point of 1,855°C and a boiling point of 4,409°C. Hafnium, due to its greater density, has an even higher melting point of 2,233°C and a boiling point of 4,603°C. These high melting points make these metals ideal for use in high-temperature applications.

20.2.2 Density and Atomic Radius

As one moves down the group, the density and atomic radius increase due to the addition of electron shells. Titanium is relatively lightweight with a density of 4.5 g/cm³, while zirconium has a density of 6.52 g/cm³, and hafnium, the heaviest of the three naturally occurring elements, has a density of 13.31 g/cm³. This increase in density corresponds to the increase in atomic number and the number of protons and neutrons within the nucleus.

20.3 CHEMICAL PROPERTIES AND REACTIVITY OF GROUP 4 ELEMENTS

Group 4 elements are typically reactive with oxygen and other nonmetals, forming stable oxides and halides.

The primary oxidation state for all elements in this group is +4, which reflects the stability of their compounds and the inertness of the fully filled 4f and 5d orbitals.

20.3.1 Oxidation States

The primary oxidation state exhibited by Group 4 elements is +4, allowing them to form a variety of compounds, including oxides, halides, and sulfides. In most of their compounds, they form tetravalent ions (M^{4+}). For instance, titanium forms titanium dioxide (TiO_2), zirconium forms zirconium dioxide (ZrO_2), and hafnium forms hafnium dioxide (HfO_2), all of which are highly stable and refractory.

20.3.2 Reactivity with Oxygen and Water

Group 4 metals react with oxygen to form oxides that are usually protective, meaning they form a passivation layer that prevents further oxidation. Titanium, for example, forms a thin layer of titanium dioxide (TiO_2) on its surface when exposed to air, providing excellent corrosion resistance.

However, these metals are generally resistant to attack by water and dilute acids, although they can be dissolved by hydrofluoric acid due to the formation of complex fluorides.

20.3.3 Reactivity with Halogens

Group 4 elements react with halogens at high temperatures to form halides, such as titanium tetrachloride ($TiCl_4$) and zirconium tetrafluoride (ZrF_4). These halides are usually volatile and are used in various industrial applications, including the production of high-purity metals.

20.4 OCCURRENCE AND EXTRACTION OF GROUP 4 ELEMENTS

Titanium, zirconium, and hafnium are relatively abundant in the Earth's crust and are typically found in mineral forms. Their extraction, however, can be complex due to the need for extensive purification to isolate the metals from ores.

20.4.1 Titanium

Titanium is primarily obtained from minerals such as ilmenite ($FeTiO_3$) and rutile (TiO_2). The most common extraction process is the Kroll process, which involves the reduction of titanium tetrachloride ($TiCl_4$) with magnesium to produce metallic titanium.

20.4.2 Zirconium and Hafnium

Zirconium is typically found in the mineral zircon ($ZrSiO_4$), while hafnium is often present in zirconium ores as an impurity. The separation of zirconium and hafnium is challenging due to their similar chemical properties and requires complex processes like solvent extraction. Both elements are essential in nuclear reactors, where they are used for their neutron-absorbing properties.

20.5 INDUSTRIAL AND TECHNOLOGICAL APPLICATIONS OF GROUP 4 ELEMENTS

Group 4 elements have diverse applications due to their unique properties, particularly their high strength, corrosion resistance, and thermal stability.

20.5.1 Aerospace and Biomedical Applications

Titanium is widely used in aerospace due to its lightweight yet strong nature. Its biocompatibility also makes it ideal for medical implants, such as joint replacements and dental implants. Titanium's resistance to corrosion and its ability to integrate with bone tissue have made it a popular choice for medical devices.

20.5.2 Nuclear Power

Zirconium is crucial in nuclear reactors due to its low neutron absorption, which minimizes interference with the nuclear reaction. Hafnium, on the other hand, is a powerful neutron absorber and is used as a control material in reactors, enabling the regulation of nuclear reactions.

20.5.3 Electronics and High-Temperature Alloys

Hafnium has found applications in the electronics industry, particularly in semiconductor manufacturing. Its high melting point and electrical properties make it an excellent material for gate insulators in field-effect transistors. Additionally, Group 4 elements are widely used in high-temperature alloys due to their resistance to thermal deformation.

20.6 ENVIRONMENTAL IMPACT AND HEALTH CONSIDERATIONS

While Group 4 elements are generally considered to be of low toxicity, their mining and refining processes can have environmental impacts. The extraction of titanium, for instance, requires the use of chemicals that can contribute to environmental pollution if not managed correctly. In terms of health, titanium is biocompatible and safe for medical use, but zirconium and hafnium compounds should be handled with care to prevent inhalation or prolonged skin contact.

20.7 CONCLUSION

Group 4 elements, including titanium, zirconium, and hafnium, are indispensable to various industries due to their unique physical and chemical properties. Their applications range from aerospace and medical technologies to nuclear reactors and electronics, underscoring their significance in modern technology. As extraction techniques improve and new uses are discovered, the importance of these elements is likely to grow, making them critical components of technological advancement in the 21st century.

CHAPTER 21

GROUP 5—THE VANADIUM GROUP

Group 5 of the periodic table, also known as the Vanadium Group, includes the elements vanadium (V), niobium (Nb), tantalum (Ta), and dubnium (Db). These elements are transition metals known for their high melting points, hardness, and resistance to corrosion. They play crucial roles in various industries, particularly in alloy production, electronics, and the medical field.

21.1 OVERVIEW OF GROUP 5 ELEMENTS

Group 5 elements exhibit typical transition metal characteristics, such as forming multiple oxidation states and complex compounds. They also share similar physical properties, such as hardness, high melting points, and good resistance to corrosion. The general electron configuration for these elements is $(n\text{-}1)d^3\ ns^2$. Moving down the group from vanadium to dubnium, the elements show increasing atomic size, density, and stability.

21.1.1 Vanadium (V)

Vanadium, with atomic number 23, is a soft, silver-gray metal with good ductility and malleability. It has an electron configuration of [Ar] $3d^3\ 4s^2$, and its most common oxidation states are +2, +3, +4, and +5. Vanadium is primarily used in steel alloys for strengthening and enhancing corrosion resistance.

21.1.2 Niobium (Nb)

Niobium, atomic number 41, is a soft, gray, ductile metal with an electron configuration of [Kr] $4d^4\ 5s^1$. It has similar properties to vanadium and is widely used in alloys, particularly in the production of superconducting materials. Niobium is also resistant to oxidation, which makes it valuable in high-temperature applications.

21.1.3 Tantalum (Ta)

Tantalum, atomic number 73, is a dense, hard, and corrosion-resistant metal. With an electron configuration of [Xe] $4f^{14}\ 5d^3\ 6s^2$, tantalum is highly resistant to chemical attack, making it suitable for medical implants and high-performance electronic components. Its primary oxidation state is +5, and it shares similar properties with niobium.

21.1.4 Dubnium (Db)

Dubnium, with atomic number 105, is a synthetic, radioactive element that has no stable isotopes. Its electron configuration is predicted to be [Rn] $5f^{14}$ $6d^3$ $7s^2$, and it is typically studied for its theoretical properties due to its limited existence. Dubnium has no practical applications and is primarily used in scientific research to understand the properties of heavy elements.

21.2 PHYSICAL PROPERTIES OF GROUP 5 ELEMENTS

Group 5 elements share several physical characteristics, such as high melting and boiling points, hardness, and density. These properties are a result of the metallic bonding and filled d-orbitals typical of transition metals.

21.2.1 Melting and Boiling Points

The elements in Group 5 have high melting and boiling points, indicating strong metallic bonds. Vanadium melts at 1,910°C and boils at 3,407°C. Niobium has a melting point of 2,468°C and a boiling point of 4,927°C, while tantalum has a melting point of 3,017°C and a boiling point of 5,458°C. These high temperatures make these metals ideal for high-temperature environments and applications.

21.2.2 Density and Hardness

Density and hardness increase down the group, with vanadium being relatively lightweight compared to niobium and tantalum. Vanadium has a density of 6.11 g/cm^3, niobium has a density of 8.57 g/cm^3, and tantalum, the densest of the three, has a density of 16.69 g/cm^3. These characteristics make tantalum highly desirable for applications requiring materials with high strength and durability.

21.3 CHEMICAL PROPERTIES AND REACTIVITY OF GROUP 5 ELEMENTS

Group 5 elements exhibit a range of oxidation states, but the most stable and common state for each element is +5. This oxidation state allows these metals to form various compounds and participate in numerous chemical reactions.

21.3.1 Oxidation States

The most common oxidation states for Group 5 elements are +3, +4, and +5, with +5 being the most stable for niobium and tantalum. Vanadium can exhibit all three states (+2, +3, +4, and +5), while niobium and tantalum are primarily found in the +5 oxidation state. This versatility in oxidation states allows Group 5 elements to form complex compounds, particularly oxides, halides, and oxyanions.

21.3.2 Reactivity with Oxygen

Group 5 metals react with oxygen to form stable oxides. For instance, vanadium forms several oxides, such as V_2O_3, V_2O_4, and V_2O_5, with V_2O_5 being the most stable. Niobium and tantalum also form stable oxides—Nb_2O_5 and Ta_2O_5—which are highly resistant to corrosion. These oxides are widely used in electronic and optical applications due to their excellent dielectric properties.

21.3.3 Reactivity with Halogens and Other Nonmetals

Group 5 metals react with halogens to form halides. For example, vanadium reacts with chlorine to form vanadium chloride (VCl_4), while niobium and tantalum form pentahalides like $NbCl_5$ and $TaCl_5$. These compounds are generally stable and can be used as precursors for other chemical reactions.

21.4 OCCURRENCE AND EXTRACTION OF GROUP 5 ELEMENTS

Vanadium, niobium, and tantalum are naturally occurring elements found in various minerals, while dubnium is synthetic. The extraction processes for vanadium, niobium, and tantalum are specialized and tailored to their specific ore types.

21.4.1 Vanadium

Vanadium is found in minerals like vanadinite ($Pb_5(VO_4)_3Cl$) and magnetite (Fe_3O_4) ores. The extraction of vanadium typically involves roasting the ore with sodium carbonate to produce vanadium pentoxide (V_2O_5), which is then reduced to produce pure vanadium.

21.4.2 Niobium and Tantalum

Niobium and tantalum are commonly found together in minerals such as columbite and tantalite. Their extraction involves a series of chemical processes, including solvent extraction, to separate the metals due to their similar chemical properties. Niobium is often extracted through the reduction of niobium pentoxide (Nb_2O_5) using carbon or aluminum, while tantalum is obtained by reducing tantalum pentachloride ($TaCl_5$).

21.4.3 Dubnium

Dubnium does not occur naturally and is produced synthetically through nuclear reactions. Its production requires particle accelerators, and it has no significant applications outside scientific research.

21.5 INDUSTRIAL AND TECHNOLOGICAL APPLICATIONS OF GROUP 5 ELEMENTS

Group 5 elements are used in a variety of industries due to their unique properties. Vanadium, niobium, and tantalum are essential for making high-performance alloys, electronic components, and medical devices.

21.5.1 Vanadium: Steel Alloys and Catalysts

Vanadium is primarily used as an alloying element in steel to improve strength, toughness, and wear resistance. Vanadium-based catalysts, such as vanadium pentoxide (V_2O_5), are also important in the chemical industry for processes like the production of sulfuric acid and the reduction of emissions from power plants.

21.5.2 Niobium: Superconducting Materials and Alloys

Niobium is widely used in the production of superconducting materials, such as niobium-titanium (Nb-Ti) alloys, which are used in MRI machines and particle accelerators. Niobium alloys are also used in jet engines and rocket nozzles due to their high-temperature stability and corrosion resistance.

21.5.3 Tantalum: Electronics and Medical Applications

Tantalum's excellent corrosion resistance and biocompatibility make it ideal for medical implants, such as pacemakers and orthopedic devices. Tantalum is also used in capacitors and high-performance electronic components, particularly in mobile phones and other digital devices.

21.6 ENVIRONMENTAL IMPACT AND HEALTH CONSIDERATIONS

While Group 5 elements are generally considered low in toxicity, their extraction and processing can pose environmental challenges. Vanadium mining, for instance, may release pollutants that can impact air and water quality. Tantalum mining, often associated with political and ethical concerns, has led to increased focus on responsible sourcing and the regulation of "conflict minerals." Niobium and tantalum, however, are generally safe when used in medical and electronic applications.

21.7 CONCLUSION

Group 5 elements—vanadium, niobium, tantalum, and dubnium—are notable for their high melting points, hardness, and resistance to corrosion. These transition metals are widely used across various industries, from strengthening steel alloys to enabling advanced electronics and medical technologies. Their unique chemical and physical properties make them essential components in modern technology, and as extraction techniques improve, the applications of these elements are likely to expand. As concerns about environmental impact and ethical sourcing continue to grow, the responsible use and management of these valuable elements will become increasingly important for sustainable development in the 21st century.

CHAPTER 22

GROUP 6—THE CHROMIUM GROUP

Group 6 of the periodic table, also known as the Chromium Group, includes the elements chromium (Cr), molybdenum (Mo), tungsten (W), and seaborgium (Sg). These elements are transition metals characterized by their ability to form multiple oxidation states, high melting points, and significant roles in industrial applications.

22.1 OVERVIEW OF GROUP 6 ELEMENTS

Group 6 elements exhibit a range of physical and chemical properties, making them essential in various industrial processes. They share a common electron configuration of (n-1)d^5 ns^1, which contributes to their unique chemical reactivity and bonding characteristics. The most notable members of this group are chromium, molybdenum, and tungsten, which are widely used in metallurgy, electronics, and chemical industries.

22.1.1 Chromium (Cr)

Chromium, with the atomic number 24, is a shiny, silvery metal known for its high corrosion resistance and hardness. Its electron configuration is [Ar] 3d^5 4s^1. Chromium exists in multiple oxidation states, with +3 and +6 being the most common. Chromium is widely used in stainless steel production, where it provides excellent resistance to corrosion and staining. It is also used in plating, dyes, and pigments due to its vibrant colors.

22.1.2 Molybdenum (Mo)

Molybdenum, atomic number 42, is a hard, silvery metal with a high melting point. Its electron configuration is [Kr] 4d^5 5s^1. Molybdenum exhibits oxidation states of +2, +3, +4, +5, and +6, with +6 being the most stable. It is primarily used in steel alloys to enhance strength, toughness, and wear resistance. Molybdenum is also vital in the production of electrical contacts and filaments.

22.1.3 Tungsten (W)

Tungsten, with atomic number 74, is a dense, hard metal known for its high melting point, the highest of all metals. Its electron configuration is [Xe] 4f^14 5d^4 6s^2. Tungsten exhibits oxidation states of +2, +3, +4, +5, and +6. Due to its unique properties, tungsten is used in a wide variety of applications, including light bulb filaments, cutting tools, and high-temperature aerospace components.

22.1.4 Seaborgium (Sg)

Seaborgium, atomic number 106, is a synthetic, radioactive element that has no stable isotopes. Its electron configuration is predicted to be [Rn] 5f^14 6d^4 7s^2. Seaborgium is primarily studied in research laboratories, and its practical applications are currently limited due to its short half-life and lack of sufficient quantities for industrial use.

22.2 PHYSICAL PROPERTIES OF GROUP 6 ELEMENTS

The elements in Group 6 exhibit several physical properties, such as high melting and boiling points, hardness, and metallic luster. These properties result from the strong metallic bonds and the filled d-orbitals typical of transition metals.

22.2.1 Melting and Boiling Points

Group 6 elements have high melting and boiling points due to the strong metallic bonds formed between their atoms. Chromium melts at 1,907°C and boils at 2,672°C. Molybdenum has a melting point of 2,623°C and a boiling point of 4,639°C, while tungsten, the element with the highest melting point, melts at 3,422°C and boils at 5,555°C. These high temperatures make these metals ideal for high-temperature applications in industries such as aerospace and manufacturing.

22.2.2 Density and Hardness

Density and hardness are notable characteristics of Group 6 elements. Chromium has a density of 7.19 g/cm^3, molybdenum has a density of 10.28 g/cm^3, and tungsten is the densest, with a density of 19.25 g/cm^3. Tungsten's hardness makes it an essential material for tools and other applications requiring durability. The high density and hardness of these metals contribute to their suitability for various industrial applications.

22.3 CHEMICAL PROPERTIES AND REACTIVITY OF GROUP 6 ELEMENTS

Group 6 elements exhibit a range of oxidation states, with chromium and molybdenum showing the ability to form complex compounds. Their chemical behavior is influenced by their electron configurations, particularly the presence of d-electrons.

22.3.1 Oxidation States

The oxidation states of Group 6 elements vary widely. Chromium can exist in oxidation states ranging from -2 to +6, with +3 and +6 being the most stable. Molybdenum shows oxidation states from -2 to +6, with +6 being the most common in its compounds. Tungsten can also exhibit multiple oxidation states, from -2 to +6, although +6 is the most stable for tungsten as well. This versatility in oxidation states allows these metals to form a variety of compounds.

22.3.2 Reactivity with Oxygen

Group 6 metals react with oxygen to form oxides. Chromium forms several oxides, including chromium(III) oxide (Cr_2O_3) and chromium(VI) oxide (CrO_3), the latter being a powerful oxidizing agent. Molybdenum reacts with oxygen to form molybdenum trioxide (MoO_3), which is used in the production of molybdenum compounds. Tungsten, on the other hand, forms tungsten oxides such as tungsten(VI) oxide (WO_3), which is utilized in various chemical applications.

22.3.3 Reactivity with Halogens and Other Nonmetals

Group 6 metals react with halogens to form halides. Chromium reacts with chlorine to form chromium(VI) chloride ($CrCl_6$), while molybdenum and tungsten can form several halides, such as molybdenum(VI) chloride ($MoCl_6$) and tungsten(VI) chloride (WCl_6). These compounds are essential for synthesizing other chemical compounds and serve as precursors for various reactions in organic chemistry.

22.4 OCCURRENCE AND EXTRACTION OF GROUP 6 ELEMENTS

The elements in Group 6 are found in various minerals and ores. Their extraction processes often involve specialized techniques tailored to the specific ore types.

22.4.1 Chromium

Chromium is primarily obtained from the mineral chromite ($FeCr_2O_4$). The extraction process typically involves roasting chromite with sodium carbonate to produce sodium chromate (Na_2CrO_4), which is then converted into chromium trioxide (CrO_3). This compound can be reduced to obtain pure chromium metal.

22.4.2 Molybdenum

Molybdenum is commonly found in minerals such as molybdenite (MoS_2) and wulfenite ($PbMoO_4$). The extraction process usually involves crushing the ore and then roasting it to convert molybdenum sulfide to molybdenum oxide (MoO_3). The oxide is then reduced using hydrogen or carbon to yield pure molybdenum metal.

22.4.3 Tungsten

Tungsten is primarily extracted from minerals such as wolframite ($(Fe,Mn)WO_4$) and scheelite ($CaWO_4$). The extraction process involves crushing the ore, followed by chemical processing to obtain tungsten trioxide (WO_3). The oxide is then reduced to produce tungsten metal using carbon or hydrogen.

22.4.4 Seaborgium

Seaborgium is a synthetic element produced in particle accelerators through nuclear reactions involving heavier elements. Its short half-life and the difficulty of producing it in significant quantities limit its practical applications. Seaborgium is mainly of interest for research in nuclear chemistry and atomic physics.

22.5 INDUSTRIAL AND TECHNOLOGICAL APPLICATIONS OF GROUP 6 ELEMENTS

Group 6 elements play vital roles in various industries due to their unique properties and reactivity. Their applications range from metallurgy to electronics and chemical manufacturing.

22.5.1 Chromium: Stainless Steel and Coatings

Chromium is primarily used in the production of stainless steel, where it provides corrosion resistance and enhances the material's overall strength. The addition of chromium to steel improves its durability and allows it to withstand harsh environmental conditions. Additionally, chromium compounds are used in electroplating and coating applications to create protective and decorative surfaces.

22.5.2 Molybdenum: High-Performance Alloys

Molybdenum is essential in manufacturing high-performance alloys, particularly in aerospace and military applications. Its ability to withstand high temperatures and resist corrosion makes it ideal for use in turbine engines, exhaust systems, and chemical processing equipment. Molybdenum is also used in the production of electrical contacts and filaments.

22.5.3 Tungsten: Filaments and Cutting Tools

Tungsten's high melting point and hardness make it a crucial material in the production of light bulb filaments, where it provides longevity and reliability. Additionally, tungsten is used in cutting tools, such as drills and saw blades, due to its strength and wear resistance. Tungsten carbide, a compound of tungsten, is widely used in various industrial applications for its exceptional hardness.

22.6 ENVIRONMENTAL IMPACT AND HEALTH CONSIDERATIONS

While the elements in Group 6 have numerous industrial benefits, their extraction and processing can pose environmental challenges. The mining and refining processes for chromium, molybdenum, and tungsten may lead to habitat destruction, soil erosion, and water pollution. It is essential to implement sustainable practices and regulations to minimize these impacts.

22.6.1 Chromium and Health Concerns

Chromium, particularly in its hexavalent form (Cr(VI)), is a known carcinogen and poses health risks to workers in industries where it is used. Proper handling and safety measures must be implemented to protect workers and the environment from exposure. Monitoring and regulation of chromium emissions are crucial to mitigate health risks.

22.6.2 Molybdenum and Tungsten

Molybdenum and tungsten are generally considered low in toxicity, but exposure to dust or fumes can cause respiratory issues. Adhering to safety guidelines during extraction and processing is essential to protect workers' health. Environmental regulations should also be enforced to prevent contamination of air and water sources during mining operations.

22.7 CONCLUSION

Group 6 elements—chromium, molybdenum, tungsten, and seaborgium—are essential transition metals with unique properties and numerous industrial applications.

From stainless steel production to high-performance alloys, these elements play a critical role in modern technology and manufacturing processes. While their extraction and use come with environmental and health considerations, responsible practices can help mitigate these challenges. As industries continue to advance, the importance of Group 6 elements in technology and materials science will remain significant, underscoring their value in a wide array of applications.

CHAPTER 23

GROUP 7—THE MANGANESE GROUP

Group 7 of the periodic table, often referred to as the manganese group, contains the elements manganese (Mn), technetium (Tc), rhenium (Re), and bohrium (Bh). These elements are transition metals and share similar physical and chemical properties due to their position in the periodic table. Manganese is the most abundant and widely studied member of this group, while technetium and rhenium have more specialized roles. Bohrium, being a synthetic and highly radioactive element, is primarily of interest in scientific research rather than practical applications.

23.1 PHYSICAL AND CHEMICAL PROPERTIES

Group 7 elements are characterized by their metallic properties. They are hard, high-density metals with high melting points, a characteristic feature of transition metals. The electron configuration of these elements, especially the filling of their d-orbitals, plays a significant role in their behavior.

Manganese, for example, has an atomic number of 25 and an electron configuration of $[Ar]3d^5 4s^2$. This configuration makes it highly reactive and able to form a variety of oxidation states, ranging from -3 to +7, with +2, +3, +4, and +7 being the most common.

Technetium and rhenium also show multiple oxidation states, but technetium tends to exhibit fewer oxidation states than manganese, typically ranging from +1 to +7. Rhenium is unique in that it can form compounds in very high oxidation states, such as +7 in rhenium heptoxide (Re_2O_7), similar to manganese, but it also has some very stable low oxidation states.

Bohrium is the least studied due to its synthetic and extremely short-lived nature, which limits the amount of information available. As a superheavy element, it has a predicted electron configuration of $[Rn]\ 5f^{14}6d^5 7s^2$, making it analogous to rhenium in terms of its expected chemical properties.

23.2 OCCURRENCE AND ABUNDANCE

Manganese is the most abundant element in Group 7, and it is the 12th most abundant element in the Earth's crust, where it occurs primarily in the form of manganese dioxide (MnO_2) and other manganese oxides. Manganese ores, such as pyrolusite, are mined extensively due to the wide range of uses for manganese in industry, particularly in steel production.

Technetium, on the other hand, is a rare and unique element. It is the first element in the periodic table that does not have any stable isotopes. Technetium-99, a radioactive isotope with a half-life of approximately 211,000 years, is produced as a byproduct of uranium fission in nuclear reactors.

Because technetium is not found in significant amounts in the Earth's crust, its occurrence is largely artificial, though trace amounts have been detected in certain stars.

Rhenium is another rare element, ranking as one of the least abundant elements in the Earth's crust. It is usually found in small quantities in ores of other elements, such as molybdenite (MoS_2), from which it is extracted as a byproduct. Rhenium is more abundant in some regions, particularly in certain copper-molybdenum mining areas, but its overall abundance remains low compared to other metals.

Bohrium is a synthetic element, first produced in 1981 by bombarding bismuth-209 with chromium-54 ions. It exists only in laboratory settings and has no natural occurrence due to its highly unstable and radioactive nature.

23.3 INDUSTRIAL USES AND APPLICATIONS

23.3.1 Manganese

Manganese is an essential metal in various industrial applications, most notably in steel production. In fact, around 90% of all manganese produced is used in steelmaking.

Manganese improves the hardness, toughness, and strength of steel while also acting as a deoxidizing and desulfurizing agent. The metal is also used in the production of aluminum alloys, dry cell batteries (particularly alkaline batteries), and in chemical processes as a catalyst. Manganese compounds, such as potassium permanganate ($KMnO_4$), are used as strong oxidizing agents in laboratories and industrial settings.

23.3.2 Technetium

Technetium's primary application is in the field of medicine, specifically in nuclear medicine. The isotope technetium-99m (Tc-99m) is widely used in diagnostic imaging due to its short half-life and the ability to emit gamma radiation without beta radiation, which minimizes damage to surrounding tissues. Tc-99m is employed in radiopharmaceuticals for imaging organs such as the heart, lungs, and bones. Despite its limited natural occurrence, technetium has become critical in medical diagnostics, improving the ability to detect and monitor various conditions, including cancers, heart disease, and infections.

23.3.3 Rhenium

Rhenium is highly valuable due to its exceptional resistance to heat and corrosion. It is used in high-temperature superalloys, particularly in jet engines and gas turbine blades, where it improves strength and heat resistance.

The use of rhenium in nickel-based superalloys has revolutionized the aerospace industry by allowing engines to operate at higher temperatures and efficiencies. In addition to aerospace applications, rhenium is also used as a catalyst in the petrochemical industry, where it plays a role in the production of high-octane gasoline and in the reforming of petroleum. Its ability to withstand harsh chemical environments and high temperatures makes it ideal for these catalytic processes.

23.3.4 Bohrium

Bohrium, due to its synthetic nature and short half-life, has no commercial or industrial applications. Its primary use is in scientific research, where it is studied for its nuclear properties and to better understand the behavior of superheavy elements.

23.4 BIOLOGICAL ROLE AND TOXICITY

Manganese is an essential trace element for all living organisms. In humans, it plays a role in enzyme function, bone formation, and metabolism. It is a cofactor for several enzymes, including those involved in the synthesis of cartilage and bone, and it also has antioxidant properties.

However, excessive exposure to manganese, particularly in industrial settings, can lead to manganism, a neurological disorder with symptoms similar to Parkinson's disease.

Technetium has no known biological role, and its radioactivity makes it potentially hazardous if not handled properly. However, its use in medical diagnostics is considered safe when administered in controlled doses.

Rhenium, like technetium, has no known biological role and is considered non-toxic in its elemental form. Its compounds, however, may pose health risks if inhaled or ingested in large amounts, although exposure is uncommon due to its limited industrial use outside specialized applications.

Bohrium is not found in nature and has no known biological role. Its extreme radioactivity would make it highly toxic, but it is produced in such minute quantities that its biological impact is negligible.

23.5 CONCLUSION

The elements in Group 7—manganese, technetium, rhenium, and bohrium—each possess unique properties and play diverse roles in industry and scientific research. Manganese is an essential industrial metal with widespread applications, particularly in steel production. Technetium, though rare and radioactive, is critical in medical diagnostics, while rhenium is prized for its heat resistance and catalytic properties in aerospace and petrochemical industries. Bohrium, being synthetic, remains largely a subject of nuclear research.

As we continue to explore the properties and applications of these elements, they remain important in various technological and scientific advancements.

CHAPTER 24

GROUP 8—THE IRON GROUP

Group 8 of the periodic table, also known as the iron group, consists of the elements iron (Fe), ruthenium (Ru), osmium (Os), and hassium (Hs). These elements are transition metals and share several physical and chemical characteristics, including high melting points, the ability to form complex ions, and variable oxidation states. The iron group plays a significant role in various industrial, technological, and biological applications.

24.1 PHYSICAL AND CHEMICAL PROPERTIES

The elements in Group 8 are characterized by their metallic nature. These metals are hard, high-density materials with high melting and boiling points, and they are good conductors of electricity and heat, which is typical of transition metals. They have partially filled d-orbitals, which contribute to their ability to exhibit a variety of oxidation states.

24.1.1 Iron (Fe)

Iron is the most common and well-known member of Group 8. It has an atomic number of 26 and an electron configuration of [Ar] $3d^6 4s^2$. Its most stable oxidation states are +2 and +3, although it can also form compounds in other states, such as +4 and +6 in specific environments. The versatility of iron's oxidation states makes it highly reactive and important in chemical and biological processes. Iron is also magnetic, which contributes to its wide usage in various applications, particularly in manufacturing and engineering.

24.1.2 Ruthenium (Ru)

Ruthenium, atomic number 44, is a rare transition metal that belongs to the platinum group of elements. It has an electron configuration of [Kr] $4d^7 5s^1$ and typically exhibits oxidation states of +2, +3, and +4, although it can also reach higher states, such as +8. Ruthenium is known for its hardness, high resistance to corrosion, and ability to form stable compounds with other elements.

24.1.3 Osmium (Os)

Osmium, with an atomic number of 76, is one of the densest elements in the periodic table. It has an electron configuration of [Xe] $4f^{14} 5d^6 6s^2$, and like ruthenium, it belongs to the platinum group of metals. Osmium can exhibit a variety of oxidation states, most commonly +4 and +8. It is known for its extreme hardness, high melting point, and resistance to corrosion. Due to its density and hardness, osmium is used in specialized industrial applications.

24.1.4 Hassium (Hs)

Hassium is the least known and studied member of Group 8. With an atomic number of 108, hassium is a synthetic element that was first produced in 1984 by bombarding lead with iron ions.

It is highly radioactive and exists only in laboratory settings, with no practical applications outside of scientific research. Due to its short half-life, its properties are not well understood, although it is predicted to share some characteristics with osmium based on its position in the periodic table.

24.2 OCCURRENCE AND ABUNDANCE

24.2.1 Iron

Iron is the fourth most abundant element in the Earth's crust and is found in a variety of ores, including hematite (Fe_2O_3), magnetite (Fe_3O_4), and limonite ($FeO(OH)\cdot nH_2O$). It is also a key component of the Earth's core, where it exists in a molten state due to the high temperatures and pressures. Iron is essential for life, playing a critical role in biological processes, particularly in oxygen transport within hemoglobin in the blood of animals.

24.2.2 Ruthenium

Ruthenium is much rarer than iron, with an average abundance of about 0.001 parts per million in the Earth's crust. It is usually found in ores alongside other platinum group metals, such as platinum and palladium. It is primarily sourced from regions with large deposits of platinum ores, including South Africa and Russia.

24.2.3 Osmium

Osmium is one of the rarest naturally occurring elements, with an abundance of only 0.002 parts per million in the Earth's crust. It is typically found in platinum ores, and like ruthenium, it is extracted as a byproduct during the mining and refining of platinum group metals. Major producers of osmium include Russia, South Africa, and North America.

24.2.4 Hassium

Hassium is a synthetic element and does not occur naturally. It is produced in minute amounts through nuclear reactions, and because of its extreme radioactivity and short half-life, it does not have any natural occurrence on Earth. Its production requires sophisticated equipment and is conducted only in specialized laboratories.

24.3 INDUSTRIAL USES AND APPLICATIONS

24.3.1 Iron

Iron is one of the most important metals in human history, with extensive applications across various industries. The primary use of iron is in the production of steel, an alloy of iron and carbon that forms the backbone of modern infrastructure. Steel is used in the construction of buildings, bridges, automobiles, machinery, and countless other products. Iron is also used in the manufacturing of magnets, tools, and industrial machinery.

In addition to its structural uses, iron compounds have applications in chemistry and medicine. Iron(III) chloride ($FeCl_3$) is used as a coagulant in water treatment, while iron(II) sulfate ($FeSO_4$) is used in agriculture as a soil amendment and in medicine as a supplement to treat iron deficiency anemia.

24.3.2 Ruthenium

Ruthenium's main industrial use is as a catalyst in chemical reactions. It is used in the production of ammonia in the Haber process, where it helps facilitate the conversion of nitrogen and hydrogen into ammonia. Ruthenium is also used in the electronics industry, particularly in the production of resistors and electrical contacts due to its excellent conductivity and resistance to wear. Additionally, ruthenium is used in the coating of platinum and palladium to increase their durability in harsh environments, such as in catalytic converters for automobiles. In recent years, ruthenium has gained attention in the field of nanotechnology and electronics, where its unique properties are being utilized in the development of advanced materials.

24.3.3 Osmium

Osmium has several specialized applications, primarily due to its hardness and high density. It is used in the production of hard alloys, such as those used in fountain pen tips, electrical contacts, and surgical implants. Osmium tetroxide (OsO_4) is a volatile compound of osmium that is used as a staining agent in electron microscopy and as a reagent in organic synthesis. Osmium's extreme density also makes it useful in scientific instruments that require heavy, durable materials. Osmium's scarcity and toxicity, however, limit its widespread use. Handling osmium tetroxide requires great care, as it is highly toxic and can cause severe damage to the respiratory system, eyes, and skin.

24.3.4 Hassium

Hassium, being a synthetic element with a very short half-life, has no known industrial applications. Its primary use is in scientific research, particularly in the study of superheavy elements and nuclear physics. Researchers produce hassium in extremely small quantities to study its properties and to expand our understanding of the periodic table and nuclear structure.

24.4 BIOLOGICAL ROLE AND TOXICITY

24.4.1 Iron

Iron is an essential element for all known forms of life. In humans and animals, it is a crucial component of hemoglobin, the protein responsible for transporting oxygen in the blood. Iron is also involved in various enzymatic processes, including energy production and DNA synthesis. A deficiency in iron can lead to anemia, a condition characterized by fatigue, weakness, and impaired cognitive function. While iron is necessary for life, excessive amounts can be toxic, leading to a condition known as iron overload, which can damage organs such as the liver and heart. Proper regulation of iron intake is essential for maintaining health.

24.4.2 Ruthenium

Ruthenium has no known biological role, and its toxicity is relatively low compared to other heavy metals. However, some ruthenium compounds, particularly those in high oxidation states, can be toxic if ingested or inhaled in large amounts. Due to its limited use in everyday applications, exposure to ruthenium is uncommon.

24.4.3 Osmium

Osmium, like ruthenium, has no known biological function, and its compounds can be toxic. Osmium tetroxide is particularly hazardous, as it is highly volatile and can cause severe damage to tissues upon contact. Prolonged exposure to osmium tetroxide can lead to respiratory issues, skin burns, and even blindness. Because of these dangers, osmium compounds are handled with extreme caution in laboratory environments.

24.4.4 Hassium

Hassium, being a synthetic element, has no biological role and is not encountered outside of research laboratories. Due to its extreme radioactivity, it would be highly toxic if exposure were possible, but its short half-life and the small quantities produced make this risk negligible.

24.5 CONCLUSION

Group 8 elements—iron, ruthenium, osmium, and hassium—each contribute in unique ways to industry, technology, and scientific research. Iron, the most abundant and widely used, is a cornerstone of modern civilization, while ruthenium and osmium find specialized applications in catalysis, electronics, and material science. Hassium, as a synthetic element, provides valuable insights into the properties of superheavy elements.

CHAPTER 25

GROUP 9—THE COBALT GROUP

Group 9 of the periodic table, commonly referred to as the cobalt group, consists of the elements cobalt (Co), rhodium (Rh), iridium (Ir), and meitnerium (Mt). These transition metals share several similarities in terms of their physical and chemical properties, such as high melting points, variable oxidation states, and the ability to form complex compounds. Group 9 metals play crucial roles in industry, technology, and scientific research, with cobalt being the most well-known and widely used. Rhodium and iridium, while less common, are vital in various high-tech applications. Meitnerium, a synthetic element, remains primarily a subject of research.

25.1 PHYSICAL AND CHEMICAL PROPERTIES

The metals in Group 9 share the characteristic properties of transition metals, including high density, high melting and boiling points, and good electrical and thermal conductivity. Their electron configurations lead to complex bonding behaviors and the ability to form compounds with a wide range of oxidation states.

25.1.1 Cobalt (Co)

Cobalt, with an atomic number of 27, has an electron configuration of [Ar] $3d^7$ $4s^2$. It is a hard, lustrous, silvery-gray metal. Cobalt can exist in various oxidation states, with +2 and +3 being the most common. Its chemical reactivity allows it to form stable complexes, particularly with oxygen, sulfur, and halogens. Cobalt is magnetic and shares this property with iron and nickel, which are located adjacent to it on the periodic table.

25.1.2 Rhodium (Rh)

Rhodium, atomic number 45, is a rare, silver-white metal that belongs to the platinum group of elements. It has an electron configuration of [Kr] $4d^8 5s^1$ and is characterized by its remarkable resistance to corrosion, even at high temperatures. Rhodium typically exhibits oxidation states ranging from 0 to +4, with +3 being the most common. It is a noble metal, meaning it is highly unreactive under normal conditions, making it ideal for certain industrial applications.

25.1.3 Iridium (Ir)

Iridium, with an atomic number of 77, is one of the densest elements in the periodic table. Its electron configuration is [Xe] $4f^{14}5d^7\ 6s^2$. Iridium shares many properties with rhodium, including its resistance to corrosion and its status as a noble metal. Iridium exhibits oxidation states from 0 to +6, although the +3 and +4 states are the most common in its compounds. Iridium is known for its extraordinary hardness and high melting point.

25.1.4 Meitnerium (Mt)

Meitnerium, atomic number 109, is a synthetic element that was first produced in 1982 by bombarding bismuth with iron ions. Its electron configuration is predicted to be [Rn] $5f^{14}\ 6d^7\ 7s^2$, placing it below iridium in the periodic table. Meitnerium is highly radioactive and exists only in minute, short-lived quantities. Its properties remain largely theoretical due to its rarity and the difficulty of producing it in significant amounts.

25.2 OCCURRENCE AND ABUNDANCE

25.2.1 Cobalt

Cobalt is a moderately abundant element in the Earth's crust, typically found in association with other metals such as nickel, copper, and iron. It is rarely found in its pure elemental form and is commonly extracted from minerals such as cobaltite (CoAsS), erythrite ($Co_3(AsO_4)_2 \cdot 8H_2O$), and linnaeite ($Co_3S_4$). Significant cobalt reserves are located in the Democratic Republic of the Congo, which accounts for over half of global production. Cobalt is also found in meteorites and plays a role in the composition of the Earth's core.

25.2.2 Rhodium

Rhodium is one of the rarest elements in the Earth's crust, with an average abundance of about 1 part per billion. It is typically found in platinum or nickel ores and is primarily sourced from South Africa, which has large deposits of platinum group metals. Rhodium is also recovered as a byproduct of platinum and nickel mining and refining processes.

25.2.3 Iridium

Iridium is another extremely rare element, with an abundance of around 0.4 parts per billion in the Earth's crust. It is typically found in alluvial deposits along with platinum and other platinum group metals. Like rhodium, iridium is mostly sourced from South Africa and Russia.

Interestingly, iridium has been found in higher concentrations in the boundary layer between the Cretaceous and Paleogene geological periods, suggesting that an asteroid impact contributed to the mass extinction that occurred 66 million years ago.

25.2.4 Meitnerium

Meitnerium does not occur naturally and can only be synthesized in laboratories through nuclear reactions. It is produced in extremely small quantities and exists for only fractions of a second before decaying into lighter elements. As a result, it has no significant presence in the Earth's crust or in any natural environment.

25.3 INDUSTRIAL USES AND APPLICATIONS

25.3.1 Cobalt

Cobalt is a highly versatile metal with numerous industrial applications. One of its primary uses is in the production of superalloys, which are materials that can withstand extreme temperatures and pressures. These superalloys are critical in the aerospace industry, particularly in turbine blades for jet engines. Cobalt is also used in rechargeable batteries, especially lithium-ion batteries, which are essential for portable electronics, electric vehicles, and renewable energy storage.

In addition, cobalt is an important component in permanent magnets, which are used in electric motors, generators, and various other devices. Cobalt compounds, such as cobalt blue ($CoAl_2O_4$), are also used as pigments in ceramics, glass, and paints. Moreover, cobalt-based catalysts are employed in the petroleum and chemical industries to facilitate hydrogenation reactions.

25.3.2 Rhodium

Rhodium's primary industrial application is in catalytic converters for automobiles. As a catalyst, rhodium helps reduce harmful emissions by converting nitrogen oxides (NOx) into nitrogen and oxygen. Due to its high resistance to corrosion and excellent catalytic properties, rhodium is an essential component in this process, especially in reducing smog-causing emissions.

Rhodium is also used in the jewelry industry, where it is applied as a plating material for white gold and platinum to enhance their appearance and prevent tarnishing. In addition, rhodium is used in the production of thermocouples, which are devices that measure temperature in extreme environments.

25.3.3 Iridium

Iridium is highly valued for its resistance to corrosion and its ability to withstand high temperatures, making it ideal for use in high-performance industrial equipment.

One of its most important applications is in the production of spark plugs for aviation engines, where reliability in extreme conditions is critical. Iridium is also used in the manufacture of crucibles for growing high-purity crystals, such as those used in electronics and optics.

Another notable use of iridium is in the field of space exploration. Due to its durability and resistance to radiation, iridium is used in spacecraft, including in the construction of re-entry shields and as part of deep-space probes. In addition, iridium is used in the medical field for radiation therapy in cancer treatment and in the production of high-precision surgical instruments.

25.3.4 Meitnerium

Meitnerium has no known industrial applications due to its short half-life and synthetic nature. Its production is limited to research laboratories, where it is used to study the properties of superheavy elements and to expand our understanding of nuclear physics. Research on meitnerium and other synthetic elements helps scientists explore the limits of the periodic table and investigate the behavior of atomic nuclei under extreme conditions.

25.4 BIOLOGICAL ROLE AND TOXICITY

25.4.1 Cobalt

Cobalt is an essential trace element for humans and animals, as it is a key component of vitamin B12 (cobalamin), which is necessary for red blood cell production and neurological function. In plants, cobalt plays a role in nitrogen fixation, which is critical for soil fertility and agriculture.

However, excessive exposure to cobalt can be harmful. In industrial settings, inhaling cobalt dust or fumes can lead to respiratory problems and other health issues, including lung disease. Cobalt toxicity can also result from chronic exposure, particularly in industries that involve cobalt mining, refining, or the production of cobalt-containing products.

25.4.2 Rhodium

Rhodium has no known biological role, and its toxicity is relatively low compared to other heavy metals. However, rhodium compounds should still be handled with care, as exposure to rhodium dust or fumes can cause irritation to the skin, eyes, and respiratory system.

25.4.3 Iridium

Iridium is not essential for biological systems, and it is considered relatively non-toxic in its elemental form. Iridium compounds, however, can be hazardous if ingested or inhaled in large quantities. Due to the limited exposure to iridium in everyday life, cases of iridium toxicity are rare.

25.4.4 Meitnerium

Meitnerium has no biological role and is highly radioactive. Given that it is produced in such small quantities and decays almost immediately, there is no known risk of exposure to meitnerium outside of specialized research environments.

25.5 CONCLUSION

The Group 9 elements—cobalt, rhodium, iridium, and meitnerium—are integral to various industries and scientific fields. Cobalt is essential for modern technology, with applications in batteries, superalloys, and magnets, while rhodium and iridium find specialized uses in catalytic converters, high-performance equipment, and space exploration.

Although meitnerium has no practical applications, its study provides valuable insights into the nature of superheavy elements. The cobalt group demonstrates the diversity and importance of transition metals in both everyday applications and cutting-edge research.

CHAPTER 26
GROUP 10—THE NICKEL GROUP

Group 10 of the periodic table consists of the elements nickel (Ni), palladium (Pd), platinum (Pt), and darmstadtium (Ds). These elements are transition metals, well-known for their corrosion resistance, high melting points, and catalytic properties. Nickel, palladium, and platinum are critical to various industries due to their versatile applications, while darmstadtium, being a synthetic and highly radioactive element, has no practical uses beyond research.

26.1 PHYSICAL AND CHEMICAL PROPERTIES

Group 10 elements share the fundamental characteristics of transition metals, such as the ability to form complex compounds, exhibit variable oxidation states, and have partially filled d-orbitals, which contribute to their reactivity and chemical bonding behavior.

26.1.1 Nickel (Ni)

Nickel, with an atomic number of 28, has an electron configuration of $[Ar]3d^8 4s^2$. It is a hard, ductile, and silvery-white metal with high corrosion resistance and excellent mechanical strength. Nickel's most common oxidation state is +2, although it can also form compounds in the +3 and +4 states. It is ferromagnetic, meaning it can be magnetized, which makes it valuable for certain technological applications.

26.1.2 Palladium (Pd)

Palladium, atomic number 46, is a rare and lustrous metal that has an electron configuration of $[Kr]\ 4d^{10}$. Unlike nickel, palladium is not magnetic, but it shares nickel's resistance to corrosion. Palladium's most common oxidation states are +2 and +4. It is highly malleable and ductile, and one of its key chemical properties is its ability to absorb large amounts of hydrogen, making it useful in hydrogen storage and purification.

26.1.3 Platinum (Pt)

Platinum, atomic number 78, has an electron configuration of $[Xe]\ 4f^{14}\ 5d^9\ 6s^1$. It is a dense, malleable, and highly unreactive metal. Known for its remarkable resistance to corrosion, platinum's most common oxidation states are +2 and +4, although it can exhibit other states as well. Platinum's inert nature makes it invaluable in chemical reactions where reactivity needs to be minimized. This metal is also highly conductive, making it useful in electronics and industrial processes.

26.1.4 Darmstadtium (Ds)

Darmstadtium, with an atomic number of 110, is a synthetic element with an electron configuration predicted to be [Rn] $5f^{14}$ $6d^8$ $7s^2$. It was first synthesized in 1994 by bombarding lead with nickel ions. Being highly radioactive, darmstadtium has no stable isotopes, and its half-life is extremely short, making it impossible to study its chemical and physical properties in detail. Its behavior is expected to be similar to that of platinum, given its position in the periodic table, but its properties remain largely speculative.

26.2 OCCURRENCE AND ABUNDANCE

26.2.1 Nickel

Nickel is the fifth most abundant element on Earth, constituting about 0.008% of the Earth's crust. It is commonly found in minerals such as pentlandite ($(Ni, Fe)_9S_8$), garnierite (a mixture of nickel and magnesium silicates), and laterites, which are weathered ores found in tropical regions. Major sources of nickel are Canada, Russia, Australia, and Indonesia. Nickel is also a significant component of meteorites, which contain iron-nickel alloys.

26.2.2 Palladium

Palladium is much rarer than nickel, with an abundance of around 0.015 parts per million in the Earth's crust. It is primarily extracted from ores containing platinum group metals, such as cooperite (PdS), sperrylite ($PtAs_2$), and nickel-copper deposits. Major producers of palladium include Russia, South Africa, Canada, and the United States, often as a byproduct of mining platinum, nickel, or copper ores.

26.2.3 Platinum

Platinum is another rare metal, with an abundance of about 0.005 parts per million in the Earth's crust. It is typically found in alluvial deposits and is mined alongside palladium and other platinum group metals. Platinum occurs naturally in the mineral sperrylite ($PtAs_2$) and as native platinum alloyed with other metals. South Africa accounts for nearly 70% of the world's platinum production, with Russia and Canada also being significant producers.

26.2.4 Darmstadtium

Darmstadtium does not occur naturally and can only be synthesized in particle accelerators. It is produced by bombarding heavy elements with ions, and only a few atoms of darmstadtium have ever been observed. Given its short half-life, it has no natural abundance or commercial mining prospects.

26.3 INDUSTRIAL USES AND APPLICATIONS

26.3.1 Nickel

Nickel's primary use is in the production of stainless steel, an alloy of iron, chromium, and nickel, which is prized for its corrosion resistance and strength. Stainless steel is used in construction, transportation, kitchenware, and medical devices.

Nickel is also crucial in the production of superalloys, which can withstand extreme temperatures and pressures, making them ideal for jet engines, gas turbines, and other high-stress environments.

In addition to its role in alloys, nickel is used in rechargeable batteries, particularly in nickel-metal hydride (NiMH) and nickel-cadmium (NiCd) batteries, which power many portable electronic devices. Nickel is also a key component in electroplating, where it is used to coat other metals, improving their durability and resistance to corrosion.

26.3.2 Palladium

Palladium is widely used in the automotive industry as a catalyst in catalytic converters, which reduce harmful emissions from internal combustion engines by converting toxic gases into less harmful substances. Its ability to absorb hydrogen also makes palladium essential in hydrogen fuel cells and purification processes.

In electronics, palladium is used in multilayer ceramic capacitors (MLCCs), which are integral components in smartphones, laptops, and other digital devices. Palladium's resistance to corrosion and tarnishing also makes it valuable in jewelry, where it is used as an alloy in white gold to enhance its durability and luster.

26.3.3 Platinum

Platinum is highly valued for its catalytic properties, particularly in the automotive industry, where it is used in catalytic converters alongside palladium. Platinum is also employed in the chemical industry as a catalyst in processes such as the production of fertilizers, explosives, and synthetic fibers.

In the medical field, platinum compounds such as cisplatin are used in chemotherapy to treat various types of cancer. Platinum's inertness and biocompatibility also make it suitable for medical implants, such as pacemakers and stents. Additionally, platinum is used in the jewelry industry, where it is prized for its rarity, beauty, and resistance to tarnishing.

Platinum is also used in the petroleum industry for refining crude oil and in fuel cells, which generate electricity through the chemical reaction between hydrogen and oxygen, producing water as the only byproduct.

26.3.4 Darmstadtium

Due to its extremely short half-life and the difficulty of producing it in meaningful quantities, darmstadtium has no known industrial applications. Its production is limited to scientific research, where it is used to study the properties of superheavy elements and to expand our understanding of atomic structure and nuclear reactions.

26.4 BIOLOGICAL ROLE AND TOXICITY

26.4.1 Nickel

Nickel plays a minor biological role in some organisms, including bacteria and plants, where it is involved in enzymatic processes. However, it is not considered essential for human health. In fact, exposure to nickel, particularly in large amounts, can lead to allergic reactions, respiratory issues, and even cancer in cases of long-term exposure to nickel compounds. Nickel is also a known skin allergen, and prolonged contact with nickel-containing objects, such as jewelry, can cause nickel dermatitis, a condition characterized by itching and skin irritation.

26.4.2 Palladium

Palladium has no known biological role and is considered relatively non-toxic in small amounts. However, prolonged exposure to palladium dust or compounds can cause allergic reactions, particularly skin rashes and respiratory issues. As palladium is often used in dental materials and jewelry, it is important to monitor potential allergic responses, although such occurrences are rare.

26.4.3 Platinum

Platinum is not biologically active, and its compounds are generally considered to be non-toxic. However, some platinum compounds, particularly those used in chemotherapy, are highly reactive and can cause side effects such as nausea, vomiting, and kidney damage. These side effects are carefully managed in medical treatments to ensure that the benefits of platinum-based therapies outweigh the risks.

26.4.4 Darmstadtium

Darmstadtium has no biological role, and its radioactivity makes it extremely hazardous. However, due to its fleeting existence and the fact that it is only produced in trace amounts in laboratory settings, there is no risk of exposure to darmstadtium outside of research facilities.

26.5 ENVIRONMENTAL IMPACT

The mining and processing of Group 10 elements, particularly nickel, palladium, and platinum, have environmental impacts. Nickel mining, for instance, can result in deforestation, soil degradation, and the release of harmful byproducts such as sulfur dioxide, which contributes to acid rain. Efforts are being made to reduce the environmental footprint of nickel mining through more sustainable practices, such as recycling nickel-containing products.

Palladium and platinum are both recovered from recycling automotive catalytic converters, reducing the need for additional mining and lowering their environmental impact. The recovery of these metals from electronic waste is also becoming increasingly important as the demand for palladium and platinum in digital devices grows.

26.6 CONCLUSION

Group 10 elements—nickel, palladium, platinum, and darmstadtium—are integral to modern industry and technology. Nickel's versatility makes it essential in stainless steel production, batteries, and superalloys, while palladium and platinum are key players in catalysis, electronics, and jewelry. Darmstadtium, although largely a scientific curiosity, contributes to the ongoing exploration of superheavy elements. As society continues to innovate, the role of these elements will likely expand, making them even more vital to the global economy and technological advancement.

CHAPTER 27

GROUP 11—THE COPPER GROUP: COINAGE METALS

Group 11 of the periodic table, often referred to as the coinage metals, consists of copper (Cu), silver (Ag), and gold (Au), with roentgenium (Rg) added as a synthetic element. These elements share similar physical and chemical properties, including high thermal and electrical conductivity, malleability, and a tendency to resist corrosion. Historically, copper, silver, and gold have been used for coinage, hence their group name. Despite their distinct applications today, their significance in economic, industrial, and cultural contexts remains profound.

27.1 PHYSICAL AND CHEMICAL PROPERTIES

27.1.1 Copper (Cu)

Copper, with an atomic number of 29, has the electron configuration [Ar] $3d^{10}$ $4s^1$. It is a soft, malleable metal with excellent electrical and thermal conductivity, second only to silver among pure metals. Copper's characteristic reddish-brown color distinguishes it from most other metals, and its high corrosion resistance makes it valuable in various industries. Copper typically exists in the +1 and +2 oxidation states, although its most common state in compounds is +2. It readily forms alloys, such as bronze (with tin) and brass (with zinc), which have enhanced mechanical properties.

27.1.2 Silver (Ag)

Silver, atomic number 47, has the electron configuration [Kr] $4d^{10}$ $5s^1$. It is a soft, white, and lustrous metal known for its unparalleled electrical and thermal conductivity, as well as its reflectivity. Like copper, silver is highly malleable and ductile, making it easy to shape and use in various applications. Silver exhibits oxidation states of +1 and, less commonly, +2. It forms compounds with halogens, sulfur, and oxygen, and is widely used in alloys for improved strength.

27.1.3 Gold (Au)

Gold, with an atomic number of 79, has the electron configuration [Xe] $4f^{14}$ $5d^{10}$ $6s^1$. Known for its distinctive yellow luster, gold is the most malleable and ductile of all metals. It resists corrosion and tarnishing, making it ideal for long-term use in artifacts and jewelry. Gold is chemically less reactive than copper and silver, existing predominantly in the +1 and +3 oxidation states. It readily forms alloys with other metals, improving its hardness for practical applications in electronics, jewelry, and dentistry.

27.1.4 Roentgenium (Rg)

Roentgenium, atomic number 111, is a synthetic, highly radioactive element with an electron configuration predicted to be [Rn] $5f^{14}$ $6d^9$ $7s^2$. As a superheavy element, its physical and chemical properties remain largely speculative due to its short half-life, which limits experimentation. Scientists predict that it would behave similarly to gold based on its position in the periodic table, but it has no known uses outside of research.

27.2 OCCURRENCE AND ABUNDANCE

27.2.1 Copper

Copper is one of the most abundant metals in the Earth's crust, constituting approximately 0.0068% by weight. It occurs in a variety of mineral forms, most commonly in sulfide ores such as chalcopyrite ($CuFeS_2$), bornite (Cu_5FeS_4), and covellite (CuS). Significant copper deposits are found in Chile, Peru, China, and the United States. Native copper (metallic copper) also occurs in nature but is relatively rare.

The high recyclability of copper means that much of the copper in use today is derived from recycled materials, particularly from electrical wiring and plumbing systems.

27.2.2 Silver

Silver is much rarer than copper, with an average concentration of 0.08 parts per million in the Earth's crust. It is often found in association with other metals, such as lead, zinc, and copper, in ores like argentite (Ag_2S) and galena (PbS). The largest producers of silver include Mexico, Peru, China, and Australia. Silver is also recovered as a byproduct of mining for other metals, particularly in polymetallic ore deposits. Additionally, silver is highly recyclable, especially from old jewelry, coins, and industrial waste.

27.2.3 Gold

Gold is one of the rarest elements in the Earth's crust, with an average concentration of 0.004 parts per million. It is typically found in veins of quartz or other minerals, often associated with sulfur, in hydrothermal deposits. Gold is extracted from ores such as native gold, tellurides, and gold-bearing pyrites. The top producers of gold include China, Australia, Russia, and the United States. Gold has been mined for thousands of years and has retained its value and significance throughout human history. Like silver and copper, gold is highly recyclable, particularly from jewelry, electronics, and old coins.

27.2.4 Roentgenium

Roentgenium does not occur naturally and can only be synthesized in laboratories by bombarding heavy elements like bismuth with nickel ions. As a result, it exists only in trace amounts and has no natural abundance. Roentgenium is produced in particle accelerators, and its production is limited to scientific research on superheavy elements and the investigation of nuclear structure.

27.3 INDUSTRIAL USES AND APPLICATIONS

27.3.1 Copper

Copper is integral to modern industry, with its primary use being in electrical wiring due to its excellent conductivity. Copper is also used extensively in plumbing, roofing, and building construction. Its high thermal conductivity makes it ideal for use in heat exchangers, radiators, and refrigeration systems.

Beyond electrical applications, copper is a key component of many alloys. Brass, an alloy of copper and zinc, is used in decorative items and musical instruments, while bronze, a copper-tin alloy, is used for statues, medals, and bearings. Copper's antimicrobial properties also make it useful in medical settings for surfaces that prevent the spread of infections.

27.3.2 Silver

Silver's unmatched electrical conductivity makes it essential in the electronics industry, where it is used in switches, contacts, and conductors. Its reflective properties are employed in mirrors, solar panels, and coatings for energy-efficient windows. Silver is also used in photography due to its light-sensitive properties, although this application has declined with the rise of digital photography.

Silver's antimicrobial and non-toxic nature has long made it valuable in medicine. Silver compounds are used in wound dressings, medical devices, and water purification systems. In addition, silver remains a popular material for jewelry, tableware, and decorative items due to its aesthetic appeal and tarnish resistance.

27.3.3 Gold

Gold's primary use is in the jewelry industry, where its malleability, ductility, and luster make it highly desirable. Gold is also used as a standard for wealth and currency, with many countries maintaining reserves in gold bullion. In technology, gold is essential for its electrical conductivity and resistance to corrosion, making it invaluable in electronics, particularly in connectors, contacts, and microchips.

Gold is also used in dentistry for fillings, crowns, and other dental work due to its biocompatibility. Additionally, gold nanoparticles are used in medical research for drug delivery, diagnostics, and imaging.

27.3.4 Roentgenium

Due to its short half-life and synthetic nature, roentgenium has no known practical applications. It is used exclusively for scientific research, primarily in studying the properties of superheavy elements and the boundaries of the periodic table.

27.4 HISTORICAL AND CULTURAL SIGNIFICANCE

27.4.1 Copper

Copper was one of the first metals used by humans, with evidence of its use dating back over 10,000 years. It played a crucial role in the development of early human civilizations, particularly during the Bronze Age, when it was alloyed with tin to create bronze, a much harder material than copper alone. Copper has been used in coinage for millennia, with many early currencies made from copper or copper alloys. Today, copper remains culturally significant, often symbolizing wealth, prosperity, and durability.

27.4.2 Silver

Silver has been used as a form of currency and in trade for thousands of years. Ancient civilizations, including the Greeks and Romans, minted silver coins, and the metal continued to play a critical role in global economies well into the modern era. Silver has also held religious and cultural significance, often being associated with purity and protection. Its use in sacred objects, ceremonial artifacts, and religious offerings underscores its historical importance.

27.4.3 Gold

Gold has long been associated with wealth, power, and divinity. Ancient civilizations, including the Egyptians, Greeks, and Romans, valued gold for its rarity and beauty, using it to create ornate jewelry, religious artifacts, and currency. Gold has also been a symbol of purity and immortality in various cultures. The discovery of gold has spurred economic booms, most notably during the Gold Rushes of the 19th century, which helped shape the development of countries like the United States and Australia.

27.5 BIOLOGICAL ROLE AND TOXICITY

27.5.1 Copper

Copper is an essential trace element in human nutrition, playing a key role in the production of red blood cells, the maintenance of healthy nerves, and the immune system. However, excessive exposure to copper, particularly through drinking water or industrial processes, can lead to copper toxicity, which can cause liver and kidney damage.

27.5.2 Silver

Silver has no known biological role, but its antimicrobial properties make it useful in medical treatments. Silver is generally considered safe, although excessive exposure can cause argyria, a condition where the skin takes on a bluish-gray tint due to the accumulation of silver particles.

27.5.3 Gold

Gold is biologically inert, meaning it does not react with body tissues, making it safe for use in medical implants and dentistry. Gold compounds are used in some treatments for rheumatoid arthritis, although such treatments are becoming less common due to the development of more effective drugs.

27.5.4 Roentgenium

Roentgenium has no biological role, and its radioactivity makes it highly toxic, though its rarity prevents any real risk of exposure.

27.6 CONCLUSION

The Group 11 elements—copper, silver, and gold—are essential to both historical and modern industries. Their unique properties, including electrical and thermal conductivity, corrosion resistance, and malleability, have made them invaluable in a wide range of applications, from electrical wiring to jewelry and medicine. Roentgenium, while not practical due to its short half-life, contributes to the study of atomic structure. As society continues to evolve, the importance of copper, silver, and gold will remain significant, driven by their versatility, historical value, and enduring economic relevance.

CHAPTER 28
GROUP 12—THE ZINC GROUP

Group 12 of the periodic table, also known as the zinc group, consists of four elements: zinc (Zn), cadmium (Cd), mercury (Hg), and copernicium (Cn). These elements are categorized as post-transition metals and share several common physical and chemical properties. They are characterized by their relatively low melting and boiling points, as well as their tendency to form oxides and sulfides. While zinc, cadmium, and mercury have various practical applications in industries such as metallurgy, electronics, and medicine, copernicium remains a synthetic element with limited research opportunities.

28.1 PHYSICAL AND CHEMICAL PROPERTIES

28.1.1 Zinc (Zn)

Zinc, with an atomic number of 30, has an electron configuration of [Ar] $3d^{10}$ $4s^2$. It is a bluish-white metal known for its moderate strength, malleability, and ability to resist corrosion. Zinc has a relatively low melting point of 419.5°C (787.1°F) and a boiling point of 907°C (1,665°F). Chemically, zinc primarily exhibits a +2 oxidation state, forming compounds like zinc oxide (ZnO) and zinc sulfide (ZnS). It reacts readily with acids, releasing hydrogen gas, and can also react with bases and other nonmetals.

28.1.2 Cadmium (Cd)

Cadmium, with an atomic number of 48, has the electron configuration [Kr] $4d^{10}$ $5s^2$. It is a soft, bluish-white metal that is malleable and ductile. Cadmium has a melting point of 321.1°C (609.98°F) and a boiling point of 767°C (1,413°F). It commonly exhibits a +2 oxidation state in its compounds. Cadmium is known for its corrosion resistance and ability to form stable complexes with various anions. Its chemical reactivity is lower than that of zinc, and it does not react with air or water at room temperature.

28.1.3 Mercury (Hg)

Mercury, atomic number 80, has an electron configuration of [Xe] $4f^{10}$ $5d^{10}$ $6s^2$. It is a unique metal that is liquid at room temperature, characterized by its silvery appearance and high density. Mercury has a melting point of -38.83°C (-37.89°F) and a boiling point of 356.73°C (674.11°F). Like zinc and cadmium, mercury typically exhibits a +2 oxidation state, forming compounds such as mercury(II) chloride ($HgCl_2$) and mercury(II) sulfide (HgS). Mercury is known for its toxicity and ability to vaporize at room temperature, which poses significant health and environmental risks.

28.1.4 Copernicium (Cn)

Copernicium, with an atomic number of 112, is a synthetic element that has only been produced in trace amounts in laboratory conditions. Its electron configuration is predicted to be $[Rn]5f^{14}6d^{10}7s^2$. Due to its instability and short half-life, copernicium does not have practical applications and is primarily of interest for research into superheavy elements and nuclear physics. The predicted properties of copernicium suggest it may exhibit some behavior similar to that of mercury, but extensive research is limited due to its rapid decay.

28.2 OCCURRENCE AND ABUNDANCE

28.2.1 Zinc

Zinc is the fourth most widely used metal and is abundant in the Earth's crust, constituting approximately 0.0075% by weight. It occurs naturally in minerals such as sphalerite (ZnS), smithsonite ($ZnCO_3$), and hemimorphite ($Zn_4Si_2O_7(OH)_2{\cdot}H_2O$). Major producers of zinc include China, Australia, and Peru, which account for a significant portion of global zinc production. The metal is primarily obtained through mining and is often recovered from ores in the form of zinc concentrate before being smelted and refined.

28.2.2 Cadmium

Cadmium is less abundant than zinc, with an average concentration of approximately 0.0006% in the Earth's crust. It is usually found in association with zinc ores, primarily as a byproduct of zinc extraction.

Cadmium is most commonly obtained from ores such as sphalerite, where it occurs in trace amounts. Significant sources of cadmium include Canada, China, and Russia. While cadmium is recycled, its primary production still relies heavily on mining operations.

28.2.3 Mercury

Mercury is a relatively rare element in the Earth's crust, with an average concentration of about 0.00007%. It is primarily found in the mineral cinnabar (HgS), where it occurs in significant concentrations. Major producers of mercury include China, Spain, and Mexico. Mercury extraction often involves mining cinnabar and heating it to release the elemental mercury. Due to its toxicity, the use of mercury has declined in recent years, leading to reduced production levels.

28.2.4 Copernicium

Copernicium is a synthetic element that does not occur naturally and is produced only in trace amounts through nuclear reactions in laboratories. The first synthesis of copernicium was achieved by bombarding lead with zinc ions, leading to its classification as a member of Group 12.

Due to its short half-life, copernicium has no natural abundance and is solely used for research purposes.

28.3 INDUSTRIAL USES AND APPLICATIONS

28.3.1 Zinc

Zinc is primarily used for galvanization, a process that involves coating iron or steel to prevent corrosion. This application is crucial in construction, automotive manufacturing, and the production of various metal products. Additionally, zinc is an essential alloying agent in producing brass and bronze, which are used in plumbing fittings, musical instruments, and decorative items.

Zinc compounds also have significant industrial applications. Zinc oxide (ZnO) is widely used in rubber manufacturing, paints, and cosmetics, while zinc sulfide (ZnS) is used in phosphorescent materials and as a pigment. Zinc is also an essential nutrient in agriculture, added to fertilizers to promote plant growth.

28.3.2 Cadmium

Cadmium is used primarily in the production of nickel-cadmium (NiCd) batteries, which are employed in portable electronics, power tools, and renewable energy systems. Despite the growing popularity of lithium-ion batteries, NiCd batteries remain relevant due to their ability to withstand deep discharges and extreme temperatures.

Cadmium is also used in various industrial applications, such as the production of pigments, coatings, and stabilizers in plastics. Cadmium telluride ($CdTe$) is a key material used in thin-film solar cells, contributing to the development of renewable energy technologies.

28.3.3 Mercury

Mercury has historically been used in various applications, including thermometers, barometers, and electrical switches. Its unique properties, such as high density and ability to conduct electricity, made it suitable for these uses. However, due to the health risks associated with mercury exposure, many of these applications have been phased out or replaced with safer alternatives.

Mercury is still used in some industrial processes, such as the production of chlorine and caustic soda through the mercury cell process. Additionally, mercury compounds are utilized in certain medical applications, such as dental amalgams, although their use is declining due to health concerns.

28.3.4 Copernicium

Copernicium does not have any practical applications due to its synthetic nature and instability. Its primary relevance lies in scientific research, particularly in the study of nuclear reactions and the properties of superheavy elements. As research on copernicium progresses, scientists may uncover new information about its potential behavior and interactions.

28.4 BIOLOGICAL ROLE AND TOXICITY

28.4.1 Zinc

Zinc is an essential trace element for human health, playing a vital role in numerous biological processes. It is involved in enzyme function, immune response, protein synthesis, and DNA synthesis. Zinc deficiency can lead to various health issues, including weakened immune function, impaired wound healing, and growth retardation in children. However, excessive zinc intake can result in toxicity, leading to nausea, vomiting, and impaired copper absorption.

28.4.2 Cadmium

Cadmium is not an essential element for human health, and exposure to cadmium can lead to serious health effects. Cadmium toxicity can occur through inhalation or ingestion, leading to kidney damage, respiratory issues, and bone disease. Due to its potential health risks, the use of cadmium in many products has been restricted or banned in several countries. Occupational exposure, particularly in industries dealing with batteries or cadmium pigments, poses significant health risks.

28.4.3 Mercury

Mercury is highly toxic and poses serious health risks to humans and the environment. Exposure to mercury can occur through inhalation, ingestion, or skin contact. Mercury can accumulate in the body, leading to neurological and developmental damage, particularly in fetuses and young children. Chronic exposure to mercury can result in symptoms such as tremors, memory problems, and kidney damage. Due to its toxicity, the use of mercury in many applications has been significantly reduced, and regulations are in place to limit exposure.

28.4.4 Copernicium

Copernicium has no known biological role, and its radioactivity poses potential health hazards. However, due to its rarity and synthetic nature, it is unlikely to pose any direct risk to human health or the environment outside of laboratory conditions.

28.5 ENVIRONMENTAL IMPACT

The extraction and use of Group 12 elements can have significant environmental impacts. Zinc mining can lead to habitat destruction, soil degradation, and water pollution due to the release of heavy metals and other contaminants. However, zinc is highly recyclable, and efforts are underway to improve the sustainability of zinc production through better waste management and recycling practices.

Cadmium poses environmental risks, particularly in agriculture, where its accumulation in soil can lead to contamination of crops and water supplies. The use of cadmium in batteries and pigments has raised concerns about its disposal and potential leaching into the environment. Regulatory measures have been implemented in many countries to limit cadmium emissions and promote safer alternatives.

Mercury contamination is a major environmental issue, particularly in aquatic ecosystems. Mercury can bioaccumulate in fish and other organisms, posing health risks to wildlife and humans who consume contaminated seafood. Efforts to reduce mercury emissions from industrial processes and phase out mercury-containing products are essential to mitigate its environmental impact.

28.6 CONCLUSION

Group 12 elements—zinc, cadmium, mercury, and copernicium—play significant roles in various industrial applications and human health. Zinc is widely used for its corrosion resistance and as an essential nutrient, while cadmium and mercury are associated with both industrial utility and significant health risks. Copernicium, though currently a subject of scientific inquiry, lacks practical applications due to its synthetic nature and instability. Understanding the properties, occurrence, and implications of these elements is crucial for their sustainable use and management in both industrial and environmental contexts. As society continues to advance, balancing the benefits and risks associated with these elements will be vital for future technological and ecological sustainability.

PART FIVE

THE F-BLOCK ELEMENTS

CHAPTER 29
THE LANTHANIDE SERIES

The lanthanide or lanthanoid series of chemical elements comprises at least the 14 metallic chemical elements with atomic numbers 57–70, from lanthanum through ytterbium. In the periodic table, they fill the 4f orbitals. Lutetium (element 71) is also sometimes considered a lanthanide, despite being a d-block element and a transition metal.

The informal chemical symbol Ln is used in general discussions of lanthanide chemistry to refer to any lanthanide. All but one of the lanthanides are f-block elements, corresponding to the filling of the 4f electron shell. Lutetium is a d-block element (thus also a transition metal), and on this basis its inclusion has been questioned; however, like its congeners scandium and yttrium in group 3, it behaves similarly to the other 14. The term rare-earth element or rare-earth metal is often used to include the stable group 3 elements Sc, Y, and Lu in addition to the 4f elements. All lanthanide elements form trivalent cations, Ln^{3+}, whose chemistry is largely determined by the ionic radius, which decreases steadily from lanthanum (La) to lutetium (Lu).

These elements are called lanthanides because the elements in the series are chemically similar to lanthanum. Because "lanthanide" means "like lanthanum", it has been argued that lanthanum cannot logically be a lanthanide, but the International Union of Pure and Applied Chemistry (IUPAC) acknowledges its inclusion based on common usage.

In presentations of the periodic table, the f-block elements are customarily shown as two additional rows below the main body of the table. This convention is entirely a matter of aesthetics and formatting practicality; a rarely used wide-formatted periodic table inserts the 4f and 5f series in their proper places, as parts of the table's sixth and seventh rows (periods), respectively.

29.1 CHEMISTRY AND COMPOUNDS

29.1.1 Effect of 4f Orbitals

Viewing the lanthanides from left to right in the periodic table, the seven 4f atomic orbitals become progressively more filled. The electronic configuration of most neutral gas-phase lanthanide atoms is $[Xe]6s^24f^n$, where n is 56 less than the atomic number Z. Exceptions are La, Ce, Gd, and Lu, which have $4f^{n-1}5d^1$ (though even then $4f^n$ is a low-lying excited state for La, Ce, and Gd; for Lu, the 4f shell is already full, and the fifteenth electron has no choice but to enter 5d). With the exception of lutetium, the 4f orbitals are chemically active in all lanthanides and produce profound differences between lanthanide chemistry and transition metal chemistry. The 4f orbitals penetrate the [Xe] core and are isolated, and thus they do not participate much in bonding. This explains why crystal field effects are small and why they do not form π bonds. As there are seven 4f orbitals, the number of unpaired electrons can be as high as 7, which gives rise to the large magnetic moments observed for lanthanide compounds.

Measuring the magnetic moment can be used to investigate the 4f electron configuration, and this is a useful tool in providing an insight into the chemical bonding. The lanthanide contraction, i.e., the reduction in size of the Ln^{3+} ion from La^{3+} (103 pm) to Lu^{3+} (86.1 pm), is often explained by the poor shielding of the 5s and 5p electrons by the 4f electrons.

The chemistry of the lanthanides is dominated by the +3 oxidation state, and in Ln^{III} compounds the 6s electrons and (usually) one 4f electron are lost and the ions have the configuration $[Xe]4f^{(n-1)}$. All the lanthanide elements exhibit the oxidation state +3. In addition, Ce^{3+} can lose its single f electron to form Ce^{4+} with the stable electronic configuration of xenon. Also, Eu^{3+} can gain an electron to form Eu^{2+} with the f^7 configuration that has the extra stability of a half-filled shell. Other than Ce(IV) and Eu(II), none of the lanthanides are stable in oxidation states other than +3 in aqueous solution.

In terms of reduction potentials, the $Ln^{0/3+}$ couples are nearly the same for all lanthanides, ranging from -1.99 (for Eu) to -2.35 V (for Pr). Thus these metals are highly reducing, with reducing power similar to alkaline earth metals such as Mg (-2.36 V).

29.1.2 Lanthanide Oxidation States

The ionization energies for the lanthanides can be compared with aluminium. In aluminium the sum of the first three ionization energies is 5139 $kJ{\cdot}mol^{-1}$, whereas the lanthanides fall in the range 3455 – 4186 $kJ{\cdot}mol^{-1}$. This correlates with the highly reactive nature of the lanthanides.

The sum of the first two ionization energies for europium, 1632 $kJ{\cdot}mol^{-1}$ can be compared with that of barium 1468.1 $kJ{\cdot}mol^{-1}$ and europium's third ionization energy is the highest of the lanthanides. The sum of the first two ionization energies for ytterbium are the second lowest in the series and its third ionization energy is the second highest. The high third ionization energy for Eu and Yb correlate with the half filling $4f^7$ and complete filling $4f^{14}$ of the 4f subshell, and the stability afforded by such configurations due to exchange energy. Europium and ytterbium form salt like compounds with Eu^{2+} and Yb^{2+}, for example the salt like dihydrides. Both europium and ytterbium dissolve in liquid ammonia forming solutions of $Ln^{2+}(NH_3)_x$ again demonstrating their similarities to the alkaline earth metals.

The relative ease with which the 4th electron can be removed in cerium and (to a lesser extent praseodymium) indicates why Ce(IV) and Pr(IV) compounds can be formed, for example CeO_2 is formed rather than Ce_2O_3 when cerium reacts with oxygen. Also Tb has a well-known IV state, as removing the 4th electron in this case produces a half-full $4f^7$ configuration.

The additional stable valences for Ce and Eu mean that their abundances in rocks sometimes varies significantly relative to the other rare earth elements.

29.1.3 Separation of Lanthanides

The similarity in ionic radius between adjacent lanthanide elements makes it difficult to separate them from each other in naturally occurring ores and other mixtures. Historically, the very laborious processes of cascading and fractional crystallization were used. Because the lanthanide ions have slightly different radii, the lattice energy of their salts and hydration energies of the ions will be slightly different, leading to a small difference in solubility. Salts of the formula $Ln(NO_3)_3 \cdot 2NH_4NO_3 \cdot 4H_2O$ can be used. Industrially, the elements are separated from each other by solvent extraction. Typically an aqueous solution of nitrates is extracted into kerosene containing tri-*n*-butylphosphate. The strength of the complexes formed increases as the ionic radius decreases, so solubility in the organic phase increases. Complete separation can be achieved continuously by use of countercurrent exchange methods. The elements can also be separated by ion-exchange chromatography, making use of the fact that the stability constant for formation of EDTA complexes increases for log K ≈ 15.5 for $[La(EDTA)]^-$ to log K ≈ 19.8 for $[Lu(EDTA)]^-$.

29.1.4 Coordination Chemistry and Catalysis

When in the form of coordination complexes, lanthanides exist overwhelmingly in their +3 oxidation state, although particularly stable 4f configurations can also give +4 (Ce, Tb) or +2 (Eu, Yb) ions. All of these forms are strongly electropositive and thus lanthanide ions are hard Lewis acids. The oxidation states are also very stable; with the exceptions of SmI_2 and cerium(IV) salts, lanthanides are not used for redox chemistry. 4f electrons have a high probability of being found close to the nucleus and are thus strongly affected as the nuclear charge increases across the series; this results in a corresponding decrease in ionic radii referred to as the lanthanide contraction.

The low probability of the 4f electrons existing at the outer region of the atom or ion permits little effective overlap between the orbitals of a lanthanide ion and any binding ligand. Thus lanthanide complexes typically have little or no covalent character and are not influenced by orbital geometries. The lack of orbital interaction also means that varying the metal typically has little effect on the complex (other than size), especially when compared to transition metals. Complexes are held together by weaker electrostatic forces which are omni-directional and thus the ligands alone dictate the symmetry and coordination of complexes. Steric factors therefore dominate, with coordinative saturation of the metal being balanced against inter-ligand repulsion. This results in a diverse range of coordination geometries, many of which are irregular, and also manifests itself in the highly fluxional nature of the complexes. As there is no energetic reason to be locked into a single geometry, rapid intramolecular and intermolecular ligand exchange will take place. This typically results in complexes that rapidly fluctuate between all possible configurations.

Many of these features make lanthanide complexes effective catalysts. Hard Lewis acids are able to polarise bonds upon coordination and thus alter the electrophilicity of compounds, with a classic example being the Luche reduction. The large size of the ions coupled with their labile ionic bonding allows even bulky coordinating species to bind and dissociate rapidly, resulting in very high turnover rates; thus excellent yields can often be achieved with loadings of only a few mol%. The lack of orbital interactions combined with the lanthanide contraction means that the lanthanides change in size across the series but that their chemistry remains much the same. This allows for easy tuning of the steric environments and examples exist where this has been used to improve the catalytic activity of the complex and change the nuclearity of metal clusters.

Despite this, the use of lanthanide coordination complexes as homogeneous catalysts is largely restricted to the laboratory and there are currently few examples them being used on an industrial scale. Lanthanides exist in many forms other than coordination complexes and many of these are industrially useful. In particular lanthanide metal oxides are used as heterogeneous catalysts in various industrial processes.

Ln(III) Compounds

The trivalent lanthanides mostly form ionic salts. The trivalent ions are hard acceptors and form more stable complexes with oxygen-donor ligands than with nitrogen-donor ligands. The larger ions are 9-coordinate in aqueous solution, $[Ln(H_2O)_9]^{3+}$ but the smaller ions are 8-coordinate, $[Ln(H_2O)_8]^{3+}$. There is some evidence that the later lanthanides have more water molecules in the second coordination sphere. Complexation with monodentate ligands is generally weak because it is difficult to displace water molecules from the first coordination sphere. Stronger complexes are formed with chelating ligands because of the chelate effect, such as the tetra-anion derived from 1,4,7,10-tetraazacyclododecane-1,4,7,10-tetraacetic acid (DOTA).

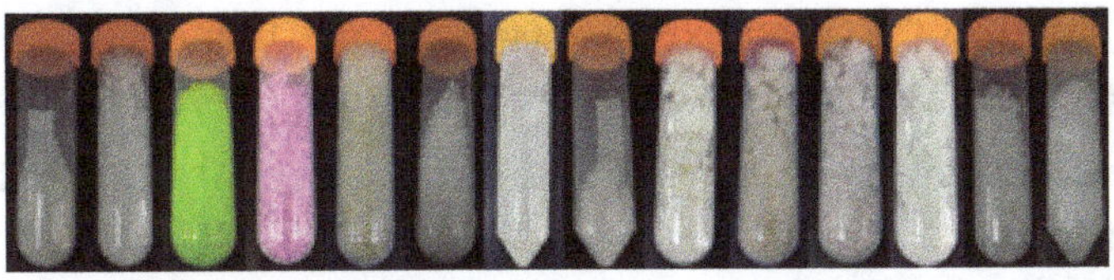

Fig 1: Samples of Lanthanide Nitrates in their Hexahydrate form. From Left to Right: La, Ce, Pr, Nd, Sm, Eu, Gd, Tb, Dy, Ho, Er, Tm, Yb, Lu

Ln(II) and Ln(IV) Compounds

The most common divalent derivatives of the lanthanides are for Eu(II), which achieves a favorable f^7 configuration. Divalent halide derivatives are known for all of the lanthanides. They are either conventional salts or are Ln(III) "electride"-like salts. The simple salts include YbI_2, EuI_2, and SmI_2. The electride-like salts, described as Ln^{3+}, $2I^-$, e^-, include LaI_2, CeI_2 and GdI_2. Many of the iodides form soluble complexes with ethers, e.g. TmI_2(dimethoxyethane)$_3$. Samarium(II) iodide is a useful reducing agent. Ln(II) complexes can be synthesized by transmetalation reactions. The normal range of oxidation states can be expanded via the use of sterically bulky cyclopentadienyl ligands, in this way many lanthanides can be isolated as Ln(II) compounds.

Ce(IV) in ceric ammonium nitrate is a useful oxidizing agent. The Ce(IV) is the exception owing to the tendency to form an unfilled f shell. Otherwise tetravalent lanthanides are rare. However, recently Tb(IV) and Pr(IV) complexes have been shown to exist.

Hydrides

Lanthanide metals react exothermically with hydrogen to form LnH_2, dihydrides. With the exception of Eu and Yb, which resemble the Ba and Ca hydrides (non-conducting, transparent salt-like compounds), they form black, pyrophoric, conducting compounds where the metal sub-lattice is face centred cubic and the H atoms occupy tetrahedral sites. Further hydrogenation produces a trihydride which is non-stoichiometric, non-conducting, more salt like. The formation of trihydride is associated with and increase in 8–10% volume and this is linked to greater localization of charge on the hydrogen atoms which become more anionic (H^- hydride anion) in character.

Halides

The only tetrahalides known are the tetrafluorides of cerium, praseodymium, terbium, neodymium and dysprosium, the last two known only under matrix isolation conditions. All of the lanthanides form trihalides with fluorine, chlorine, bromine and iodine. They are all high melting and predominantly ionic in nature. The fluorides are only slightly soluble in water and are not sensitive to air, and this contrasts with the other halides which are air sensitive, readily soluble in water and react at high temperature to form oxohalides.

The trihalides were important as pure metal can be prepared from them. In the gas phase the trihalides are planar or approximately planar, the lighter lanthanides have a lower % of dimers, the heavier lanthanides a higher proportion. The dimers have a similar structure to Al_2Cl_6.

Some of the dihalides are conducting while the rest are insulators. The conducting forms can be considered as Ln^{III} electride compounds where the electron is delocalised into a conduction band, $Ln^{3+} (X^-)_2(e^-)$. All of the diiodides have relatively short metal-metal separations. The $CuTi_2$ structure of the lanthanum, cerium and praseodymium diiodides along with HP-NdI_2 contain 4^4 nets of metal and iodine atoms with short metal-metal bonds (393-386 La-Pr). these compounds should be considered to be two-dimensional metals (two-dimensional in the same way that graphite is). The salt-like dihalides include those of Eu, Dy, Tm, and Yb. The formation of a relatively stable +2 oxidation state for Eu and Yb is usually explained by the stability (exchange energy) of half filled (f^7) and fully filled f^{14}. GdI_2 possesses the layered MoS_2 structure, is ferromagnetic and exhibits colossal magnetoresistance.

The sesquihalides Ln_2X_3 and the Ln_7I_{12} compounds listed in the table contain metal clusters, discrete Ln_6I_{12} clusters in Ln_7I_{12} and condensed clusters forming chains in the sesquihalides. Scandium forms a similar cluster compound with chlorine, Sc_7Cl_{12} Unlike many transition metal clusters these lanthanide clusters do not have strong metal-metal interactions and this is due to the low number of valence electrons involved, but instead are stabilised by the surrounding halogen atoms.

LaI and TmI are the only known monohalides. LaI, prepared from the reaction of LaI_3 and La metal, it has a NiAs type structure and can be formulated $La^{3+} (I^-)(e^-)_2$. TmI is a true Tm(I) compound, however it is not isolated in a pure state.

Oxides and Hydroxides

All of the lanthanides form sesquioxides, Ln_2O_3. The lighter/larger lanthanides adopt a hexagonal 7-coordinate structure while the heavier/smaller ones adopt a cubic 6-coordinate "C-M_2O_3" structure. All of the sesquioxides are basic, and absorb water and carbon dioxide from air to form carbonates, hydroxides and hydroxycarbonates. They dissolve in acids to form salts.

Cerium forms a stoichiometric dioxide, CeO_2, where cerium has an oxidation state of +4. CeO_2 is basic and dissolves with difficulty in acid to form Ce^{4+} solutions, from which Ce^{IV} salts can be isolated, for example the hydrated nitrate $Ce(NO_3)_4.5H_2O$.

CeO_2 is used as an oxidation catalyst in catalytic converters. Praseodymium and terbium form non-stoichiometric oxides containing Ln^{IV}, although more extreme reaction conditions can produce stoichiometric (or near stoichiometric) PrO_2 and TbO_2.

Europium and ytterbium form salt-like monoxides, EuO and YbO, which have a rock salt structure. EuO is ferromagnetic at low temperatures, and is a semiconductor with possible applications in spintronics. A mixed Eu^{II}/Eu^{III} oxide Eu_3O_4 can be produced by reducing Eu_2O_3 in a stream of hydrogen. Neodymium and samarium also form monoxides, but these are shiny conducting solids, although the existence of samarium monoxide is considered dubious.

All of the lanthanides form hydroxides, $Ln(OH)_3$. With the exception of lutetium hydroxide, which has a cubic structure, they have the hexagonal UCl_3 structure. The hydroxides can be precipitated from solutions of Ln^{III}. They can also be formed by the reaction of the sesquioxide, Ln_2O_3, with water, but although this reaction is thermodynamically favorable it is kinetically slow for the heavier members of the series. Fajans' rules indicate that the smaller Ln^{3+} ions will be more polarizing and their salts correspondingly less ionic. The hydroxides of the heavier lanthanides become less basic, for example $Yb(OH)_3$ and $Lu(OH)_3$ are still basic hydroxides but will dissolve in hot concentrated NaOH.

Chalcogenides (S, Se, Te)

All of the lanthanides form Ln_2Q_3 (Q= S, Se, Te). The sesquisulfides can be produced by reaction of the elements or (with the exception of Eu_2S_3) sulfidizing the oxide (Ln_2O_3) with H_2S.

The sesquisulfides, Ln_2S_3 generally lose sulfur when heated and can form a range of compositions between Ln_2S_3 and Ln_3S_4. The sesquisulfides are insulators but some of the Ln_3S_4 are metallic conductors (e.g. Ce_3S_4) formulated $(Ln^{3+})_3 (S^{2-})_4 (e^-)$, while others (e.g. Eu_3S_4 and Sm_3S_4) are semiconductors. Structurally the sesquisulfides adopt structures that vary according to the size of the Ln metal. The lighter and larger lanthanides favoring 7-coordinate metal atoms, the heaviest and smallest lanthanides (Yb and Lu) favoring 6 coordination and the rest structures with a mixture of 6 and 7 coordination.

Polymorphism is common amongst the sesquisulfides. The colors of the sesquisulfides vary metal to metal and depend on the polymorphic form. The colors of the γ-sesquisulfides are La_2S_3, white/yellow; Ce_2S_3, dark red; Pr_2S_3, green; Nd_2S_3, light green; Gd_2S_3, sand; Tb_2S_3, light yellow and Dy_2S_3, orange. The shade of γ-Ce_2S_3 can be varied by doping with Na or Ca with hues ranging from dark red to yellow, and Ce_2S_3 based pigments are used commercially and are seen as low toxicity substitutes for cadmium based pigments.

All of the lanthanides form monochalcogenides, LnQ, (Q= S, Se, Te). The majority of the monochalcogenides are conducting, indicating a formulation $Ln^{III}Q^{2-}$(e-) where the electron is in conduction bands. The exceptions are SmQ, EuQ and YbQ which are semiconductors or insulators but exhibit a pressure induced transition to a conducting state. Compounds LnQ_2 are known but these do not contain Ln^{IV} but are Ln^{III} compounds containing polychalcogenide anions.

Oxysulfides Ln_2O_2S are well known, they all have the same structure with 7-coordinate Ln atoms, and 3 sulfur and 4 oxygen atoms as near neighbours. Doping these with other lanthanide elements produces phosphors. As an example, gadolinium oxysulfide, Gd_2O_2S doped with Tb^{3+} produces visible photons when irradiated with high energy X-rays and is used as a scintillator in flat panel detectors. When mischmetal, an alloy of lanthanide metals, is added to molten steel to remove oxygen and sulfur, stable oxysulfides are produced that form an immiscible solid.

Pnictides (Group 15)

All of the lanthanides form a mononitride, LnN, with the rock salt structure. The mononitrides have attracted interest because of their unusual physical properties. SmN and EuN are reported as being "half metals". NdN, GdN, TbN and DyN are ferromagnetic, SmN is antiferromagnetic. Applications in the field of spintronics are being investigated. CeN is unusual as it is a metallic conductor, contrasting with the other nitrides also with the other cerium pnictides. A simple description is $Ce^{4+}N^{3-}$ (e–) but the interatomic distances are a better match for the trivalent state rather than for the tetravalent state. A number of different explanations have been offered. The nitrides can be prepared by the reaction of lanthanum metals with nitrogen. Some nitride is produced along with the oxide, when lanthanum metals are ignited in air. Alternative methods of synthesis are a high temperature reaction of lanthanide metals with ammonia or the decomposition of lanthanide amides, $Ln(NH_2)_3$. Achieving pure stoichiometric compounds, and crystals with low defect density has proved difficult. The lanthanide nitrides are sensitive to air and hydrolyse producing ammonia.

The other pnictides phosphorus, arsenic, antimony and bismuth also react with the lanthanide metals to form monopnictides, LnQ, where Q = P, As, Sb or Bi. Additionally a range of other compounds can be produced with varying stoichiometries, such as LnP_2, LnP_5, LnP_7, Ln_3As, Ln_5As_3 and $LnAs_2$.

Carbides

Carbides of varying stoichiometries are known for the lanthanides. Non-stoichiometry is common. All of the lanthanides form LnC_2 and Ln_2C_3 which both contain C_2 units. The dicarbides with exception of EuC_2, are metallic conductors with the calcium carbide structure and can be formulated as $Ln^{3+}C_2^{2-}(e-)$. The C-C bond length is longer than that in CaC_2, which contains the C_2^{2-} anion, indicating that the antibonding orbitals of the C_2^{2-} anion are involved in the conduction band. These dicarbides hydrolyse to form hydrogen and a mixture of hydrocarbons. EuC_2 and to a lesser extent YbC_2 hydrolyse differently producing a higher percentage of acetylene (ethyne). The sesquicarbides, Ln_2C_3 can be formulated as $Ln_4(C_2)_3$.

These compounds adopt the Pu_2C_3 structure which has been described as having C_2^{2-} anions in bisphenoid holes formed by eight near Ln neighbours. The lengthening of the C-C bond is less marked in the sesquicarbides than in the dicarbides, with the exception of Ce_2C_3. Other carbon rich stoichiometries are known for some lanthanides. Ln_3C_4 (Ho-Lu) containing C, C_2 and C_3 units; Ln_4C_7 (Ho-Lu) contain C atoms and C_3 units and Ln_4C_5 (Gd-Ho) containing C and C_2 units. Metal rich carbides contain interstitial C atoms and no C_2 or C_3 units. These are Ln_4C_3 (Tb and Lu); Ln_2C (Dy, Ho, Tm) and Ln_3C (Sm-Lu).

Borides

All of the lanthanides form a number of borides. The "higher" borides (LnB_x where $x > 12$) are insulators/semiconductors whereas the lower borides are typically conducting. The lower borides have stoichiometries of LnB_2, LnB_4, LnB_6 and LnB_{12}. Applications in the field of spintronics are being investigated. The range of borides formed by the lanthanides can be compared to those formed by the transition metals. The boron rich borides are typical of the lanthanides (and groups 1–3) whereas for the transition metals tend to form metal rich, "lower" borides. The lanthanide borides are typically grouped together with the group 3 metals with which they share many similarities of reactivity, stoichiometry and structure. Collectively these are then termed the rare earth borides.

Many methods of producing lanthanide borides have been used, amongst them are direct reaction of the elements; the reduction of Ln_2O_3 with boron; reduction of boron oxide, B_2O_3, and Ln_2O_3 together with carbon; reduction of metal oxide with boron carbide, B_4C. Producing high purity samples has proved to be difficult. Single crystals of the higher borides have been grown in a low melting metal (e.g. Sn, Cu, Al).

Diborides, LnB_2, have been reported for Sm, Gd, Tb, Dy, Ho, Er, Tm, Yb and Lu. All have the same, AlB_2, structure containing a graphitic layer of boron atoms. Low temperature ferromagnetic transitions for Tb, Dy, Ho and Er. TmB_2 is ferromagnetic at 7.2 K.

Tetraborides, LnB_4 have been reported for all of the lanthanides except EuB_4, all have the same UB_4 structure. The structure has a boron sub-lattice consists of chains of octahedral B_6 clusters linked by boron atoms. The unit cell decreases in size successively from LaB_4 to LuB_4. The tetraborides of the lighter lanthanides melt with decomposition to LnB_6. Attempts to make EuB_4 have failed. The LnB_4 are good conductors and typically antiferromagnetic.

Hexaborides, LnB_6 have been reported for all of the lanthanides. They all have the CaB_6 structure, containing B_6 clusters. They are non-stoichiometric due to cation defects. The hexaborides of the lighter lanthanides (La – Sm) melt without decomposition, EuB_6 decomposes to boron and metal and the heavier lanthanides decompose to LnB_4 with exception of YbB_6 which decomposes forming YbB_{12}.

The stability has in part been correlated to differences in volatility between the lanthanide metals. In EuB_6 and YbB_6 the metals have an oxidation state of +2 whereas in the rest of the lanthanide hexaborides it is +3.

This rationalises the differences in conductivity, the extra electrons in the Ln^{III} hexaborides entering conduction bands. EuB_6 is a semiconductor and the rest are good conductors. LaB_6 and CeB_6 are thermionic emitters, used, for example, in scanning electron microscopes.

Dodecaborides, LnB_{12}, are formed by the heavier smaller lanthanides, but not by the lighter larger metals, La – Eu. With the exception YbB_{12} (where Yb takes an intermediate valence and is a Kondo insulator), the dodecaborides are all metallic compounds. They all have the UB_{12} structure containing a 3 dimensional framework of cubooctahedral B_{12} clusters.

The higher boride LnB_{66} is known for all lanthanide metals. The composition is approximate as the compounds are non-stoichiometric. They all have similar complex structure with over 1600 atoms in the unit cell. The boron cubic sub lattice contains super icosahedra made up of a central B_{12} icosahedra surrounded by 12 others, $B_{12}(B_{12})_{12}$. Other complex higher borides LnB_{50} (Tb, Dy, Ho Er Tm Lu) and LnB_{25} are known (Gd, Tb, Dy, Ho, Er) and these contain boron icosahedra in the boron framework.

Organometallic Compounds

Lanthanide-carbon σ bonds are well known; however as the 4f electrons have a low probability of existing at the outer region of the atom there is little effective orbital overlap, resulting in bonds with significant ionic character. As such organo-lanthanide compounds exhibit carbanion-like behavior, unlike the behavior in transition metal organometallic compounds.

Because of their large size, lanthanides tend to form more stable organometallic derivatives with bulky ligands to give compounds such as $Ln[CH(SiMe_3)_3]$. Analogues of uranocene are derived from dilithiocyclooctatetraene, $Li_2C_8H_8$. Organic lanthanide(II) compounds are also known, such as Cp^*_2Eu.

29.2 PHYSICAL PROPERTIES

29.2.1 Magnetic and Spectroscopic

All the trivalent lanthanide ions, except lanthanum and lutetium, have unpaired f electrons. (Ligand-to-metal charge transfer can nonetheless produce a nonzero f-occupancy even in La(III) compounds.) However, the magnetic moments deviate considerably from the spin-only values because of strong spin–orbit coupling. The maximum number of unpaired electrons is 7, in Gd^{3+}, with a magnetic moment of 7.94 B.M., but the largest magnetic moments, at 10.4–10.7 B.M., are exhibited by Dy^{3+} and Ho^{3+}. However, in Gd^{3+} all the electrons have parallel spin and this property is important for the use of gadolinium complexes as contrast reagent in MRI scans.

Crystal field splitting is rather small for the lanthanide ions and is less important than spin–orbit coupling in regard to energy levels. Transitions of electrons between f orbitals are forbidden by the Laporte rule. Furthermore, because of the "buried" nature of the f orbitals, coupling with molecular vibrations is weak. Consequently, the spectra of lanthanide ions are rather weak and the absorption bands are similarly narrow.

Glass containing holmium oxide and holmium oxide solutions (usually in perchloric acid) have sharp optical absorption peaks in the spectral range 200–900 nm and can be used as a wavelength calibration standard for optical spectrophotometers, and are available commercially.

As f-f transitions are Laporte-forbidden, once an electron has been excited, decay to the ground state will be slow. This makes them suitable for use in lasers as it makes the population inversion easy to achieve. The Nd:YAG laser is one that is widely used. Europium-doped yttrium vanadate was the first red phosphor to enable the development of color television screens. Lanthanide ions have notable luminescent properties due to their unique 4f orbitals. Laporte forbidden f-f transitions can be activated by excitation of a bound "antenna" ligand. This leads to sharp emission bands throughout the visible, NIR, and IR and relatively long luminescence lifetimes.

29.3 OCCURRENCE

Samarskite and similar minerals contain lanthanides in association with the elements such as tantalum, niobium, hafnium, zirconium, vanadium, and titanium, from group 4 and group 5, often in similar oxidation states. Monazite is a phosphate of numerous group 3 + lanthanide + actinide metals and mined especially for the thorium content and specific rare earths, especially lanthanum, yttrium and cerium. Cerium and lanthanum as well as other members of the rare-earth series are often produced as a metal called mischmetal containing a variable mixture of these elements with cerium and lanthanum predominating; it has direct uses such as lighter flints and other spark sources which do not require extensive purification of one of these metals.

There are also lanthanide-bearing minerals based on group-2 elements, such as yttrocalcite, yttrocerite and yttrofluorite, which vary in content of yttrium, cerium, lanthanum and others. Other lanthanide-bearing minerals include bastnäsite, florencite, chernovite, perovskite, xenotime, cerite, gadolinite, lanthanite, fergusonite, polycrase, blomstrandine, håleniusite, miserite, loparite, lepersonnite, euxenite, all of which have a range of relative element concentration and may be denoted by a predominating one, as in monazite-(Ce). Group 3 elements do not occur as native-element minerals in the fashion of gold, silver, tantalum and many others on Earth, but may occur in lunar soil. Very rare halides of cerium, lanthanum, and presumably other lanthanides, feldspars and garnets are also known to exist.

The lanthanide contraction is responsible for the great geochemical divide that splits the lanthanides into light and heavy-lanthanide enriched minerals, the latter being almost inevitably associated with and dominated by yttrium.

This divide is reflected in the first two "rare earths" that were discovered: yttria (1794) and ceria (1803). The geochemical divide has put more of the light lanthanides in the Earth's crust, but more of the heavy members in the Earth's mantle. The result is that although large rich ore-bodies are found that are enriched in the light lanthanides, correspondingly large ore-bodies for the heavy members are few. The principal ores are monazite and bastnäsite.

Monazite sands usually contain all the lanthanide elements, but the heavier elements are lacking in bastnäsite. The lanthanides obey the Oddo–Harkins rule – odd-numbered elements are less abundant than their even-numbered neighbors.

Three of the lanthanide elements have radioactive isotopes with long half-lives (^{138}La, ^{147}Sm and ^{176}Lu) that can be used to date minerals and rocks from Earth, the Moon and meteorites. Promethium is effectively a man-made element, as all its isotopes are radioactive with half-lives shorter than 20 years.

29.4 APPLICATIONS

29.4.1 Industrial

Lanthanide elements and their compounds have many uses but the quantities consumed are relatively small in comparison to other elements. About 15000 ton/year of the lanthanides are consumed as catalysts and in the production of glasses. This 15000 tons corresponds to about 85% of the lanthanide production. From the perspective of value, however, applications in phosphors and magnets are more important.

The devices lanthanide elements are used in include superconductors, samarium-cobalt and neodymium-iron-boron high-flux rare-earth magnets, magnesium alloys, electronic polishers, refining catalysts and hybrid car components (primarily batteries and magnets). Lanthanide ions are used as the active ions in luminescent materials used in optoelectronics applications, most notably the Nd:YAG laser. Erbium-doped fiber amplifiers are significant devices in optical-fiber communication systems. Phosphors with lanthanide dopants are also widely used in cathode-ray tube technology such as television sets. The earliest color television CRTs had a poor-quality red; europium as a phosphor dopant made good red phosphors possible. Yttrium iron garnet (YIG) spheres can act as tunable microwave resonators.

Lanthanide oxides are mixed with tungsten to improve their high temperature properties for TIG welding, replacing thorium, which was mildly hazardous to work with. Many defense-related products also use lanthanide elements such as night-vision goggles and rangefinders. The SPY-1 radar used in some Aegis equipped warships, and the hybrid propulsion system of *Arleigh Burke*-class destroyers all use rare earth magnets in critical capacities. The price for lanthanum oxide used in fluid catalytic cracking has risen from $5 per kilogram in early 2010 to $140 per kilogram in June 2011.

Most lanthanides are widely used in lasers, and as (co-)dopants in doped-fiber optical amplifiers; for example, in Er-doped fiber amplifiers, which are used as repeaters in the terrestrial and submarine fiber-optic transmission links that carry internet traffic. These elements deflect ultraviolet and infrared radiation and are commonly used in the production of sunglass lenses. Other applications are summarized in the following table:

Application	Percentage
Catalytic converters	45%
Petroleum refining catalysts	25%
Permanent magnets	12%
Glass polishing and ceramics	7%
Metallurgical	7%
Phosphors	3%
Other	1%

The complex Gd(DOTA) is used in magnetic resonance imaging.

29.4.2 Life Science

Lanthanide complexes can be used for optical imaging. Applications are limited by the lability of the complexes.

Some applications depend on the unique luminescence properties of lanthanide chelates or cryptates. These are well-suited for this application due to their large Stokes shifts and extremely long emission lifetimes (from microseconds to milliseconds) compared to more traditional fluorophores (e.g., fluorescein, allophycocyanin, phycoerythrin, and rhodamine).

The biological fluids or serum commonly used in these research applications contain many compounds and proteins which are naturally fluorescent. Therefore, the use of conventional, steady-state fluorescence measurement presents serious limitations in assay sensitivity. Long-lived fluorophores, such as lanthanides, combined with time-resolved detection (a delay between excitation and emission detection) minimizes prompt fluorescence interference.

Time-resolved fluorometry (TRF) combined with Förster resonance energy transfer (FRET) offers a powerful tool for drug discovery researchers: Time-Resolved Förster Resonance Energy Transfer or TR-FRET. TR-FRET combines the low background aspect of TRF with the homogeneous assay format of FRET. The resulting assay provides an increase in flexibility, reliability and sensitivity in addition to higher throughput and fewer false positive/false negative results.

This method involves two fluorophores: a donor and an acceptor. Excitation of the donor fluorophore (in this case, the lanthanide ion complex) by an energy source (e.g. flash lamp or laser) produces an energy transfer to the acceptor fluorophore if they are within a given proximity to each other (known as the Förster's radius). The acceptor fluorophore in turn emits light at its characteristic wavelength.

The two most commonly used lanthanides in life science assays are shown below along with their corresponding acceptor dye as well as their excitation and emission wavelengths and resultant Stokes shift (separation of excitation and emission wavelengths).

Donor	Excitation⇒Emission λ (nm)	Acceptor	Excitation⇒Emission λ (nm)	Stoke's Shift (nm)
Eu^{3+}	340⇒615	Allophycocyanin	615⇒660	320
Tb^{3+}	340⇒545	Phycoerythrin	545⇒575	235

29.4.3 Possible Medical Uses

Currently there is research showing that lanthanide elements can be used as anticancer agents. The main role of the lanthanides in these studies is to inhibit proliferation of the cancer cells. Specifically cerium and lanthanum have been studied for their role as anti-cancer agents.

One of the specific elements from the lanthanide group that has been tested and used is cerium (Ce). There have been studies that use a protein-cerium complex to observe the effect of cerium on the cancer cells. The hope was to inhibit cell proliferation and promote cytotoxicity. Transferrin receptors in cancer cells, such as those in breast cancer cells and epithelial cervical cells, promote the cell proliferation and malignancy of the cancer. Transferrin is a protein used to transport iron into the cells and is needed to aid the cancer cells in DNA replication. Transferrin acts as a growth factor for the cancerous cells and is dependent on iron. Cancer cells have much higher levels of transferrin receptors than normal cells and are very dependent on iron for their proliferation. In the field of magnetic resonance imaging (MRI), compounds containing gadolinium are utilized extensively.

The photobiological characteristics, anticancer, anti-leukemia, and anti-HIV activities of the lanthanides with coumarin and its related compounds are demonstrated by the biological activities of the complex. Cerium has shown results as an anti-cancer agent due to its similarities in structure and biochemistry to iron. Cerium may bind in the place of iron on to the transferrin and then be brought into the cancer cells by transferrin-receptor mediated endocytosis. The cerium binding to the transferrin in place of the iron inhibits the transferrin activity in the cell. This creates a toxic environment for the cancer cells and causes a decrease in cell growth.

This is the proposed mechanism for cerium's effect on cancer cells, though the real mechanism may be more complex in how cerium inhibits cancer cell proliferation. Specifically in HeLa cancer cells studied in vitro, cell viability was decreased after 48 to 72 hours of cerium treatments. Cells treated with just cerium had decreases in cell viability, but cells treated with both cerium and transferrin had more significant inhibition for cellular activity. Another specific element that has been tested and used as an anti-cancer agent is lanthanum, more specifically lanthanum chloride ($LaCl_3$). The lanthanum ion is used to affect the levels of let-7a and microRNAs miR-34a in a cell throughout the cell cycle.

When the lanthanum ion was introduced to the cell in vivo or in vitro, it inhibited the rapid growth and induced apoptosis of the cancer cells (specifically cervical cancer cells). This effect was caused by the regulation of the let-7a and microRNAs by the lanthanum ions. The mechanism for this effect is still unclear but it is possible that the lanthanum is acting in a similar way as the cerium and binding to a ligand necessary for cancer cell proliferation. In the field of magnetic resonance imaging (MRI), compounds containing gadolinium are utilized extensively.

29.5 BIOLOGICAL EFFECTS

Due to their sparse distribution in the earth's crust and low aqueous solubility, the lanthanides have a low availability in the biosphere, and for a long time were not known to naturally form part of any biological molecules. In 2007 a novel methanol dehydrogenase that strictly uses lanthanides as enzymatic cofactors was discovered in a bacterium from the phylum Verrucomicrobiota, *Methylacidiphilum fumariolicum*.

This bacterium was found to survive only if there are lanthanides present in the environment. Compared to most other nondietary elements, non-radioactive lanthanides are classified as having low toxicity. The same nutritional requirement has also been observed in *Methylorubrum extorquens* and *Methylobacterium radiotolerans*.

CHAPTER 30
THE ACTINIDES

The actinide or actinoid series encompasses at least the 14 metallic chemical elements in the 5f series, with atomic numbers from 89 to 102, actinium through nobelium. (Number 103, lawrencium, is sometimes also included despite being part of the 6d transition series.) The actinide series derives its name from the first element in the series, actinium. The informal chemical symbol An is used in general discussions of actinide chemistry to refer to any actinide.

The 1985 IUPAC *Red Book* recommends that *actinoid* be used rather than *actinide*, since the suffix *-ide* normally indicates a negative ion. However, owing to widespread current use, *actinide* is still allowed. Since *actinoid* literally means *actinium-like* (cf. *humanoid* or *android*), it has been argued for semantic reasons that actinium cannot logically be an actinoid, but IUPAC acknowledges its inclusion based on common usage.

Actinium through nobelium are f-block elements, while lawrencium is a d-block element and a transition metal. The series mostly corresponds to the filling of the 5f electron shell, although as isolated atoms in the ground state many have anomalous configurations involving the filling of the 6d shell due to interelectronic repulsion. In comparison with the lanthanides, also mostly f-block elements, the actinides show much more variable valence. They all have very large atomic and ionic radii and exhibit an unusually large range of physical properties. While actinium and the late actinides (from curium onwards) behave similarly to the lanthanides, the elements thorium, protactinium, and uranium are much more similar to transition metals in their chemistry, with neptunium, plutonium, and americium occupying an intermediate position.

All actinides are radioactive and release energy upon radioactive decay; naturally occurring uranium and thorium, and synthetically produced plutonium are the most abundant actinides on Earth. These have been used in nuclear reactors, and uranium and plutonium are critical elements of nuclear weapons. Uranium and thorium also have diverse current or historical uses, and americium is used in the ionization chambers of most modern smoke detectors.

Of the actinides, primordial thorium and uranium occur naturally in substantial quantities. The radioactive decay of uranium produces transient amounts of actinium and protactinium, and atoms of neptunium and plutonium are occasionally produced from transmutation reactions in uranium ores. The other actinides are purely synthetic elements. Nuclear weapons tests have released at least six actinides heavier than plutonium into the environment; analysis of debris from a 1952 hydrogen bomb explosion showed the presence of americium, curium, berkelium, californium, einsteinium and fermium.

In presentations of the periodic table, the f-block elements are customarily shown as two additional rows below the main body of the table. This convention is entirely a matter of aesthetics and formatting practicality; a rarely used wide-formatted periodic table inserts the 4f and 5f series in their proper places, as parts of the table's sixth and seventh rows (periods).

30.1 DISCOVERY, ISOLATION AND SYNTHESIS

Like the lanthanides, the actinides form a family of elements with similar properties. Within the actinides, there are two overlapping groups: transuranium elements, which follow uranium in the periodic table; and transplutonium elements, which follow plutonium. Compared to the lanthanides, which (except for promethium) are found in nature in appreciable quantities, most actinides are rare. Most do not occur in nature, and of those that do, only thorium and uranium do so in more than trace quantities. The most abundant or easily synthesized actinides are uranium and thorium, followed by plutonium, americium, actinium, protactinium, neptunium, and curium.

Table 1: Synthesis of Transuranium Elements

Element	Year	Method
Neptunium	1940	Bombarding ^{238}U with neutrons
Plutonium	1941	Bombarding ^{238}U with deuterons
Americium	1944	Bombarding ^{239}Pu with neutrons
Curium	1944	Bombarding ^{239}Pu with α-particles
Berkelium	1949	Bombarding ^{241}Am with α-particles
Californium	1950	Bombarding ^{242}Cm with α-particles
Einsteinium	1952	As a product of nuclear explosion
Fermium	1952	As a product of nuclear explosion
Mendelevium	1955	Bombarding ^{253}Es with α-particles
Nobelium	1965	Bombarding ^{243}Am with ^{15}N or ^{238}U with ^{22}Ne
Lawrencium	1961 –1971	Bombarding ^{252}Cf with ^{10}B or ^{11}B and of ^{243}Am with ^{18}O

The existence of transuranium elements was suggested in 1934 by Enrico Fermi, based on his experiments. However, even though four actinides were known by that time, it was not yet understood that they formed a family similar to lanthanides. The prevailing view that dominated early research into transuranics was that they were regular elements in the 7th period, with thorium, protactinium and uranium corresponding to 6th-period hafnium, tantalum and tungsten, respectively. Synthesis of transuranics gradually undermined this point of view. By 1944, an observation that curium failed to exhibit oxidation states above 4 (whereas its supposed 6th period homolog, platinum, can reach oxidation state of 6) prompted Glenn Seaborg to formulate an "actinide hypothesis". Studies of known actinides and discoveries of further transuranic elements provided more data in support of this position, but the phrase "actinide hypothesis" (the implication being that a "hypothesis" is something that has not been decisively proven) remained in active use by scientists through the late 1950s.

At present, there are two major methods of producing isotopes of transplutonium elements: (1) irradiation of the lighter elements with neutrons; (2) irradiation with accelerated charged particles. The first method is more important for applications, as only neutron irradiation using nuclear reactors allows the production of sizeable amounts of synthetic actinides; however, it is limited to relatively light elements. The advantage of the second method is that elements heavier than plutonium, as well as neutron-deficient isotopes, can be obtained, which are not formed during neutron irradiation.

In 1962–1966, there were attempts in the United States to produce transplutonium isotopes using a series of six underground nuclear explosions. Small samples of rock were extracted from the blast area immediately after the test to study the explosion products, but no isotopes with mass number greater than 257 could be detected, despite predictions that such isotopes would have relatively long half-lives of α-decay. This non-observation was attributed to spontaneous fission owing to the large speed of the products and to other decay channels, such as neutron emission and nuclear fission.

30.1.1 From Actinium to Uranium

Uranium and thorium were the first actinides discovered. Uranium was identified in 1789 by the German chemist Martin Heinrich Klaproth in pitchblende ore. He named it after the planet Uranus, which had been discovered eight years earlier. Klaproth was able to precipitate a yellow compound (likely sodium diuranate) by dissolving pitchblende in nitric acid and neutralizing the solution with sodium hydroxide.

He then reduced the obtained yellow powder with charcoal, and extracted a black substance that he mistook for metal. Sixty years later, the French scientist Eugène-Melchior Péligot identified it as uranium oxide. He also isolated the first sample of uranium metal by heating uranium tetrachloride with metallic potassium. The atomic mass of uranium was then calculated as 120, but Dmitri Mendeleev in 1872 corrected it to 240 using his periodicity laws. This value was confirmed experimentally in 1882 by K. Zimmerman.

Thorium oxide was discovered by Friedrich Wöhler in the mineral thorianite, which was found in Norway (1827). Jöns Jacob Berzelius characterized this material in more detail in 1828. By reduction of thorium tetrachloride with potassium, he isolated the metal and named it thorium after the Norse god of thunder and lightning Thor. The same isolation method was later used by Péligot for uranium.

Actinium was discovered in 1899 by André-Louis Debierne, an assistant of Marie Curie, in the pitchblende waste left after removal of radium and polonium. He described the substance (in 1899) as similar to titanium and (in 1900) as similar to thorium. The discovery of actinium by Debierne was however questioned in 1971 and 2000, arguing that Debierne's publications in 1904 contradicted his earlier work of 1899–1900. This metal was discovered not by its own radiation but by the radiation of the daughter products. Owing to the close similarity of actinium and lanthanum and low abundance, pure actinium could only be produced in 1950. The term actinide was probably introduced by Victor Goldschmidt in 1937.

Protactinium was possibly isolated in 1900 by William Crookes. It was first identified in 1913, when Kasimir Fajans and Oswald Helmuth Göhring encountered the short-lived isotope ^{234m}Pa (half-life 1.17 minutes) during their studies of the ^{238}U decay chain. They named the new element *brevium* (from Latin *brevis* meaning brief); the name was changed to *protoactinium* in 1918 when two groups of scientists, led by the Austrian Lise Meitner and Otto Hahn of Germany and Frederick Soddy and John Arnold Cranston of Great Britain, independently discovered the much longer-lived ^{231}Pa. The name was shortened to *protactinium* in 1949. This element was little characterized until 1960, when Alfred Maddock and his co-workers in the U.K. isolated 130 grams of protactinium from 60 tonnes of waste left after extraction of uranium from its ore.

30.2 ISOTOPES

33 isotopes of actinium and eight excited isomeric states of some of its nuclides are known, ranging in mass number from 204 to 236. Three isotopes, ^{225}Ac, ^{227}Ac and ^{228}Ac, were found in nature and the others were produced in the laboratory; only the three natural isotopes are used in applications.

Actinium-225 is a member of the radioactive neptunium series; it was first discovered in 1947 as a decay product of uranium-233 and it is an α-emitter with a half-life of 10 days. Actinium-225 is less available than actinium-228, but is more promising in radiotracer applications. Actinium-227 (half-life 21.77 years) occurs in all uranium ores, but in small quantities. One gram of uranium (in radioactive equilibrium) contains only 2×10^{-10} gram of ^{227}Ac. Actinium-228 is a member of the radioactive thorium series formed by the decay of ^{228}Ra; it is a β^- emitter with a half-life of 6.15 hours. In one tonne of thorium there is 5×10^{-8} gram of ^{228}Ac. It was discovered by Otto Hahn in 1906.

There are 32 known isotopes of thorium ranging in mass number from 207 to 238. Of these, the longest-lived is ^{232}Th, whose half-life of 1.4×10^{10} years means that it still exists in nature as a primordial nuclide. The next longest-lived is ^{230}Th, an intermediate decay product of ^{238}U with a half-life of 75,400 years. Several other thorium isotopes have half-lives over a day; all of these are also transient in the decay chains of ^{232}Th, ^{235}U, and ^{238}U.

28 isotopes of protactinium are known with mass numbers 212–239 as well as three excited isomeric states. Only ^{231}Pa and ^{234}Pa have been found in nature. All the isotopes have short lifetimes, except for protactinium-231 (half-life 32,760 years). The most important isotopes are ^{231}Pa and ^{233}Pa, which is an intermediate product in obtaining uranium-233 and is the most affordable among artificial isotopes of protactinium. ^{233}Pa has convenient half-life and energy of γ-radiation, and thus was used in most studies of protactinium chemistry. Protactinium-233 is a β-emitter with a half-life of 26.97 days.

There are 27 known isotopes of uranium, having mass numbers 215–242 (except 220). Three of them, ^{234}U, ^{235}U and ^{238}U, are present in appreciable quantities in nature. Among others, the most important is ^{233}U, which is a final product of transformation of ^{232}Th irradiated by slow neutrons. ^{233}U has a much higher fission efficiency by low-energy (thermal) neutrons, compared e.g. with ^{235}U. Most uranium chemistry studies were carried out on uranium-238 owing to its long half-life of 4.4×10^{9} years.

There are 25 isotopes of neptunium with mass numbers 219–244 (except 221); they are all highly radioactive. The most popular among scientists are long-lived ^{237}Np ($t_{1/2} = 10^{6}\times2.20$ years) and short-lived ^{239}Np, ^{238}Np ($t_{1/2} \sim 2$ days).

There are 20 known isotopes of plutonium, having mass numbers 228–247. The most stable isotope of plutonium is ^{244}Pu with half-life of 8.13×10^{7} years.

Eighteen isotopes of americium are known with mass numbers from 229 to 247 (with the exception of 231). The most important are ^{241}Am and ^{243}Am, which are alpha-emitters and also emit soft, but intense γ-rays; both of them can be obtained in an isotopically pure form. Chemical properties of americium were first studied with ^{241}Am, but later shifted to ^{243}Am, which is almost 20 times less radioactive. The disadvantage of ^{243}Am is production of the short-lived daughter isotope ^{239}Np, which has to be considered in the data analysis.

Among 19 isotopes of curium, ranging in mass number from 233 to 251, the most accessible are ^{242}Cm and ^{244}Cm; they are α-emitters, but with much shorter lifetime than the americium isotopes. These isotopes emit almost no γ-radiation, but undergo spontaneous fission with the associated emission of neutrons. More long-lived isotopes of curium ($^{245-248}$Cm, all α-emitters) are formed as a mixture during neutron irradiation of plutonium or americium. Upon short irradiation, this mixture is dominated by ^{246}Cm, and then ^{248}Cm begins to accumulate. Both of these isotopes, especially ^{248}Cm, have a longer half-life (3.48×10^5 years) and are much more convenient for carrying out chemical research than ^{242}Cm and ^{244}Cm, but they also have a rather high rate of spontaneous fission. ^{247}Cm has the longest lifetime among isotopes of curium (1.56×10^7 years), but is not formed in large quantities because of the strong fission induced by thermal neutrons.

Seventeen isotopes of berkelium have been identified with mass numbers 233, 234, 236, 238, and 240–252. Only ^{249}Bk is available in large quantities; it has a relatively short half-life of 330 days and emits mostly soft β-particles, which are inconvenient for detection. Its alpha radiation is rather weak (1.45×10^{-3}% with respect to β-radiation), but is sometimes used to detect this isotope. ^{247}Bk is an alpha-emitter with a long half-life of 1,380 years, but it is hard to obtain in appreciable quantities; it is not formed upon neutron irradiation of plutonium because β-decay of curium isotopes with mass number below 248 is not known.

The 20 isotopes of californium with mass numbers 237–256 are formed in nuclear reactors; californium-253 is a β-emitter and the rest are α-emitters. The isotopes with even mass numbers (^{250}Cf, ^{252}Cf and ^{254}Cf) have a high rate of spontaneous fission, especially ^{254}Cf of which 99.7% decays by spontaneous fission. Californium-249 has a relatively long half-life (352 years), weak spontaneous fission and strong γ-emission that facilitates its identification. ^{249}Cf is not formed in large quantities in a nuclear reactor because of the slow β-decay of the parent isotope ^{249}Bk and a large cross section of interaction with neutrons, but it can be accumulated in the isotopically pure form as the β-decay product of (pre-selected) ^{249}Bk. Californium produced by reactor-irradiation of plutonium mostly consists of ^{250}Cf and ^{252}Cf, the latter being predominant for large neutron fluences, and its study is hindered by the strong neutron radiation.

Table 2: Properties of Some Transplutonium Isotope Pairs

Parent Isotope	$t_{1/2}$	Daughter Isotope	$t_{1/2}$	Time to Establish Radioactive Equilibrium
^{243}Am	7370 years	^{239}Np	2.35 days	47.3 days
^{245}Cm	8265 years	^{241}Pu	14 years	129 years
^{247}Cm	1.64×10^7 years	^{243}Pu	4.95 hours	7.2 days
^{254}Es	270 days	^{250}Bk	3.2 hours	35.2 hours
^{255}Es	39.8 days	^{255}Fm	22 hours	5 days
^{257}Fm	79 days	^{253}Cf	17.6 days	49 days

Among the 18 known isotopes of einsteinium with mass numbers from 240 to 257, the most affordable is ^{253}Es. It is an α-emitter with a half-life of 20.47 days, a relatively weak γ-emission and small spontaneous fission rate as compared with the isotopes of californium. Prolonged neutron irradiation also produces a long-lived isotope ^{254}Es ($t_{1/2}$ = 275.5 days).

Twenty isotopes of fermium are known with mass numbers of 241–260. ^{254}Fm, ^{255}Fm and ^{256}Fm are α-emitters with a short half-life (hours), which can be isolated in significant amounts. ^{257}Fm ($t_{1/2}$ = 100 days) can accumulate upon prolonged and strong irradiation. All these isotopes are characterized by high rates of spontaneous fission.

Among the 17 known isotopes of mendelevium (mass numbers from 244 to 260), the most studied is ^{256}Md, which mainly decays through electron capture (α-radiation is ≈10%) with a half-life of 77 minutes. Another alpha emitter, ^{258}Md, has a half-life of 53 days.

Both these isotopes are produced from rare einsteinium (^{253}Es and ^{255}Es respectively), that therefore limits their availability.

Long-lived isotopes of nobelium and isotopes of lawrencium (and of heavier elements) have relatively short half-lives. For nobelium, 13 isotopes are known, with mass numbers 249–260 and 262. The chemical properties of nobelium and lawrencium were studied with ^{255}No ($t_{1/2}$ = 3 min) and ^{256}Lr ($t_{1/2}$ = 35 s). The longest-lived nobelium isotope, ^{259}No, has a half-life of approximately 1 hour. Lawrencium has 14 known isotopes with mass numbers 251–262, 264, and 266. The most stable of them is ^{266}Lr with a half life of 11 hours.

Among all of these, the only isotopes that occur in sufficient quantities in nature to be detected in anything more than traces and have a measurable contribution to the atomic weights of the actinides are the primordial ^{232}Th, ^{235}U, and ^{238}U, and three long-lived decay products of natural uranium, ^{230}Th, ^{231}Pa, and ^{234}U. Natural thorium consists of 0.02(2)% ^{230}Th and 99.98(2)% ^{232}Th; natural protactinium consists of 100% ^{231}Pa; and natural uranium consists of 0.0054(5)% ^{234}U, 0.7204(6)% ^{235}U, and 99.2742(10)% ^{238}U.

30.3 FORMATION IN NUCLEAR REACTORS

The figure *buildup of actinides* is a table of nuclides with the number of neutrons on the horizontal axis (isotopes) and the number of protons on the vertical axis (elements).

The red dot divides the nuclides in two groups, so the figure is more compact. Each nuclide is represented by a square with the mass number of the element and its half-life.

Naturally existing actinide isotopes (Th, U) are marked with a bold border, alpha emitters have a yellow colour, and beta emitters have a blue colour. Pink indicates electron capture (^{236}Np), whereas white stands for a long-lasting metastable state (^{242}Am).

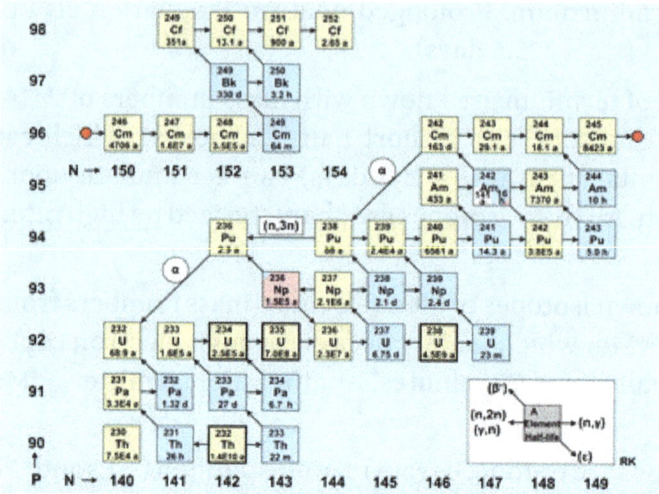

Fig 1: Table of Nuclides: Buildup of Actinides in a Nuclear Reactor, Including Radioactive Decay

The formation of actinide nuclides is primarily characterised by:

- Neutron capture reactions (n,γ), which are represented in the figure by a short right arrow.
- The (n,2n) reactions and the less frequently occurring (γ,n) reactions are also taken into account, both of which are marked by a short left arrow.
- Even more rarely and only triggered by fast neutrons, the (n,3n) reaction occurs, which is represented in the figure with one example, marked by a long left arrow.

In addition to these neutron- or gamma-induced nuclear reactions, the radioactive conversion of actinide nuclides also affects the nuclide inventory in a reactor. These decay types are marked in the figure by diagonal arrows. The beta-minus decay, marked with an arrow pointing up-left, plays a major role for the balance of the particle densities of the nuclides. Nuclides decaying by positron emission (beta-plus decay) or electron capture (e) do not occur in a nuclear reactor except as products of knockout reactions; their decays are marked with arrows pointing down-right.

Due to the long half-lives of the given nuclides, alpha decay plays almost no role in the formation and decay of the actinides in a power reactor, as the residence time of the nuclear fuel in the reactor core is rather short (a few years). Exceptions are the two relatively short-lived nuclides ^{242}Cm ($T_{1/2}$ = 163 d) and ^{236}Pu ($T_{1/2}$ = 2.9 y). Only for these two cases, the α decay is marked on the nuclide map by a long arrow pointing down-left. A few long-lived actinide isotopes, such as ^{244}Pu and ^{250}Cm, cannot be produced in reactors because neutron capture does not happen quickly enough to bypass the short-lived beta-decaying nuclides ^{243}Pu and ^{249}Cm; they can however be generated in nuclear explosions, which have much higher neutron fluxes.

30.4 DISTRIBUTION IN NATURE

Thorium and uranium are the most abundant actinides in nature with the respective mass concentrations of 16 ppm and 4 ppm. Uranium mostly occurs in the Earth's crust as a mixture of its oxides in the mineral uraninite, which is also called pitchblende because of its black color. There are several dozens of other uranium minerals such as carnotite ($KUO_2VO_4{\cdot}3H_2O$) and autunite ($Ca(UO_2)_2(PO_4)_2{\cdot}nH_2O$). The isotopic composition of natural uranium is ^{238}U (relative abundance 99.2742%), ^{235}U (0.7204%) and ^{234}U (0.0054%); of these ^{238}U has the largest half-life of 4.51×10^9 years. The worldwide production of uranium in 2009 amounted to 50,572 tonnes, of which 27.3% was mined in Kazakhstan. Other important uranium mining countries are Canada (20.1%), Australia (15.7%), Namibia (9.1%), Russia (7.0%), and Niger (6.4%).

Table 3: Content of Plutonium in Uranium and Thorium Ores

Ore	Location	Uranium Content, %	Mass Ratio ^{239}Pu/ore	Ratio ^{239}Pu/U ($\times10^{-12}$)
Uraninite	Canada	13.5	9.1×10^{-12}	7.1
Uraninite	Congo	38	4.8×10^{-12}	12
Uraninite	Colorado, US	50	3.8×10^{-12}	7.7
Monazite	Brazil	0.24	2.1×10^{-14}	8.3
Monazite	North Carolina, US	1.64	5.9×10^{-14}	3.6
Fergusonite	-	0.25	$<1\times10^{-14}$	<4
Carnotite	-	10	$<4\times10^{-14}$	<0.4

The most abundant thorium minerals are thorianite (ThO_2), thorite ($ThSiO_4$) and monazite, ((Th,Ca,Ce)PO_4). Most thorium minerals contain uranium and vice versa; and they all have significant fraction of lanthanides. Rich deposits of thorium minerals are located in the United States (440,000 tonnes), Australia and India (~300,000 tonnes each) and Canada (~100,000 tonnes).

The abundance of actinium in the Earth's crust is only about 5×10^{-15}%. Actinium is mostly present in uranium-containing, but also in other minerals, though in much smaller quantities.

The content of actinium in most natural objects corresponds to the isotopic equilibrium of parent isotope ^{235}U, and it is not affected by the weak Ac migration. Protactinium is more abundant (10^{-12}%) in the Earth's crust than actinium. It was discovered in uranium ore in 1913 by Fajans and Göhring. As actinium, the distribution of protactinium follows that of ^{235}U.

The half-life of the longest-lived isotope of neptunium, ^{237}Np, is negligible compared to the age of the Earth. Thus neptunium is present in nature in negligible amounts produced as intermediate decay products of other isotopes. Traces of plutonium in uranium minerals were first found in 1942, and the more systematic results on ^{239}Pu are summarized in the table (no other plutonium isotopes could be detected in those samples). The upper limit of abundance of the longest-living isotope of plutonium, ^{244}Pu, is $3\times10^{-20}\%$. Plutonium could not be detected in samples of lunar soil. Owing to its scarcity in nature, most plutonium is produced synthetically.

30.5 PROPERTIES

Actinides have similar properties to lanthanides. Just as the 4f electron shells are filled in the lanthanides, the 5f electron shells are filled in the actinides. Because the 5f, 6d, 7s, and 7p shells are close in energy, many irregular configurations arise; thus, in gas-phase atoms, just as the first 4f electron only appears in cerium, so the first 5f electron appears even later, in protactinium. However, just as lanthanum is the first element to use the 4f shell in compounds, so actinium is the first element to use the 5f shell in compounds. The f-shells complete their filling together, at ytterbium and nobelium. The first experimental evidence for the filling of the 5f shell in actinides was obtained by McMillan and Abelson in 1940. As in lanthanides, the ionic radius of actinides monotonically decreases with atomic number.

The shift of electron configurations in the gas phase does not always match the chemical behaviour. For example, the early-transition-metal-like prominence of the highest oxidation state, corresponding to removal of all valence electrons, extends up to uranium even though the 5f shells begin filling before that. On the other hand, electron configurations resembling the lanthanide congeners already begin at plutonium, even though lanthanide-like behaviour does not become dominant until the second half of the series begins at curium. The elements between uranium and curium form a transition between these two kinds of behaviour, where higher oxidation states continue to exist, but lose stability with respect to the +3 state. The +2 state becomes more important near the end of the series, and is the most stable oxidation state for nobelium, the last 5f element. Oxidation states rise again only after nobelium, showing that a new series of 6d transition metals has begun: lawrencium shows only the +3 oxidation state, and rutherfordium only the +4 state, making them congeners of lutetium and hafnium in the 5d row.

Actinides are typical metals. All of them are soft and have a silvery color (but tarnish in air), relatively high density and plasticity. Some of them can be cut with a knife. Their electrical resistivity varies between 15 and 150 μΩ·cm. The hardness of thorium is similar to that of soft steel, so heated pure thorium can be rolled in sheets and pulled into wire. Thorium is nearly half as dense as uranium and plutonium, but is harder than either of them. All actinides are radioactive, paramagnetic, and, with the exception of actinium, have several crystalline phases: plutonium has seven, and uranium, neptunium and californium three. The crystal structures of protactinium, uranium, neptunium and plutonium do not have clear analogs among the lanthanides and are more similar to those of the 3*d*-transition metals.

All actinides are pyrophoric, especially when finely divided, that is, they spontaneously ignite upon reaction with air at room temperature. The melting point of actinides does not have a clear dependence on the number of *f*-electrons. The unusually low melting point of neptunium and plutonium (~640 °C) is explained by hybridization of 5*f* and 6*d* orbitals and the formation of directional bonds in these metals.

30.5.1 Chemical Properties

Like the lanthanides, all actinides are highly reactive with halogens and chalcogens; however, the actinides react more easily. Actinides, especially those with a small number of 5*f*-electrons, are prone to hybridization. This is explained by the similarity of the electron energies at the 5*f*, 7*s* and 6*d* shells. Most actinides exhibit a larger variety of valence states, and the most stable are +6 for uranium, +5 for protactinium and neptunium, +4 for thorium and plutonium and +3 for actinium and other actinides.

Actinium is chemically similar to lanthanum, which is explained by their similar ionic radii and electronic structures. Like lanthanum, actinium almost always has an oxidation state of +3 in compounds, but it is less reactive and has more pronounced basic properties. Among other trivalent actinides Ac^{3+} is least acidic, i.e. has the weakest tendency to hydrolyze in aqueous solutions.

Thorium is rather active chemically. Owing to lack of electrons on 6*d* and 5*f* orbitals, tetravalent thorium compounds are colorless. At pH < 3, solutions of thorium salts are dominated by the cations $[Th(H_2O)_8]^{4+}$. The Th^{4+} ion is relatively large, and depending on the coordination number can have a radius between 0.95 and 1.14 Å. As a result, thorium salts have a weak tendency to hydrolyse. The distinctive ability of thorium salts is their high solubility both in water and polar organic solvents.

Protactinium exhibits two valence states; the +5 is stable, and the +4 state easily oxidizes to protactinium(V). Thus tetravalent protactinium in solutions is obtained by the action of strong reducing agents in a hydrogen atmosphere. Tetravalent protactinium is chemically similar to uranium(IV) and thorium(IV). Fluorides, phosphates, hypophosphates, iodates and phenylarsonates of protactinium(IV) are insoluble in water and dilute acids.

Protactinium forms soluble carbonates. The hydrolytic properties of pentavalent protactinium are close to those of tantalum(V) and niobium(V). The complex chemical behavior of protactinium is a consequence of the start of the filling of the 5*f* shell in this element.

Uranium has a valence from 3 to 6, the last being most stable. In the hexavalent state, uranium is very similar to the group 6 elements. Many compounds of uranium(IV) and uranium(VI) are non-stoichiometric, i.e., have variable composition. For example, the actual chemical formula of uranium dioxide is UO_{2+x}, where *x* varies between -0.4 and 0.32. Uranium(VI) compounds are weak oxidants. Most of them contain the linear "uranyl" group, UO_2^{2+}. Between 4 and 6 ligands can be accommodated in an equatorial plane perpendicular to the uranyl group. The uranyl group acts as a hard acid and forms stronger complexes with oxygen-donor ligands than with nitrogen-donor ligands. NpO_2^{2+} and PuO_2^{2+} are also the common form of Np and Pu in the +6 oxidation state. Uranium(IV) compounds exhibit reducing properties, e.g., they are easily oxidized by atmospheric oxygen. Uranium(III) is a very strong reducing agent. Owing to the presence of d-shell, uranium (as well as many other actinides) forms organometallic compounds, such as $U^{III}(C_5H_5)_3$ and $U^{IV}(C_5H_5)_4$.

Neptunium has valence states from 3 to 7, which can be simultaneously observed in solutions. The most stable state in solution is +5, but the valence +4 is preferred in solid neptunium compounds. Neptunium metal is very reactive. Ions of neptunium are prone to hydrolysis and formation of coordination compounds.

Plutonium also exhibits valence states between 3 and 7 inclusive, and thus is chemically similar to neptunium and uranium. It is highly reactive, and quickly forms an oxide film in air. Plutonium reacts with hydrogen even at temperatures as low as 25–50 °C; it also easily forms halides and intermetallic compounds. Hydrolysis reactions of plutonium ions of different oxidation states are quite diverse. Plutonium(V) can enter polymerization reactions.

The largest chemical diversity among actinides is observed in americium, which can have valence between 2 and 6. Divalent americium is obtained only in dry compounds and non-aqueous solutions (acetonitrile). Oxidation states +3, +5 and +6 are typical for aqueous solutions, but also in the solid state. Tetravalent americium forms stable solid compounds (dioxide, fluoride and hydroxide) as well as complexes in aqueous solutions. It was reported that in alkaline solution americium can be oxidized to the heptavalent state, but these data proved erroneous. The most stable valence of americium is 3 in aqueous solution and 3 or 4 in solid compounds.

Valence 3 is dominant in all subsequent elements up to lawrencium (with the exception of nobelium). Curium can be tetravalent in solids (fluoride, dioxide). Berkelium, along with a valence of +3, also shows the valence of +4, more stable than that of curium; the valence 4 is observed in solid fluoride and dioxide. The stability of Bk^{4+} in aqueous solution is close to that of Ce^{4+}. Only valence 3 was observed for californium, einsteinium and fermium. The divalent state is proven for mendelevium and nobelium, and in nobelium it is more stable than the trivalent state. Lawrencium shows valence 3 both in solutions and solids.

The redox potential $E_{\frac{M^{4+}}{AnO_2^{2+}}}$ increases from -0.32 V in uranium, through 0.34 V (Np) and 1.04 V (Pu) to 1.34 V in americium revealing the increasing reduction ability of the An^{4+} ion from americium to uranium. All actinides form AnH_3 hydrides of black color with salt-like properties. Actinides also produce carbides with the general formula of AnC or AnC_2 (U_2C_3 for uranium) as well as sulfides An_2S_3 and AnS_2.

30.6 COMPOUNDS

Some actinides can exist in several oxide forms such as An_2O_3, AnO_2, An_2O_5 and AnO_3. For all actinides, oxides AnO_3 are amphoteric and An_2O_3, AnO_2 and An_2O_5 are basic, they easily react with water, forming bases:

$$An_2O_3 + 3\,H_2O \rightarrow 2\,An(OH)_3.$$

These bases are poorly soluble in water and by their activity are close to the hydroxides of rare-earth metals. $Np(OH)_3$ has not yet been synthesized, $Pu(OH)_3$ has a blue color while $Am(OH)_3$ is pink and $Cm(OH)_3$ is colorless. $Bk(OH)_3$ and $Cf(OH)_3$ are also known, as are tetravalent hydroxides for Np, Pu and Am and pentavalent for Np and Am.

The strongest base is of actinium. All compounds of actinium are colorless, except for black actinium sulfide (Ac_2S_3). Dioxides of tetravalent actinides crystallize in the cubic system, same as in calcium fluoride.

Thorium reacting with oxygen exclusively forms the dioxide:

$$Th + O_2 \xrightarrow{1000\,^\circ C} \overbrace{ThO_2}^{\text{Thorium dioxide}}$$

Thorium dioxide is a refractory material with the highest melting point among any known oxide (3390°C). Adding 0.8–1% ThO_2 to tungsten stabilizes its structure, so the doped filaments have better mechanical stability to vibrations. To dissolve ThO_2 in acids, it is heated to 500–600°C; heating above 600°C produces a very resistant to acids and other reagents form of ThO_2. Small addition of fluoride ions catalyses dissolution of thorium dioxide in acids.

Two protactinium oxides have been obtained: PaO_2 (black) and Pa_2O_5 (white); the former is isomorphic with ThO_2 and the latter is easier to obtain. Both oxides are basic, and $Pa(OH)_5$ is a weak, poorly soluble base.

Decomposition of certain salts of uranium, for example $UO_2(NO_3)\cdot 6H_2O$ in air at 400°C, yields orange or yellow UO_3. This oxide is amphoteric and forms several hydroxides, the most stable being uranyl hydroxide $UO_2(OH)_2$. Reaction of uranium(VI) oxide with hydrogen results in uranium dioxide, which is similar in its properties with ThO_2. This oxide is also basic and corresponds to the uranium hydroxide $U(OH)_4$.

Plutonium, neptunium and americium form two basic oxides: An_2O_3 and AnO_2. Neptunium trioxide is unstable; thus, only Np_3O_8 could be obtained so far. However, the oxides of plutonium and neptunium with the chemical formula AnO_2 and An_2O_3 are well characterized.

30.6.1 Salts

Actinides easily react with halogens forming salts with the formulas MX_3 and MX_4 (X = halogen). So the first berkelium compound, $BkCl_3$, was synthesized in 1962 with an amount of 3 nanograms. Like the halogens of rare earth elements, actinide chlorides, bromides, and iodides are water-soluble, and fluorides are insoluble. Uranium easily yields a colorless hexafluoride, which sublimates at a temperature of 56.5°C; because of its volatility, it is used in the separation of uranium isotopes with gas centrifuge or gaseous diffusion. Actinide hexafluorides have properties close to anhydrides. They are very sensitive to moisture and hydrolyze forming AnO_2F_2. The pentachloride and black hexachloride of uranium were synthesized, but they are both unstable.

Action of acids on actinides yields salts, and if the acids are non-oxidizing then the actinide in the salt is in low-valence state:

$$U + 2\,H_2SO_4 \rightarrow U(SO_4)_2 + 2\,H_2$$

$$2\,Pu + 6\,HCl \rightarrow 2\,PuCl_3 + 3\,H_2$$

However, in these reactions the regenerating hydrogen can react with the metal, forming the corresponding hydride. Uranium reacts with acids and water much more easily than thorium.

Actinide salts can also be obtained by dissolving the corresponding hydroxides in acids. Nitrates, chlorides, sulfates and perchlorates of actinides are water-soluble. When crystallizing from aqueous solutions, these salts form hydrates, such as $Th(NO_3)_4{\cdot}6H_2O$, $Th(SO_4)_2{\cdot}9H_2O$ and $Pu_2(SO_4)_3{\cdot}7H_2O$. Salts of high-valence actinides easily hydrolyze. So, colorless sulfate, chloride, perchlorate and nitrate of thorium transform into basic salts with formulas $Th(OH)_2SO_4$ and $Th(OH)_3NO_3$. The solubility and insolubility of trivalent and tetravalent actinides is like that of lanthanide salts. So phosphates, fluorides, oxalates, iodates and carbonates of actinides are weakly soluble in water; they precipitate as hydrates, such as $ThF_4{\cdot}3H_2O$ and $Th(CrO_4)_2{\cdot}3H_2O$.

Actinides with oxidation state +6, except for the AnO_2^{2+}-type cations, form $[AnO_4]^{2-}$, $[An_2O_7]^{2-}$ and other complex anions. For example, uranium, neptunium and plutonium form salts of the Na_2UO_4 (uranate) and $(NH_4)_2U_2O_7$ (diuranate) types. In comparison with lanthanides, actinides more easily form coordination compounds, and this ability increases with the actinide valence. Trivalent actinides do not form fluoride coordination compounds, whereas tetravalent thorium forms K_2ThF_6, $KThF_5$, and even K_5ThF_9 complexes. Thorium also forms the corresponding sulfates (for example $Na_2SO_4{\cdot}Th(SO_4)_2{\cdot}5H_2O$), nitrates and thiocyanates. Salts with the general formula $An_2Th(NO_3)_6{\cdot}nH_2O$ are of coordination nature, with the coordination number of thorium equal to 12. Even easier is to produce complex salts of pentavalent and hexavalent actinides. The most stable coordination compounds of actinides – tetravalent thorium and uranium – are obtained in reactions with diketones, e.g. acetylacetone.

30.7 APPLICATIONS

While actinides have some established daily-life applications, such as in smoke detectors (americium) and gas mantles (thorium), they are mostly used in nuclear weapons and as fuel in nuclear reactors. The last two areas exploit the property of actinides to release enormous energy in nuclear reactions, which under certain conditions may become self-sustaining chain reactions.

The most important isotope for nuclear power applications is uranium-235. It is used in the thermal reactor, and its concentration in natural uranium does not exceed 0.72%. This isotope strongly absorbs thermal neutrons releasing much energy. One fission act of 1 gram of ^{235}U converts into about 1 MW·day. Of importance, is that $^{235}{}_{92}U$ emits more neutrons than it absorbs; upon reaching the critical mass, $^{235}{}_{92}U$ enters into a self-sustaining chain reaction. Typically, uranium nucleus is divided into two fragments with the release of 2–3 neutrons, for example:

$$^{235}{}_{92}U + {}^{1}{}_{0}n \longrightarrow {}^{115}{}_{45}Rh + {}^{118}{}_{47}Ag + 3{}^{1}{}_{0}n$$

Other promising actinide isotopes for nuclear power are thorium-232 and its product from the thorium fuel cycle, uranium-233.

Emission of neutrons during the fission of uranium is important not only for maintaining the nuclear chain reaction, but also for the synthesis of the heavier actinides. Uranium-239 converts via β-decay into plutonium-239, which, like uranium-235, is capable of spontaneous fission. The world's first nuclear reactors were built not for energy, but for producing plutonium-239 for nuclear weapons.

About half of produced thorium is used as the light-emitting material of gas mantles. Thorium is also added into multicomponent alloys of magnesium and zinc. Mg-Th alloys are light and strong, but also have high melting point and ductility and thus are widely used in the aviation industry and in the production of missiles. Thorium also has good electron emission properties, with long lifetime and low potential barrier for the emission. The relative content of thorium and uranium isotopes is widely used to estimate the age of various objects, including stars.

The major application of plutonium has been in nuclear weapons, where the isotope plutonium-239 was a key component due to its ease of fission and availability. Plutonium-based designs allow reducing the critical mass to about a third of that for uranium-235. The "Fat Man"-type plutonium bombs produced during the Manhattan Project used explosive compression of plutonium to obtain significantly higher densities than normal, combined with a central neutron source to begin the reaction and increase efficiency. Thus only 6.2 kg of plutonium was needed for an explosive yield equivalent to 20 kilotons of TNT. Hypothetically, as little as 4 kg of plutonium—and maybe even less—could be used to make a single atomic bomb using very sophisticated assembly designs.

Plutonium-238 is potentially more efficient isotope for nuclear reactors, since it has smaller critical mass than uranium-235, but it continues to release much thermal energy (0.56 W/g) by decay even when the fission chain reaction is stopped by control rods. Its application is limited by its high price (about US$1000/g). This isotope has been used in thermopiles and water distillation systems of some space satellites and stations. The Galileo and Apollo spacecraft (e.g. Apollo 14) had heaters powered by kilogram quantities of plutonium-238 oxide; this heat is also transformed into electricity with thermopiles. The decay of plutonium-238 produces relatively harmless alpha particles and is not accompanied by gamma rays. Therefore, this isotope (~160 mg) is used as the energy source in heart pacemakers where it lasts about 5 times longer than conventional batteries.

Actinium-227 is used as a neutron source. Its high specific energy (14.5 W/g) and the possibility of obtaining significant quantities of thermally stable compounds are attractive for use in long-lasting thermoelectric generators for remote use. ^{228}Ac is used as an indicator of radioactivity in chemical research, as it emits high-energy electrons (2.18 MeV) that can be easily detected. ^{228}Ac-^{228}Ra mixtures are widely used as an intense gamma-source in industry and medicine.

Development of self-glowing actinide-doped materials with durable crystalline matrices is a new area of actinide utilization as the addition of alpha-emitting radionuclides to some glasses and crystals may confer luminescence.

30.8 TOXICITY

Radioactive substances can harm human health via (i) local skin contamination, (ii) internal exposure due to ingestion of radioactive isotopes, and (iii) external overexposure by β-activity and γ-radiation. Together with radium and transuranium elements, actinium is one of the most dangerous radioactive poisons with high specific α-activity. The most important feature of actinium is its ability to accumulate and remain in the surface layer of skeletons. At the initial stage of poisoning, actinium accumulates in the liver. Another danger of actinium is that it undergoes radioactive decay faster than being excreted. Adsorption from the digestive tract is much smaller (~0.05%) for actinium than radium.

Protactinium in the body tends to accumulate in the kidneys and bones. The maximum safe dose of protactinium in the human body is 0.03 μCi that corresponds to 0.5 micrograms of ^{231}Pa. This isotope, which might be present in the air as aerosol, is 2.5×10^8 times more toxic than hydrocyanic acid.

Plutonium, when entering the body through air, food or blood (e.g. a wound), mostly settles in the lungs, liver and bones with only about 10% going to other organs, and remains there for decades. The long residence time of plutonium in the body is partly explained by its poor solubility in water. Some isotopes of plutonium emit ionizing α-radiation, which damages the surrounding cells. The median lethal dose (LD_{50}) for 30 days in dogs after intravenous injection of plutonium is 0.32 milligram per kg of body mass, and thus the lethal dose for humans is approximately 22 mg for a person weighing 70 kg; the amount for respiratory exposure should be approximately four times greater. Another estimate assumes that plutonium is 50 times less toxic than radium, and thus permissible content of plutonium in the body should be 5 μg or 0.3 μCi. Such amount is nearly invisible under microscope. After trials on animals, this maximum permissible dose was reduced to 0.65 μg or 0.04 μCi. Studies on animals also revealed that the most dangerous plutonium exposure route is through inhalation, after which 5–25% of inhaled substances is retained in the body. Depending on the particle size and solubility of the plutonium compounds, plutonium is localized either in the lungs or in the lymphatic system, or is absorbed in the blood and then transported to the liver and bones. Contamination via food is the least likely way. In this case, only about 0.05% of soluble and 0.01% of insoluble compounds of plutonium absorbs into blood, and the rest is excreted. Exposure of damaged skin to plutonium would retain nearly 100% of it.

Using actinides in nuclear fuel, sealed radioactive sources or advanced materials such as self-glowing crystals has many potential benefits. However, a serious concern is the extremely high radiotoxicity of actinides and their migration in the environment. Use of chemically unstable forms of actinides in MOX and sealed radioactive sources is not appropriate by modern safety standards. There is a challenge to develop stable and durable actinide-bearing materials, which provide safe storage, use and final disposal. A key need is application of actinide solid solutions in durable crystalline host phases.

PART SIX

OTHER TOPICS

CHAPTER 31
THE ATOMIC NUCLEUS

The atomic nucleus is the small, dense region consisting of protons and neutrons at the center of an atom, discovered in 1911 by Ernest Rutherford based on the 1909 Geiger–Marsden gold foil experiment. After the discovery of the neutron in 1932, models for a nucleus composed of protons and neutrons were quickly developed by Dmitri Ivanenko and Werner Heisenberg. An atom is composed of a positively charged nucleus, with a cloud of negatively charged electrons surrounding it, bound together by electrostatic force. Almost all of the mass of an atom is located in the nucleus, with a very small contribution from the electron cloud. Protons and neutrons are bound together to form a nucleus by the nuclear force.

The diameter of the nucleus is in the range of 1.70 fm (1.70×10^{-15} m) for hydrogen (the diameter of a single proton) to about 11.7 fm for uranium. These dimensions are much smaller than the diameter of the atom itself (nucleus + electron cloud), by a factor of about 26,634 (uranium atomic radius is about 156 pm (156×10^{-12} m)) to about 60,250 (hydrogen atomic radius is about 52.92 pm).

The branch of physics concerned with the study and understanding of the atomic nucleus, including its composition and the forces that bind it together, is called nuclear physics.

31.1 HISTORY

The nucleus was discovered in 1911, as a result of Ernest Rutherford's efforts to test Thomson's "plum pudding model" of the atom. The electron had already been discovered by J. J. Thomson. Knowing that atoms are electrically neutral, J. J. Thomson postulated that there must be a positive charge as well. In his plum pudding model, Thomson suggested that an atom consisted of negative electrons randomly scattered within a sphere of positive charge. Ernest Rutherford later devised an experiment with his research partner Hans Geiger and with help of Ernest Marsden, that involved the deflection of alpha particles (helium nuclei) directed at a thin sheet of metal foil. He reasoned that if J. J. Thomson's model were correct, the positively charged alpha particles would easily pass through the foil with very little deviation in their paths, as the foil should act as electrically neutral if the negative and positive charges are so intimately mixed as to make it appear neutral. To his surprise, many of the particles were deflected at very large angles.

Because the mass of an alpha particle is about 8000 times that of an electron, it became apparent that a very strong force must be present if it could deflect the massive and fast moving alpha particles. He realized that the plum pudding model could not be accurate and that the deflections of the alpha particles could only be explained if the positive and negative charges were separated from each other and that the mass of the atom was a concentrated point of positive charge. This justified the idea of a nuclear atom with a dense center of positive charge and mass.

31.2 PRINCIPLES

The nucleus of an atom consists of neutrons and protons, which in turn are the manifestation of more elementary particles, called quarks, that are held in association by the nuclear strong force in certain stable combinations of hadrons, called baryons. The nuclear strong force extends far enough from each baryon so as to bind the neutrons and protons together against the repulsive electrical force between the positively charged protons. The nuclear strong force has a very short range, and essentially drops to zero just beyond the edge of the nucleus. The collective action of the positively charged nucleus is to hold the electrically negative charged electrons in their orbits about the nucleus. The collection of negatively charged electrons orbiting the nucleus display an affinity for certain configurations and numbers of electrons that make their orbits stable. Which chemical element an atom represents is determined by the number of protons in the nucleus; the neutral atom will have an equal number of electrons orbiting that nucleus. Individual chemical elements can create more stable electron configurations by combining to share their electrons. It is that sharing of electrons to create stable electronic orbits about the nuclei that appears to us as the chemistry of our macro world.

Protons define the entire charge of a nucleus, and hence its chemical identity. Neutrons are electrically neutral, but contribute to the mass of a nucleus to nearly the same extent as the protons. Neutrons can explain the phenomenon of isotopes (same atomic number with different atomic mass). The main role of neutrons is to reduce electrostatic repulsion inside the nucleus.

31.3 COMPOSITION AND SHAPE

Protons and neutrons are fermions, with different values of the strong isospin quantum number, so two protons and two neutrons can share the same space wave function since they are not identical quantum entities. They are sometimes viewed as two different quantum states of the same particle, the *nucleon*. Two fermions, such as two protons, or two neutrons, or a proton + neutron (the deuteron) can exhibit bosonic behavior when they become loosely bound in pairs, which have integer spin.

In the rare case of a hypernucleus, a third baryon called a hyperon, containing one or more strange quarks and/or other unusual quark(s), can also share the wave function. However, this type of nucleus is extremely unstable and not found on Earth except in high-energy physics experiments.

The neutron has a positively charged core of radius ≈0.3 fm surrounded by a compensating negative charge of radius between 0.3 fm and 2 fm. The proton has an approximately exponentially decaying positive charge distribution with a mean square radius of about 0.8 fm.

The shape of the atomic nucleus can be spherical, rugby ball-shaped (prolate deformation), discus-shaped (oblate deformation), triaxial (a combination of oblate and prolate deformation) or pear-shaped.

31.4 FORCES

Nuclei are bound together by the residual strong force (nuclear force). The residual strong force is a minor residuum of the strong interaction which binds quarks together to form protons and neutrons.

This force is much weaker *between* neutrons and protons because it is mostly neutralized within them, in the same way that electromagnetic forces *between* neutral atoms (such as van der Waals forces that act between two inert gas atoms) are much weaker than the electromagnetic forces that hold the parts of the atoms together internally (for example, the forces that hold the electrons in an inert gas atom bound to its nucleus).

The nuclear force is highly attractive at the distance of typical nucleon separation, and this overwhelms the repulsion between protons due to the electromagnetic force, thus allowing nuclei to exist. However, the residual strong force has a limited range because it decays quickly with distance; thus only nuclei smaller than a certain size can be completely stable.

The largest known completely stable nucleus (i.e. stable to alpha, beta, and gamma decay) is lead-208 which contains a total of 208 nucleons (126 neutrons and 82 protons). Nuclei larger than this maximum are unstable and tend to be increasingly short-lived with larger numbers of nucleons. However, bismuth-209 is also stable to beta decay and has the longest half-life to alpha decay of any known isotope, estimated at a billion times longer than the age of the universe.

The residual strong force is effective over a very short range (usually only a few femtometres (fm); roughly one or two nucleon diameters) and causes an attraction between any pair of nucleons.

For example, between a proton and a neutron to form a deuteron [NP], and also between protons and protons, and neutrons and neutrons.

31.5 HALO NUCLEI AND NUCLEAR FORCE RANGE LIMITS

The effective absolute limit of the range of the nuclear force (also known as residual strong force) is represented by halo nuclei such as lithium-11 or boron-14, in which dineutrons, or other collections of neutrons, orbit at distances of about 10 fm (roughly similar to the 8 fm radius of the nucleus of uranium-238). These nuclei are not maximally dense. Halo nuclei form at the extreme edges of the chart of the nuclides—the neutron drip line and proton drip line—and are all unstable with short half-lives, measured in milliseconds; for example, lithium-11 has a half-life of 8.8 ms.

Halos in effect represent an excited state with nucleons in an outer quantum shell which has unfilled energy levels "below" it (both in terms of radius and energy). The halo may be made of either neutrons [NN, NNN] or protons [PP, PPP]. Nuclei which have a single neutron halo include ^{11}Be and ^{19}C. A two-neutron halo is exhibited by ^{6}He, ^{11}Li, ^{17}B, ^{19}B and ^{22}C. Two-neutron halo nuclei break into three fragments, never two, and are called *Borromean nuclei* because of this behavior (referring to a system of three interlocked rings in which breaking any ring frees both of the others). ^{8}He and ^{14}Be both exhibit a four-neutron halo. Nuclei which have a proton halo include ^{8}B and ^{26}P. A two-proton halo is exhibited by ^{17}Ne and ^{27}S. Proton halos are expected to be more rare and unstable than the neutron examples, because of the repulsive electromagnetic forces of the halo proton(s).

31.6 NUCLEAR MODELS

Although the standard model of physics is widely believed to completely describe the composition and behavior of the nucleus, generating predictions from theory is much more difficult than for most other areas of particle physics. This is due to two reasons:

- In principle, the physics within a nucleus can be derived entirely from quantum chromodynamics (QCD). In practice however, current computational and mathematical approaches for solving QCD in low-energy systems such as the nuclei are extremely limited. This is due to the phase transition that occurs between high-energy quark matter and low-energy hadronic matter, which renders perturbative techniques unusable, making it difficult to construct an accurate QCD-derived model of the forces between nucleons. Current approaches are limited to either phenomenological models such as the Argonne v18 potential or chiral effective field theory.
- Even if the nuclear force is well constrained, a significant amount of computational power is required to accurately compute the properties of nuclei *ab initio*. Developments in many-body theory have made this

possible for many low mass and relatively stable nuclei, but further improvements in both computational power and mathematical approaches are required before heavy nuclei or highly unstable nuclei can be tackled.

Historically, experiments have been compared to relatively crude models that are necessarily imperfect. None of these models can completely explain experimental data on nuclear structure.

The nuclear radius (R) is considered to be one of the basic quantities that any model must predict. For stable nuclei (not halo nuclei or other unstable distorted nuclei) the nuclear radius is roughly proportional to the cube root of the mass number (A) of the nucleus, and particularly in nuclei containing many nucleons, as they arrange in more spherical configurations:

The stable nucleus has approximately a constant density and therefore the nuclear radius R can be approximated by the following formula,

$$R = r_0 A^{1/3}$$

where A = Atomic mass number (the number of protons Z, plus the number of neutrons N) and r_0=1.25 fm=$^{15-}$10 × 1.25 m. In this equation, the constant r_0 varies by 0.2 fm, depending on the nucleus in question, but this is less than 20% change from a constant.

In other words, packing protons and neutrons in the nucleus gives *approximately* the same total size result as packing hard spheres of a constant size (like marbles) into a tight spherical or almost spherical bag (some stable nuclei are not quite spherical, but are known to be prolate).

31.6.1 Cluster Model

The cluster model describes the nucleus as a molecule-like collection of proton-neutron groups (e.g., alpha particles) with one or more valence neutrons occupying molecular orbitals.

31.6.2 Liquid Drop Model

Early models of the nucleus viewed the nucleus as a rotating liquid drop. In this model, the trade-off of long-range electromagnetic forces and relatively short-range nuclear forces, together cause behavior which resembled surface tension forces in liquid drops of different sizes.

This formula is successful at explaining many important phenomena of nuclei, such as their changing amounts of binding energy as their size and composition changes, but it does not explain the special stability which occurs when nuclei have special "magic numbers" of protons or neutrons.

The terms in the semi-empirical mass formula, which can be used to approximate the binding energy of many nuclei, are considered as the sum of five types of energies.

Then the picture of a nucleus as a drop of incompressible liquid roughly accounts for the observed variation of binding energy of the nucleus:

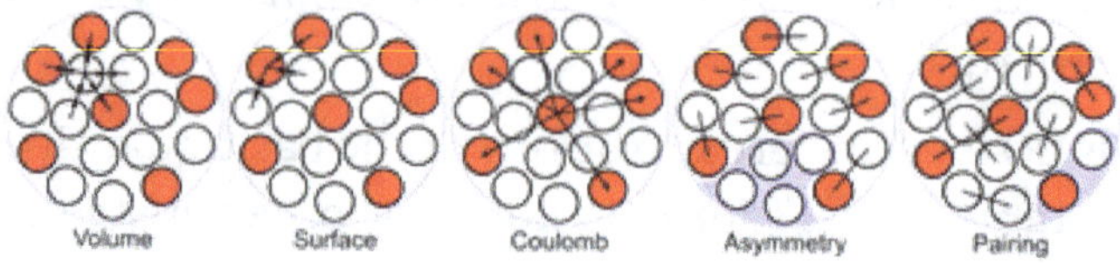

Volume Energy

When an assembly of nucleons of the same size is packed together into the smallest volume, each interior nucleon has a certain number of other nucleons in contact with it. So, this nuclear energy is proportional to the volume.

Surface Energy

A nucleon at the surface of a nucleus interacts with fewer other nucleons than one in the interior of the nucleus and hence its binding energy is less. This surface energy term takes that into account and is therefore negative and is proportional to the surface area.

Coulomb Energy

The electric repulsion between each pair of protons in a nucleus contributes toward decreasing its binding energy.

Asymmetry Energy

An energy associated with the Pauli exclusion principle. Were it not for the Coulomb energy, the most stable form of nuclear matter would have the same number of neutrons as protons, since unequal numbers of neutrons and protons imply filling higher energy levels for one type of particle, while leaving lower energy levels vacant for the other type.

Pairing Energy

An energy which is a correction term that arises from the tendency of proton pairs and neutron pairs to occur. An even number of particles is more stable than an odd number.

31.6.3 Shell Models and other Quantum Models

A number of models for the nucleus have also been proposed in which nucleons occupy orbitals, much like the atomic orbitals in atomic physics theory. These wave models imagine nucleons to be either sizeless point particles in potential wells, or else probability waves as in the "optical model", frictionlessly orbiting at high speed in potential wells.

In the above models, the nucleons may occupy orbitals in pairs, due to being fermions, which allows explanation of even/odd *Z* and *N* effects well known from experiments.

The exact nature and capacity of nuclear shells differs from those of electrons in atomic orbitals, primarily because the potential well in which the nucleons move (especially in larger nuclei) is quite different from the central electromagnetic potential well which binds electrons in atoms. Some resemblance to atomic orbital models may be seen in a small atomic nucleus like that of helium-4, in which the two protons and two neutrons separately occupy 1s orbitals analogous to the 1s orbital for the two electrons in the helium atom, and achieve unusual stability for the same reason. Nuclei with 5 nucleons are all extremely unstable and short-lived, yet, helium-3, with 3 nucleons, is very stable even with lack of a closed 1s orbital shell. Another nucleus with 3 nucleons, the triton hydrogen-3 is unstable and will decay into helium-3 when isolated. Weak nuclear stability with 2 nucleons {NP} in the 1s orbital is found in the deuteron hydrogen-2, with only one nucleon in each of the proton and neutron potential wells.

While each nucleon is a fermion, the {NP} deuteron is a boson and thus does not follow Pauli Exclusion for close packing within shells. Lithium-6 with 6 nucleons is highly stable without a closed second 1p shell orbital. For light nuclei with total nucleon numbers 1 to 6 only those with 5 do not show some evidence of stability. Observations of beta-stability of light nuclei outside closed shells indicate that nuclear stability is much more complex than simple closure of shell orbitals with magic numbers of protons and neutrons.

For larger nuclei, the shells occupied by nucleons begin to differ significantly from electron shells, but nevertheless, present nuclear theory does predict the magic numbers of filled nuclear shells for both protons and neutrons. The closure of the stable shells predicts unusually stable configurations, analogous to the noble group of nearly-inert gases in chemistry. An example is the stability of the closed shell of 50 protons, which allows tin to have 10 stable isotopes, more than any other element. Similarly, the distance from shell-closure explains the unusual instability of isotopes which have far from stable numbers of these particles, such as the radioactive elements 43 (technetium) and 61 (promethium), each of which is preceded and followed by 17 or more stable elements.

There are however problems with the shell model when an attempt is made to account for nuclear properties well away from closed shells. This has led to complex *post hoc* distortions of the shape of the potential well to fit experimental data, but the question remains whether these mathematical manipulations actually correspond to the spatial deformations in real nuclei. Problems with the shell model have led some to propose realistic two-body and three-body nuclear force effects involving nucleon clusters and then build the nucleus on this basis. Three such cluster models are the 1936 Resonating Group Structure model of John Wheeler, Close-Packed Spheron Model of Linus Pauling and the 2D Ising Model of MacGregor.

31.6.4 Consistency between Models

As with the case of superfluid liquid helium, atomic nuclei are an example of a state in which both (1) "ordinary" particle physical rules for volume and (2) non-intuitive quantum mechanical rules for a wave-like nature apply. In superfluid helium, the helium atoms have volume, and essentially "touch" each other, yet at the same time exhibit strange bulk properties, consistent with a Bose–Einstein condensation.

The nucleons in atomic nuclei also exhibit a wave-like nature and lack standard fluid properties, such as friction. For nuclei made of hadrons which are fermions, Bose-Einstein condensation does not occur, yet nevertheless, many nuclear properties can only be explained similarly by a combination of properties of particles with volume, in addition to the frictionless motion characteristic of the wave-like behavior of objects trapped in Erwin Schrödinger's quantum orbitals.

CHAPTER 32
SPECTRA

Spectra play a pivotal role in inorganic chemistry, serving as a fundamental tool for understanding the electronic structure of atoms and molecules. Spectroscopy, the study of spectra, allows chemists to analyze the interaction between matter and electromagnetic radiation. This interaction provides insights into molecular energy levels, bonding characteristics, and the identity of chemical species. The different types of spectra, including emission, absorption, and fluorescence, each provide unique information that is crucial for the interpretation of inorganic compounds and their behaviors.

32.1 TYPES OF SPECTRA

32.1.1 Emission Spectra

Emission spectra are produced when an atom or molecule absorbs energy and subsequently emits light as it returns to a lower energy state. When electrons in an atom are excited to higher energy levels, they can release energy in the form of photons when transitioning back to their ground state. The resulting emission spectrum is characterized by distinct lines or bands, each corresponding to specific wavelengths of emitted light. This spectrum is unique to each element, functioning as a "fingerprint" that can be used to identify the presence of elements in a sample.

Emission spectra can be categorized into continuous and line spectra. Continuous spectra, produced by solid, liquid, or densely packed gases, display a smooth range of colors. In contrast, line spectra arise from the emission of discrete wavelengths and are often observed in gaseous samples. The study of emission spectra is fundamental in fields such as astrophysics, where the composition of distant stars and galaxies can be inferred from their spectral lines.

32.1.2 Absorption Spectra

Absorption spectra are generated when light passes through a sample, and certain wavelengths are absorbed by the atoms or molecules present. As a result, the spectrum shows dark lines or bands superimposed on a continuous spectrum, corresponding to the wavelengths absorbed by the sample. Each element has a unique absorption spectrum, which can be used to identify the presence of specific elements or compounds.

The phenomenon of absorption is closely related to the energy levels of electrons within an atom. When a photon with the appropriate energy interacts with an electron, it can cause the electron to move to a higher energy state. The specific wavelengths absorbed depend on the energy differences between the electronic states of the atoms or molecules in the sample.

32.1.3 Fluorescence and Phosphorescence

Fluorescence and phosphorescence are types of luminescence that involve the absorption of light followed by the re-emission of light. In fluorescence, the absorbed energy is re-emitted almost immediately (within nanoseconds) as light of a longer wavelength. This rapid process is due to the electrons returning to their ground state without significant delay.

In contrast, phosphorescence involves a longer time delay before the re-emission of light. This occurs because the excited electrons transition to a triplet state, which is a metastable state.

The transition from the triplet state back to the ground state is "forbidden," resulting in a longer lifetime for the excited state and a delay in light emission. Phosphorescent materials can continue to glow for seconds or even hours after the excitation source has been removed.

32.2 THE IMPORTANCE OF SPECTRA IN INORGANIC CHEMISTRY

32.2.1 Identifying Elements and Compounds

One of the primary applications of spectroscopy in inorganic chemistry is the identification of elements and compounds. By analyzing the emission or absorption spectra of a sample, chemists can determine its composition and the presence of specific elements.

This technique is widely used in analytical chemistry, environmental monitoring, and materials science. For example, atomic absorption spectroscopy (AAS) and inductively coupled plasma optical emission spectroscopy (ICP-OES) are common methods for determining the concentration of trace elements in various matrices.

32.2.2 Understanding Electronic Transitions

Spectra provide valuable information about electronic transitions within atoms and molecules. The energy levels of electrons are quantized, meaning that electrons can only occupy specific energy states. When electrons transition between these states, they absorb or emit photons of specific energies, which correspond to the wavelengths in the spectrum. By studying the spectral lines, chemists can gain insights into the electronic structure of a substance, including the energy levels, orbital configurations, and the nature of chemical bonding.

32.2.3 Probing Molecular Environments

Inorganic chemistry often involves complex coordination compounds and metal complexes. Spectroscopy is instrumental in probing the local environment around metal ions within these compounds.

For example, ultraviolet-visible (UV-Vis) spectroscopy can be used to study the electronic transitions of d-d transitions in transition metal complexes, providing insights into their geometry, ligand field strength, and oxidation states.

Infrared (IR) spectroscopy, another important technique, allows for the investigation of vibrational transitions in molecules. The absorption of IR radiation corresponds to specific vibrational modes of chemical bonds, offering information about functional groups and molecular structure. This is particularly valuable in understanding coordination complexes and their interactions with ligands.

32.3 APPLICATIONS OF SPECTRA IN INORGANIC CHEMISTRY

32.3.1 Characterization of Coordination Compounds

Coordination compounds, consisting of metal ions bonded to ligands, are fundamental in inorganic chemistry. Spectroscopy provides crucial information for characterizing these compounds. For instance, electronic absorption spectroscopy can be employed to study the d-d transitions of transition metal complexes. The energy and intensity of these transitions depend on the ligand field around the metal center, allowing chemists to infer the geometry and electronic structure of the complex.

IR spectroscopy is also valuable for analyzing coordination compounds. Ligand vibrations and metal-ligand interactions can be observed through characteristic absorption bands. By comparing the IR spectra of different complexes, researchers can identify specific ligands and their bonding modes.

32.3.2 Determination of Reaction Mechanisms

Spectroscopic techniques can aid in elucidating reaction mechanisms in inorganic chemistry.

By monitoring the changes in the spectra of reactants and products over time, chemists can gain insights into the intermediates formed during a reaction. Time-resolved spectroscopy, for example, allows for the observation of transient species and provides information about their lifetimes and structural properties.

In studies of catalysis, spectroscopy is often employed to investigate the active sites of catalysts and the intermediates involved in catalytic cycles. Understanding these mechanisms is crucial for developing more efficient and selective catalytic processes.

32.3.3 Environmental Monitoring

Spectroscopy is widely used in environmental chemistry to monitor pollutants and assess the quality of air, water, and soil. Techniques such as atomic emission spectroscopy and mass spectrometry can detect trace levels of heavy metals and other toxic substances in environmental samples. By analyzing the spectral data, researchers can identify the sources of pollution, track changes over time, and evaluate the effectiveness of remediation efforts.

32.4 INSTRUMENTATION AND TECHNIQUES

The field of spectroscopy encompasses a variety of techniques and instrumentation, each tailored to specific applications and types of spectra.

32.4.1 UV-Visible Spectroscopy

UV-Vis spectroscopy is a powerful technique for studying electronic transitions in molecules. It employs light in the ultraviolet and visible regions of the electromagnetic spectrum. Instruments measure the absorption of light at various wavelengths, generating a spectrum that reflects the electronic structure of the sample.

32.4.2 Infrared Spectroscopy

Infrared (IR) spectroscopy focuses on vibrational transitions within molecules. It uses infrared radiation to probe the vibrational modes of chemical bonds. The resulting IR spectrum provides information about functional groups, molecular symmetry, and the presence of specific bonds.

32.4.3 Nuclear Magnetic Resonance (NMR) Spectroscopy

NMR spectroscopy exploits the magnetic properties of atomic nuclei to provide detailed information about molecular structure and dynamics. It is particularly useful for studying the arrangement of atoms within organic and inorganic compounds. While not a traditional spectroscopic technique in the context of emission or absorption, NMR plays a crucial role in understanding molecular interactions.

32.4.4 Mass Spectrometry

Mass spectrometry (MS) is a powerful analytical technique used to determine the mass-to-charge ratio of ions. While it is primarily focused on mass analysis rather than spectra in the traditional sense, mass spectra can provide insights into the composition and fragmentation patterns of inorganic compounds. Coupled with other spectroscopic techniques, mass spectrometry enhances the identification and characterization of complex mixtures.

32.5 CONCLUSION

Spectra serve as a cornerstone of analysis in inorganic chemistry, providing insights into the electronic structure, bonding characteristics, and molecular behavior of various compounds. Through techniques such as emission and absorption spectroscopy, chemists can identify elements, probe molecular environments, and elucidate reaction mechanisms.

The applications of spectroscopy extend beyond the laboratory, contributing to environmental monitoring and advancements in materials science. As technology continues to evolve, the role of spectroscopy in inorganic chemistry will remain essential, driving further discoveries and innovations in the field.

BIBLIOGRAPHY

Advanced Inorganic Chemistry: Authors: F. A. Cotton and G. Wilkinson Publisher: Wiley Year: 2020

Advanced Inorganic Chemistry: Authors: F. A. Cotton, G. Wilkinson, C. A. Murillo, and R. A. Bochmann Publisher: Wiley Year: 2015

Basic Coordination Chemistry: Author: A. G. J. P. van der Meer Publisher: Academic Press Year: 2001

Basic Inorganic Chemistry: Authors: Gary L. Miessler and Paul J. Fischer Publisher: Prentice Hall Year: 2018

Chemistry of the Elements: Authors: N.N. Greenwood and A. Earnshaw Publisher: Butterworth-Heinemann Year: 2006

Comprehensive Coordination Chemistry II: Editors: J. A. McCleverty and T. J. Meyer Publisher: Elsevier Year: 2004

Concise Coordination Chemistry: Authors: Philip A. Cox Publisher: Royal Society of Chemistry Year: 2002

Concise Inorganic Chemistry: Author: J.D. Lee Publisher: Wiley Year: 2015

Concise Inorganic Chemistry: Authors: J. Derek Woollins Publisher: Wiley Year: 2017

Concise Inorganic Chemistry: Authors: J. Derek Woollins Publisher: Wiley-Blackwell Year: 2017

Coordination Chemistry: Author: J. Derek Woollins Publisher: Wiley Year: 2019

Descriptive Inorganic Chemistry: Authors: Geoff Rayner-Canham and Tina Overton Publisher: Freeman Year: 2021

Essentials of Inorganic Chemistry: Authors: J. Derek Woollins Publisher: Royal Society of Chemistry Year: 2010

Fundamentals of Inorganic Chemistry: Authors: G. C. Rajamathi and C. K. Jayasree Publisher: New Age International Year: 2015

Inorganic Chemistry: A Textbook: Author: Jürgen O. Metzger Publisher: Springer Year: 2021

Inorganic Chemistry: Author: J. Derek Woollins Publisher: Wiley Year: 2020

Inorganic Chemistry: Authors: Gary L. Miessler and Paul J. Fischer Publisher: Pearson Year: 2021

Inorganic Chemistry: Authors: J. Derek Woollins Publisher: Wiley Year: 2012

Inorganic Chemistry: Principles of Structure and Reactivity: Authors: James Huheey, Ellen A. Keiter, and Richard L. Keiter Publisher: Pearson Year: 2013

Introduction to Inorganic Chemistry: Authors: Gary L. Miessler and Paul J. Fischer Publisher: Prentice Hall Year: 2019

Metal Complexes in Aqueous Solutions: Author: P. M. D. V. Prasad Publisher: Academic Press Year: 1990

Modern Inorganic Chemistry: Author: R. A. MacGillivray Publisher: Springer Year: 2019

Organometallic Chemistry: Authors: Gary O. Spessard and Gary A. Mills Publisher: Wiley Year: 1997

Principles of Inorganic Chemistry: Authors: B.R. Puri, L.R. Sharma, and K.C. Kalia Publisher: Vishal Publishing Year: 2017

Principles of Instrumental Analysis: Authors: Douglas A. Skoog, F. James Holler, and Timothy A. Nieman Publisher: Cengage Learning Year: 2013

The Chemistry of Organolithium Compounds: Authors: A. R. Katritzky, O. S. A. O'Reilly Publisher: Wiley Year: 2005

The Chemistry of the Actinide Elements: Authors: J. W. A. M. Van Eijk, R. G. Schiffrin Publisher: Springer Year: 1986

INDEX

J

K

L

M

N

O

P

Q

R

S

www.ingramcontent.com/pod-product-compliance
Lightning Source LLC
LaVergne TN
LVHW062220200926
848484LV00006B/1439